AF429431

Computer Methods in Power Systems
Analysis with MATLAB

Computer Methods in Power Systems
Analysis with MATLAB

P. CHANDRA SEKHAR
Associate Professor
Department of Electrical & Electronics Engineering
Mahatma Gandhi Institute of Technology,
Hyderabad, T.S

P. SATISH KUMAR
Professor of Electrical Engineering
University College of Engineering (Autonomous)
Osmania University, Hyderabad, T.S

BSP **BS Publications**
A unit of **BSP Books Pvt. Ltd.**
4-4-309/316, Giriraj Lane,
Sultan Bazar, Hyderabad - 500 095 - T.S.

Computer Methods in Power Systems *'Analysis with MATLAB'*
By P. Chandra Sekhar, P. Satish Kumar

Published by:

BSP **BS Publications**

A unit of **BSP Books Pvt., Ltd.**
4-4-309/316, Giriraj Lane, Sultan Bazar,
Hyderabad - 500 095, T.S.
Phone : 040 - 23445688
e-mail : info@bspbooks.net
Website : www.bspbooks.net

ISBN: 978-93-90211-49-4

Dedicated

to

All Our Teachers

PREFACE

This book **'Computer Methods in Power Systems – *Analysis with MATLAB'*** caters the academic requirements of under graduate students and post graduate students of electrical and electronics engineering of all universities across India. The computer methods in power systems subject is very important for the students of electrical engineering as it deals with key practical and real time aspects of power systems like load flow analysis, short circuit analysis, power system stability analysis and voltage control analysis. The contents of the book are presented in lucid style so that an average student can grasp the subject. More number of simple and complex problems are worked out to strengthen the theory and also objective questions are given at the end of each chapter to enable the students for competitive examinations. MATLAB programming is also included. This text book provides full pledged in-depth knowledge including all types of numerical problems. The organization of book is as follows:

- **First chapter** deals with transmission line analysis, analysis of short and medium transmission lines, loading and interconnection of transmission lines.
- **Second chapter** presents per unit representation of power system components.
- **Third chapter** deals with graph theory, incidence matrices and augmented loop matrix.
- **Fourth and Fifth chapters** presents admittance bus matrix and impedance bus matrix, formation of network matrices by singular transformation and for an interconnected network, calculation of impedance matrix without mutual coupling between the lines and mutual coupling between the elements.
- **Sixth chapter** presents load flow analysis, static load flow equations, power transmission in the load flow analysis and numerical methods.
- **Seventh chapter** deals with fault analysis, reactance control of short circuit currents and short circuit capacity.
- **Eighth chapter** presents symmetrical components, classification of unbalanced systems, zero sequence networks, sequence networks of an alternator operating under no load condition.
- **Ninth and Tenth chapters** presents unsymmetrical and symmetrical fault analysis.
- **Eleventh chapter** concentrates on power system stability analysis, methods to determine stability, transient stability analysis, steady state stability analysis.
- **Twelfth chapter** presents voltage control and power factor improvement methods.
- **Thirtieth chapter** includes MATLAB programming for the formulation of Y_{bus} by inspection method, singular transformation method with and without mutual coupling, formation of Y_{bus} and Z_{bus} matrix using building algorithm, calculation of ABCD parameters, power angle diagram, swing curve, jacobian matrix calculations. MATLAB programming to solve load flow equations, load flow analysis and solution.
- Each chapter consist of objective, solved and unsolved problems at the end.

Authors

ACKNOWLEDGEMENTS

Authors are grateful to Prof. D. Ravinder, Vice-Chancellor, Prof. P. Laxminarayana, Registrar, Osmania University, Hyderabad and the management of Mahatma Gandhi Institute of Technology, Hyderabad for their constant support and encouragement in this endeavor.

Authors extent their gratitude to the Principal, University College of Engineering, Dean, Faculty of Engineering, Head, Department of Electrical Engineering, Osmania Universityand the Principal, Head, Department of Electrical and Electronics, MGIT and all colleaguesfor their constant encouragement.

Authors

CONTENTS

Preface .... (vii)

Acknowledgements .... (ix)

Chapter 1: Transmission Line Analysis

1.1   Introduction 1

1.2   Concepts of a Transmission Line 1

1.3   Performance of Transmission Lines.. 1

1.4   Classification of Transmission Lines 2

1.5   Analysis of Short Transmission Lines 3

1.6   Analysis of Medium Transmission Lines.. 4

      1.6.1   Load Condenser Method.. 5

      1.6.2   Source Condenser Method... 6

      1.6.3   Nominal T–Method.. 7

      1.6.4   Nominal–π Method 10

1.7   Impedance of Transmission Line for
Open Circuit Condition & Short Circuit Condition 13

1.8   Conditions for Zero Voltage Regulation and
Maximum Voltage Regulation of a Transmission Line 14

1.9   Relation between Actual Loading of Transmission
Line and Surge Impedance Loading... 17

1.10 Interconnection of Transmission Line... 23

      1.10.1   Transmission Lines Connected in Cascade......................... 23

      1.10.2   Parallel Connection of Transmission Lines 24

Objective Questions..**25**

Chapter 2: Per Unit Quantities

2.1   Introduction 27

2.2   Per Unit Representation of Power System Components 27

      2.2.1   Absolute System of Analysis... 27

2.3   Expressing Per Unit Reactance in Terms of
Base Voltage and Base Current.. 28

2.4   Expressing Per Unit Reactance in Terms of
Base Volt Ampere and Base Voltage 29

2.5 Expressing Base Reactance (Ω) in Terms of
Base VA and Base Current 30

2.6 Expressing $X_{p.u,\,new}$ in terms of $X_{p.u,\,old}$ 30

Objective Questions **35**

Chapter 3: Graph Theory

3.1 Introduction 39

3.2 Definitions 40

3.3 Incidence Matrices 41

    3.3.1 Element-Node Incident Matrix $[\overline{A}]$ 41

    3.3.2 Bus Incidence Matrix (A) 41

    3.3.3 Branch Path Incidence Matrix (K) 42

    3.3.4 Basic Cutset Incidence Matrix [B] 44

    3.3.5 Augmented Cut-Set Incidence Matrix $[\overline{B}\ or\ \widehat{B}]$ 45

    3.3.6 Basic Loop Incidence Matrix (C) 46

3.3.7 Augmented Loop Incidence Matrix $(\overline{C})$ 47

Objective Questions **67**

Chapter 4: Admittance Bus Matrix

4.1 Introduction 71

4.2 Representation of Network Components 72

4.3 Formation of Network Matrices by Singular Transformation 73

    4.3.1 Network Performance Equations 73

    4.3.2 Bus Admittance Matrix 74

4.4 Formulation of $[Y_{Bus}]$ for an Interconnected Network 75

4.5 Properties of $[Y_{Bus}]$ 77

Objective Questions **91**

Chapter 5: Impedance Bus Matrix

5.1 Introduction 93

5.2 Calculation of Impedance Matrix without
Mutual Coupling between the Transmission Lines 93

5.3 Performance Equations of Partial Networks 102

5.4 Determination of Z_{bus} by Considering Mutual Coupling between the Elements 103
    5.4.1 Addition of a Branch 103
    5.4.2 Addition of Link 106

Objective Questions **115**

Chaper 6: Load Flow Analysis

6.1 Introduction 119
6.2 Classification of Buses 120
6.3 Determination of Static Load Flow Equations 122
6.4 Voltage at i^{th} Bus 123
6.5 Determination of Power Transmission in the Load Flow Analysis 124
6.6 Numerical Methods 125
    6.6.1 Gauss Seidal Method 125
    6.6.2 Newton Raphson Method 132
    6.6.3 Decoupled Load Flow Method 141
    6.6.4 Fast Decoupled Load Flow Method 141

Objective Questions **150**

Chapter 7: Fault Analysis

7.1 Introduction 157
7.2 Reactance Control of Short Circuit Currents 159
    7.2.1 Generator Reactors 159
    7.2.2 Feeder Reactors 160
    7.2.3 Bus Bar Reactors 160
7.3 Short Circuit Capacity or Fault Level of a Bus 161

Objective Questions **178**

Chapter 8: Symmetrical Components

8.1 Introduction 183
8.2 Classification of Unbalanced Systems 184
8.3 Significance of Operator 'k' or 'a' or 'α' 184
8.4 Per Unit Neutral Current 187
8.5 Method of Connections between Neutral and
Ground of a Three Phase Star Connected Alternator 188

8.6 Zero Sequence Networks of Transformers 188

8.7 Sequence Networks of an Alternator Operating under No Load Condition 192

8.8 Reactance offered by Alternator during Fault 193

Objective Questions **196**

Chapter 9: Unsymmetrical Fault Analysis

9.1 Introduction 199

9.2 Classification of Faults 199

9.3 Line to Ground Fault Analysis (L-G Fault) 201

9.4 Line to Line Fault Analysis (L-L Fault) 205

9.5 Line to Line to Ground Fault Analysis (L-L-G Fault) 209

Objective Questions **226**

Unsolved Problems **227**

Chapter 10: Symmetrical Fault Analysis

10.1 Introduction 231

10.2 Line–Line–Line Fault (LLL) 231

10.3 Representation of Positive X_{1eq}, Negative X_{2eq} and Zero Sequence X_{oeq} Reactances in terms of Self-Reactance, X_s and Mutual Reactance, X_m of Transmission Line 234

Solved Questions **236**

Objective Questions **245**

Chapter 11: Power System Stability Analysis

11.1 Introduction 247

    11.1.1 Power System Stability 248

    11.1.2 Classification of Power System Stability 248

11.2 Methods to Determine Steady State Stability 249

    11.2.1 Static Model 249

    11.2.2 ABCD Constants 252

    11.2.3 Methods to Improve Steady State Stability 257

    11.2.4 Determination of Transfer Reactance 258

11.3 Analysis of Transient Stability 258

    11.3.1 Conditions for Which Power Transmitted during Fault in Zero 258

    11.3.2 Dynamic Model 259

11.4   Analysis of Steady State Stability using Swing Equation 261

11.5   Analysis of Transient Stability Based on number of Synchronous Machines 263

11.6   Swinging of Synchronous Machines 265

    11.6.1   Coherent Swinging of Synchronous Machines 265

    11.6.2   Non-Coherent Swinging of Synchronous Machines 266

11.7   Equal Area Criterion Method 268

11.8   Applications of Equal Area Criterion Method 269

    11.8.1   Sudden Increase in Mechanical Input to a Synchronous Generator 270

    11.8.2   Sudden Increase in Mechanical Output of a Synchronous Motor 272

    11.8.3   Removal of one of the Parallel Transmission Lines
    Forcibly using Fast Acting Circuit Breakers 274

    11.8.4   Occurrence of Fault at the Middle of
    One of the Parallel Transmission Line 277

    11.8.5   Occurrence of Fault on one of the Parallel Transmission
    Lines near Bus Bar 281

    11.8.6   Occurrence of Fault on Bus Bar 284

    11.8.7   Critical Clearing Time, t_c 286

11.9   Step by Step Method or Point by Point Method 287

Solved Problems **289**

Unsolved Problems **300**

Objective Questions **301**

Chapter 12: Voltage Control and Power Factor Improvement Methods

12.1   Introduction 303

    12.1.1   Sign of the Net Reactive Power 304

    12.1.2   Approximate Expression for Voltage at the
    Receiving End of Transmission Line 306

    12.1.3   Reactive Power Balance Equation 307

12.2   Shunt Capacitor / Power Factor Improvement Device 308

12.3   Shunt Inductor 312

12.4   Series Capacitor / Power System Stability Device 316

12.5   Comparison between Shunt Capacitor and Series Capacitor 320

12.6   Synchronous Inductor and Synchronous Capacitor 321

12.7   Synchronous Phase Modifier / Phase Advancers 322

Key Points 324

Solved Problems 325

Unsolved Problems 333

Objective Questions 333

Chapter 13: MATLAB Programming

13.1 Y_{Bus} Formation by Inspection Method 335

13.2 Y_{Bus} Formation by Singular Transformation Method without Mutual Coupling 336

13.3 Y_{Bus} Formation by Singular Transformation Method with Mutual Coupling 339

13.4   Formation of Y-Bus Matrix 341

13.5   Formation of Y-Bus Matrix using Bus Building Algorithm 342

13.6   Formation of Z-Bus Matrix using Bus Building Algorithm – 1 344

13.7   Formation of Z-Bus Matrix using Bus Building Algorithm – 2 345

13.8   Determination of Bus Currents, Bus Power and Line Flows for a
       Specified Bus System Profile 349

13.9   Calculation of ABCD Parameters 351

13.10 Determination of Power Angle Diagram for MATLAB Program 357

13.11 Determination of Swing Curve 360

13.12 Jacobian Matrix Calculation 362

13.13 MATLAB Program to Solve Load Flow Equations using
       Gauss-Seidel Method with PQ Buses 364

13.14 Load Flow Analysis by Gauss-Seidal Method 367

13.15 Newton Raphson Load Flow Method using MATLAB 369

13.16 Fast Decoupled Power Flow Model using MATLAB 372

13.17 LG, LL, LLG Fault Analysis 374

13.18 Solution of Swing Equation by Point by Point Method 394

Solutions 397

CHAPTER

1   Transmission Line Analysis

1.1   INTRODUCTION

This chapter deals with the analysis of a transmission line. The factors analyzing the performance of a transmission line are dealt. Further, short, medium and long transmissions are analyzed. Further the analysis is performed with respect to ABCD constants, open circuit and short circuit conditions. The concept of travelling waves in power systems has been dealt.

1.2   CONCEPTS OF A TRANSMISSION LINE

An electric transmission line can be represented by a series combination of resistance, inductance and shunt combination of conductance and capacitance. These parameters are symbolized as R, L, G and C respectively. Among these R and G are least important as they do not affect much the total equivalent impedance of the line and hence the transmission capacity.

1.3   PERFORMANCE OF TRANSMISSION LINES

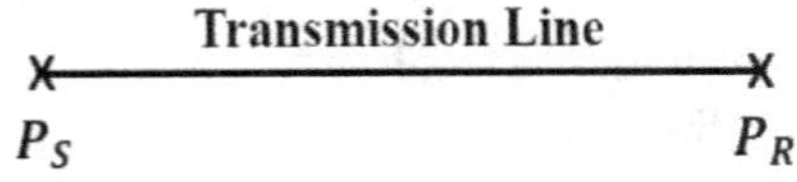

Figure 1.1 Transmission line to determine % Efficiency

The performance of a transmission line can be calculated using: (i) % Efficiency, (ii) % Regulation

(i) **% Efficiency:** Consider a transmission line as shown in Figure 1.1.

P_s is sending end power and P_R is the receiving end power

% Efficiency of a transmission line is

$$\% \; \eta \; \frac{P_R}{P_S} \times 100 = \frac{P_R}{P_R + P_{Losses}} \times 100$$

P_{losses} : For a single phase system, power loss $= I^2 R$

For a three phase system, power loss $= 3I^2 R$

where 'I' is the phase current and 'R' is the resistance per phase.

(ii) **%Regulation:** As the transmission line is a stationary device, % voltage regulation is calculated.

Consider a transmission line as shown in Figure 1.2.

V_S is sending end voltage and V_R is the receiving end voltage.

Figure 1.2 Transmission line to determine % voltage regulation

Voltage drop in transmission line $= V_S - V_R$

$$\% \in = \frac{V_S - V_R}{V_R} \times 100$$

Case: Assume that the receiving end of transmission line is operating under no–load condition,

As the receiving end is operating at no-load condition, $I = 0$

Voltage drop in transmission line $= 0$

Sending end voltage, V_S = No load-receiving end voltage, V_{R_o}

$$\% \in = \frac{V_S - V_R}{V_R} \times 100$$

The percentage regulation of a transmission line is defined as the change in receiving end voltage from no load to full load, expressed as % full load receiving end voltage.

Note: The percentage efficiency of a transmission line must be high and percentage regulation of a transmission line must be low for better performance of transmission lines.

1.4   CLASSIFICATION OF TRANSMISSION LINES

(a)   Transmission lines are primarily classified based on wavelength.

(b)   The term power transmission means travel of voltage wave and current wave from sending end to receiving end of the transmission line.

(c)   Voltage wave & current wave travels at velocity of light from sending end to receiving end of the transmission line.

Consider a voltage wave travelling from sending end to the receiving end of transmission line as shown in figure 1.3.

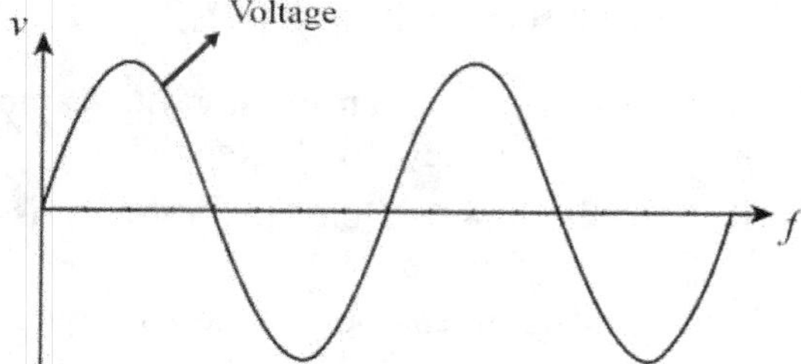

Figure 1.3 Voltage travelling wave

$$\text{Wavelength}(\lambda) = \frac{\text{Velocity of travelling wave}}{\text{Frequency}}$$

$$= \frac{3 \times 10^8 \dfrac{m}{sec}}{50 \dfrac{cycles}{sec}}$$

$$\lambda = 6000 \frac{km}{cycle}$$

i.e., voltage (or) current wave travels an angular distance of 6000 km to complete one cycle, but due to complexity of power system network, transmission lines are further classified based on

1. Physical length of the line
2. Operating voltage
3. Effect of capacitance

and is tabulated as follows.

Transmission Line	Physical Length	Operating Voltage	Effect of Capacitance
Short line	0–80km	0–20kV	Neglected
Medium–line	80–120km	20–100kV	Lumped or concentrated
Long–line	> 120 km	>100kV	uniformly distributed

1.5   ANALYSIS OF SHORT TRANSMISSION LINES

1. **Equivalent Circuit:** A short transmission line consists of resistance and inductance connected in series.

   R is the resistance and jX_L is the inductive reactance.

   Consider the equivalent circuit as shown in the Figure 1.4.

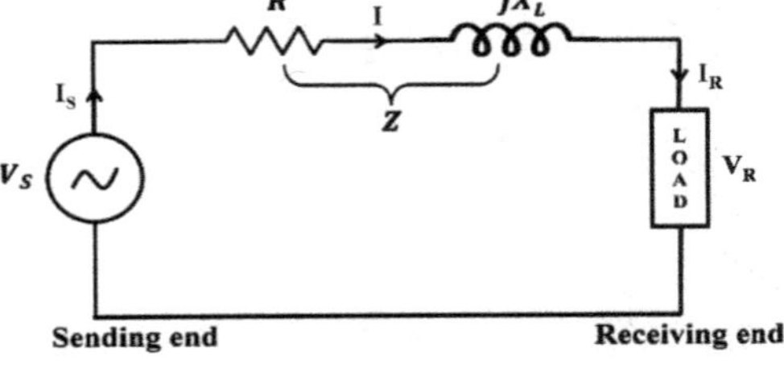

Figure 1.4 Short Transmission Line

2. **Mathematical Relations:**

$$I_S = I_R = I \text{ (say)}$$

   Resistive voltage drop = IR

   Reactive voltage drop = $j\,IX_L$

   Total voltage drop = $IR + I(jX_L) = I(R + jX_L) = IZ$

   Sending end voltage, $V_S = V_R + I_R + I\,(jX_L)$

$$V_S = V_R + I(R + jX_L)$$

$$V_S = V_R + IZ$$

3. **Vector Diagram:** Consider receiving end voltage, V_R as the reference vector.

   Assume R-L Load

   Receiving end current, I_R lags receiving end voltage, V_R by ϕ_R

   where $0 < \phi_R \leq 90$, lag

   The vector diagram is shown in Figure 1.5.

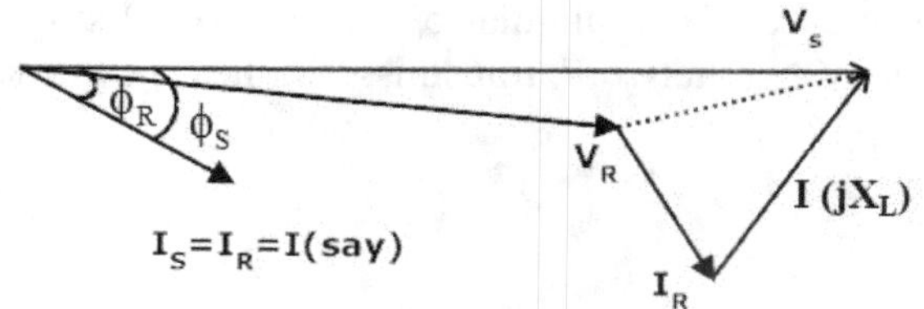

Figure 1.5 Vector diagram of short transmission line

For lagging power factor loads: $\phi_R < \phi_S$

Note: For leading power factor loads: $\phi_R > \phi_S$

4. **ABCD Constants:** For a two-port network,

$$V_S = AV_R + BI_R \qquad \qquad .....(1.1)$$

$$I_S = CV_R + DI_R \qquad \qquad .....(1.2)$$

$$V_S = V_R + Z\,I_R \qquad \qquad .....(1.3)$$

Compare (1.1) & (1.3)

$\qquad$ A = 1 and B = Z

$\qquad I_S = I_R$

$\qquad I_S = 0.\,V_R + 1.I_R \qquad \qquad .....(1.4)$

Compare (1.2) & (1.4)

C = 0 and D = 1

A = D = 1, i.e., short line is symmetrical

AD − BC = 1, i.e., short line is reciprocal

5. **Conclusion:**

1. A = 1, D = 1 either for series branch (or) shunt branch
2. Constant 'B' determines impedance (Z) of series branch
3. Constant 'C' determines admittance (Y) of shunt branch

1.6 ANALYSIS OF MEDIUM TRANSMISSION LINES

Medium lines are classified based on the location of the capacitance in the equivalent circuit into four methods.

(a) **Load condenser method:** Capacitor is connected across load
(b) **Source condenser method:** Capacitor is connected across source
(c) **Nominal-π method:** Capacitance is equally split across source and load
(d) **Nominal-T method:** Capacitor is at the middle of the line

1.6.1 Load Condenser Method

1. **Equivalent Circuit:** Consider the equivalent circuit as shown in the Figure 1.6. R is the resistance and jX_L is the inductive reactance.

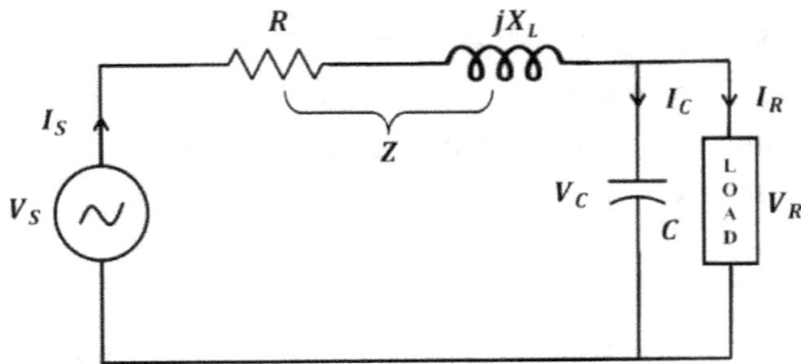

Figure 1.6 Medium Transmission Line by Load condenser method

2. **Mathematical Relations:** Sending end current, $I_S = I_R + I_C$

Resistive voltage drop $= I_S R$

Reactive voltage drop $= I_S(jX_L)$

Total voltage drop $= I_S R + I_S(jX_L)$

$$= I_S(R + jX_L)$$

$$= I_S Z$$

Sending end voltage, $V_S = V_R + I_S R + I_S(jX_L)$

$$V_S = V_R + I_S(R + jX_L)$$

$$V_S = V_R + I_S Z$$

3. **Vector Diagram:** Let V_R be the Reference vector

Assume R-L Load.

Receiving end current, I_R lags receiving end voltage, V_R by ϕ_R

where $\qquad\qquad 0 < \phi_R \leq 90$, lag

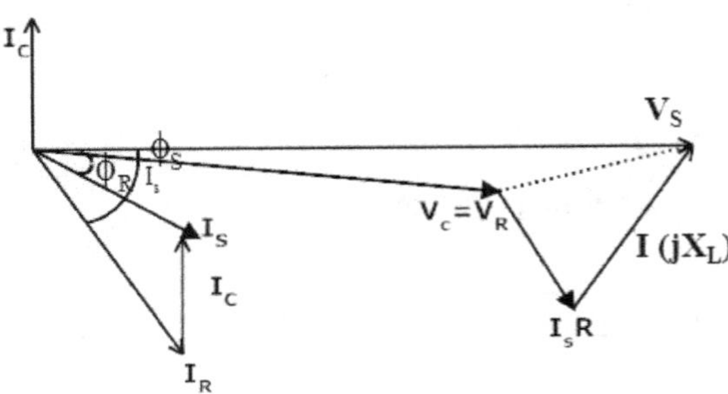

Figure 1.7 Vector diagram of Medium Transmission Line by Load condenser method

4. **ABCD Constants:**

$$\begin{bmatrix} A & B \\ C & D \end{bmatrix} = \begin{bmatrix} 1 & Z \\ 0 & 1 \end{bmatrix}\begin{bmatrix} 1 & 0 \\ Y & 1 \end{bmatrix} = \begin{bmatrix} 1+YZ & Z \\ Y & 1 \end{bmatrix}$$

$A \neq D$ i.e., Load condenser method is unsymmetrical

$AD - BC = 1$ i.e., Load condenser method is reciprocal

Note: As capacitor is unsymmetrically located in the equivalent circuit, load condenser method is unsymmetrical.

1.6.2 Source Condenser Method

1. **Equivalent Circuit:** Consider the equivalent circuit as shown in the figure 1.8.

   R is the resistance and jX_L is the inductive reactance.

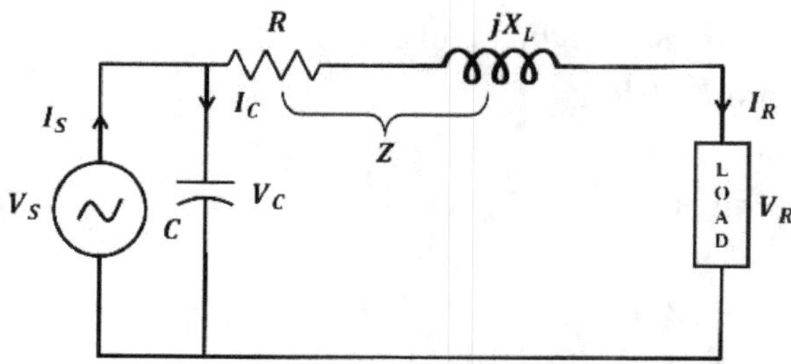

Figure 1.8 Medium Transmission Line by Source condenser method

2. **Mathematical Relations:** Sending end current, $I_S = I_R + I_C$

   Resistive voltage drop $= I_R R$

   Reactive voltage drop $= I_R(jX_L)$

   Total voltage drop $= I_R R + I_R(jX_L) = I_R (R + jX_L) = I_R Z$

   Sending end voltage, $V_S = V_R + I_R R + I_R(jX_L)$

   $$= V_R + I_R(R + jX_L)$$

   $$V_S = V_R + I_R Z$$

3. **Vector Diagram:** Let V_R be the reference vector

   Assume R-L Load.

   Receiving end current, I_R lags receiving end voltage, V_R by ϕ_R

   where $\qquad 0° < \phi_R \leq 90°$, lag

   The vector diagram is shown in Figure 1.9.

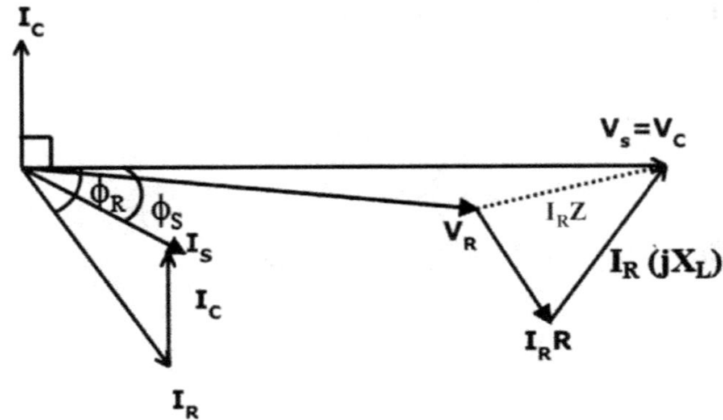

Figure 1.9 Vector diagram of Medium Transmission Line by Source condenser method

4. **ABCD–Constants:**

$$\begin{bmatrix} A & B \\ C & D \end{bmatrix} = \begin{bmatrix} 1 & 0 \\ Y & I \end{bmatrix} = \begin{bmatrix} 1 & Z \\ 0 & 1 \end{bmatrix} = \begin{bmatrix} 1 & Z \\ Y & 1+YZ \end{bmatrix}$$

i.e., $A \neq D$ i.e., source condenser method is unsymmetrical

$$AD - BC = 1 + YZ - YZ = 1$$

$AD - BC = 1$ i.e., source condenser method is reciprocal

Note: As capacitor is unsymmetrically located in the equivalent circuit, source condenser method is unsymmetrical.

1.6.3 Nominal T–Method

As capacitor is located exactly at the middle of the line, Nominal T–method is also known as middle condenser method.

The term 'nominal' represents 'rated voltage'

1. **Equivalent Circuit:** Consider the equivalent circuit as shown in the Figure 1.10.

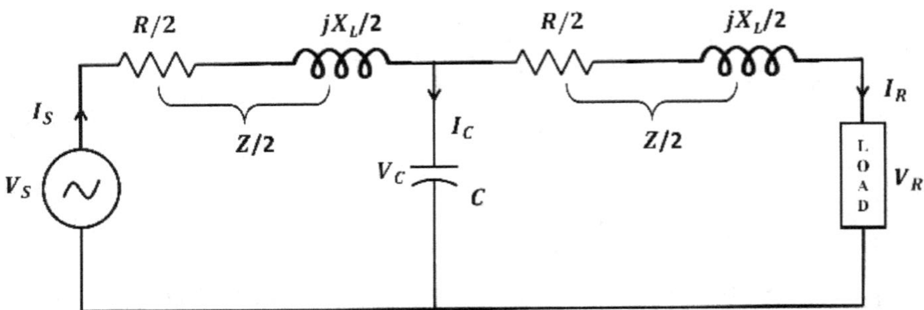

Figure 1.10 Medium Transmission Line by Nominal–T method

2. **Mathematical Relation:** Sending end current, $I_S = I_R + I_C$

   Voltage across capacitor, $V_C = V_R + I_R \dfrac{R}{2} + I_R \left(\dfrac{jX_L}{2} \right)$

   Sending end voltage, $V_S = V_C + I_S \dfrac{R}{2} + I_S \left(\dfrac{jX_L}{2} \right)$

3. **Vector Diagram:** Let V_R be the reference vector

   Assume R-L Load.

   Receiving end current, I_R lags receiving end voltage, V_R by ϕ_R

   where $\qquad 0° < \phi_R \leq 90°,$ lag

   The vector diagram is shown in Figure 1.11.

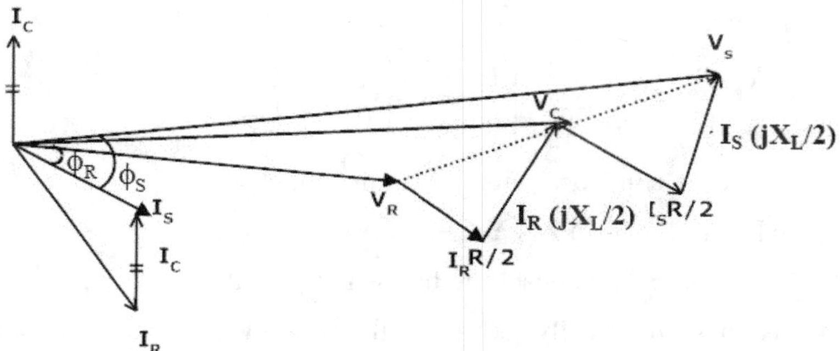

Figure 1.11 Vector diagram of Medium Transmission Line by Nominal–T method

4. **ABCD – Constants**

$$\begin{bmatrix} A & B \\ C & D \end{bmatrix} = \begin{bmatrix} 1 & \dfrac{Z}{2} \\ 0 & 1 \end{bmatrix} \begin{bmatrix} 1 & 0 \\ Y & 1 \end{bmatrix} \begin{bmatrix} 1 & \dfrac{Z}{2} \\ 0 & 1 \end{bmatrix}$$

$$\begin{bmatrix} A & B \\ C & D \end{bmatrix} = \begin{bmatrix} 1 + \dfrac{YZ}{2} & Z\left(1 + \dfrac{YZ}{4}\right) \\ 0 & 1 + \dfrac{YZ}{2} \end{bmatrix}$$

i.e., $A = D$, i.e., Nominal T–method is symmetrical

$AD - BC = 1$, i.e., Nominal T–method is reciprocal

Note: As capacitor is located symmetrically in the equivalent circuit Nominal–T method is symmetrical.

Case - With receiving end of transmission line operating under no–load condition:

1. **Equivalent Circuit:** Consider the equivalent circuit as shown in Figure 1.12.

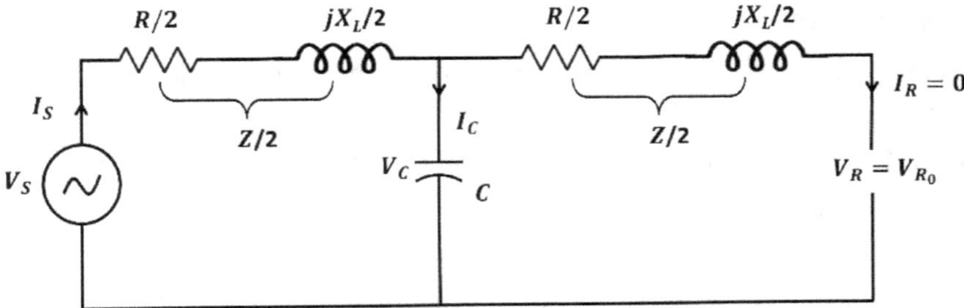

Figure 1.12 Medium Transmission Line by Nominal–T method with open circuited receiving end

2. **Mathematical Relations:** As $I_R = 0$, voltage drop due to $I_R = 0$

$$\therefore \quad V_C = V_R = V_{R_0}$$

$$V_S = V_C + I_C \frac{R}{2} + I_S\left(\frac{jX_L}{2}\right)$$

$$I_S = I_C$$

$$V_S = V_C + I_C \frac{R}{2} + I_C\left(\frac{jX_L}{2}\right)$$

3. **Vector Diagram:** Let V_R be the Reference vector

   The vector diagram is shown in Figure 1.13.

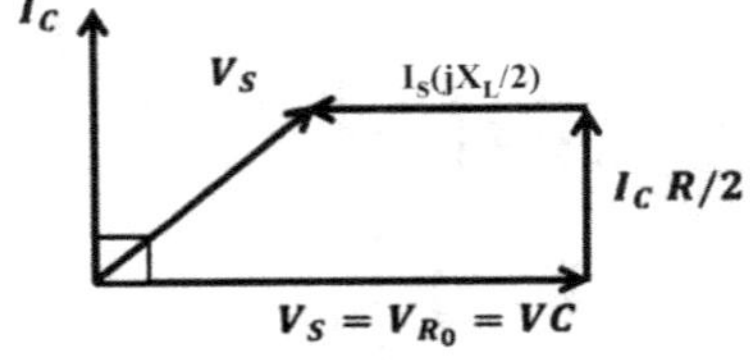

Figure 1.13 Vector diagram of Ferranti Effect

Ferranti Effect

- The magnitude of sending end voltage is less than the magnitude of receiving end voltage at no–load condition
- Ferranti effect is due to capacitive current (or) capacitance in the equivalent circuit.
- As short line do not have capacitance, Ferranti effect does not occur in short line.
- Therefore, Ferranti effect occurs in medium line and long line at no load conditions.
- % Voltage rise in transmission line due to Ferranti effect

$$= \frac{\omega^2 l^2}{18} \times 10^{-8}\, V$$

   where 'l' is the length of transmission line in Km.

- Receiving end voltage at no–load condition is

$$V_{RO} = V_C = I_C\,(-jX_C)$$

$$= \left[\frac{V_S}{\dfrac{R}{2} + j\dfrac{X_L}{2} + (-jX_C)}\right](-jX_C) = \left[\frac{V_S}{\dfrac{R}{2} + j\dfrac{\omega_L}{2} + (-\dfrac{j}{\omega C})}\right]\left(-j\frac{1}{\omega C}\right)$$

1.6.4 Nominal–π Method

*As the capacitor is split into two equal parts across source and load, nominal–π method is also referred as "split condenser method".

1. **Equivalent Circuit:** Consider the equivalent circuit as shown in the Figure 1.14.

   R is the resistance and jX_L is the inductive reactance.

   Vc_s = Voltage across capacitor at the sending end.

   Vc_R = Voltage across capacitor at the receiving end.

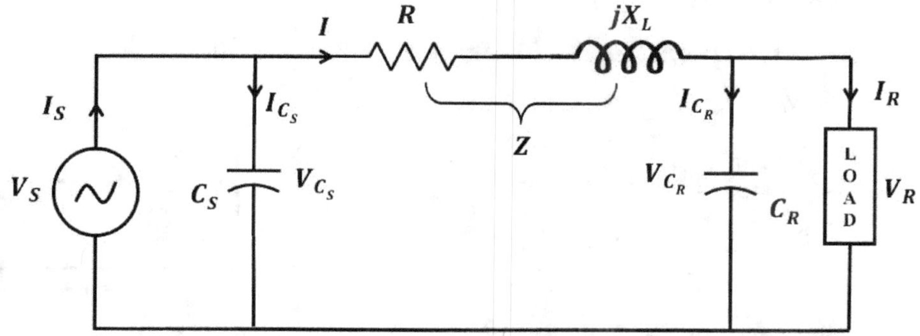

Figure 1.14 Medium Transmission Line by Nominal–π method

2. **Mathematical Relations:**

$$C_s = C_R = \frac{C}{2}$$

$$V_{C_R} = V_R$$

Resistive voltage drop = IR

Reactive voltage drop = IjX_L

Total voltage drop = $I\,R + I\,jX_L$

$$= I\,(R + jX_L)$$

$$= IZ$$

Sending end voltage, $V_S = V_R + I_R + I\,(jX_L)$

$$V_S = V_R + I\,(R + jX_L)$$

$$V_S = V_R + IZ$$

3. **Vector Diagram:** Let V_R be the reference vector.

   Assume R-L Load

   Receiving end current, I_R lags receiving end voltage, V_R by ϕ_R

$$0 < \phi_R \leq 90, \text{ lag}$$

The vector diagram is shown in Figure 1.15.

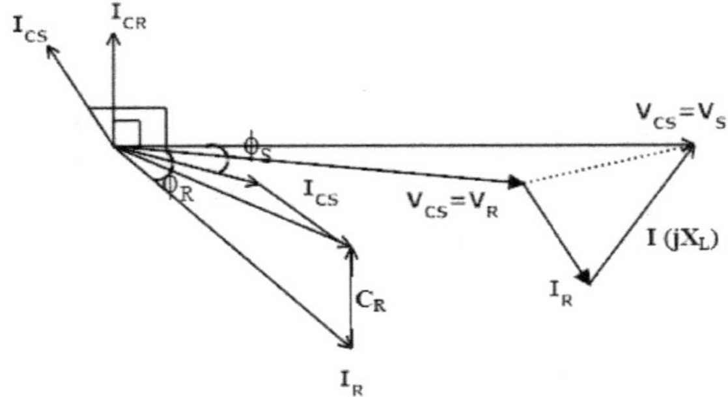

Figure 1.15 Vector diagram of Medium Transmission Line by Nominal– π method

4. ABCD – Constants:

$$\begin{bmatrix} A & B \\ C & D \end{bmatrix} = \begin{bmatrix} 1 & 0 \\ \dfrac{Y}{2} & 1 \end{bmatrix}\begin{bmatrix} 1 & Z \\ 0 & 1 \end{bmatrix}\begin{bmatrix} 1 & 0 \\ \dfrac{Y}{2} & 1 \end{bmatrix}$$

$$\begin{bmatrix} A & B \\ C & D \end{bmatrix} = \begin{bmatrix} 1+\dfrac{YZ}{2} & Z \\ Y\left[1+\dfrac{YZ}{4}\right] & 1+\dfrac{YZ}{2} \end{bmatrix}$$

i.e., A = D i.e., Nominal π– method is symmetrical

AD – BC = 1 i.e., Nominal π– method is reciprocal

Note: As Equal capacitance $\left(\dfrac{c}{2}\right)$ is across source and load, nominal π method is symmetrical.

Case - With receiving end of transmission line operating at no–load condition

1. **Equivalent Circuit:** Consider the equivalent circuit as shown in figure 1.16.

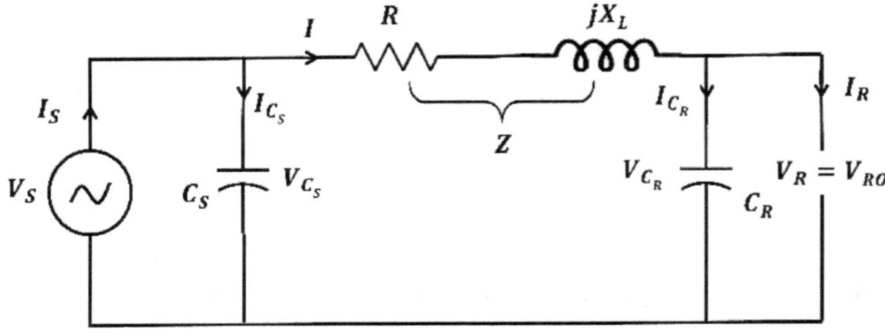

Figure 1.16 Medium Transmission Line by Nominal– π method with open circuited receiving end

2. **Mathematical Relations:** As $I_R = 0$

$$V_{C_R} = V_R = V_{R_0}$$

$$I = I_{C_R} \; (\because I_R = 0)$$

$$V_S = V_{R_0} + IR + I(jX_L)$$

$$V_S = V_{C_R} + I_{C_R}R + I_{C_R}(jX_L)$$

3. **Vector Diagram:**

Let V_{R_0} be the Reference vector

The vector diagram is shown in Figure 1.17.

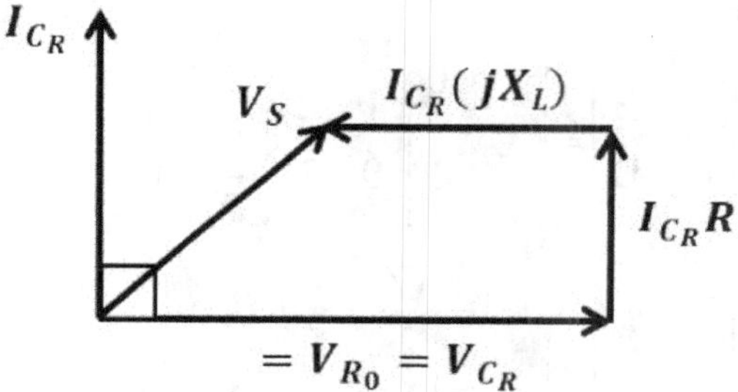

Figure 1.17 Vector diagram of Ferranti Effect for Nominal–π method

$*$ $|V_S| < |V_{R_0}|$ i.e., Ferranti effect

4. **No–Load Receiving End Voltage:**

$$V_{R_0} + V_R = V_{C_R} = I_{C_R}(-jX_{C_R})$$

$$= \left\{ \frac{V_S}{R + jX_L(-jX_{C_R})} \right\}$$

$$= \left\{ \frac{V_S}{R + j\omega_L - \dfrac{j}{\omega C_R}} \right\} = \left\{ \frac{V_S}{R + j\omega L - \dfrac{j}{\omega \dfrac{C}{2}}} \right\} \left(-\dfrac{j}{\omega \dfrac{C}{2}} \right)$$

Note: Operator 'j' rotates a vector in anticlockwise direction by 90^0.

1.7 IMPEDANCE OF TRANSMISSION LINE FOR OPEN CIRCUIT CONDITION & SHORT CIRCUIT CONDITION

Case 1: With receiving end open–circuited

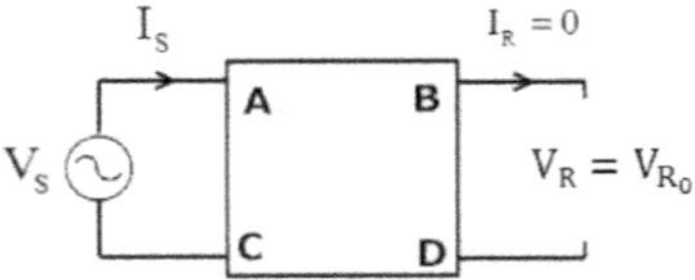

Figure 1.18 With receiving end open–circuited

As receiving end is open circuited, $I_R = 0$ and $V_R = V_{R0}$

From the basic two port equations,

$$V_S = AV_R + BI_R$$

$$V_S = A\ V_{R_0} \quad (\because I_R = 0) \qquad\qquad \dots..(1.5)$$

$$A = \frac{V_S}{V_{R_0}}$$

$$I_S = CV_R + DI_R$$

$$I_S = C\ V_{R_0} \qquad\qquad \dots..(1.6)$$

$$C = \frac{I_S}{V_{R_0}}$$

Divide (1.5) and (1.6):

$$\frac{V_S}{V_S} = \frac{A}{C} \qquad\qquad \dots..(1.7)$$

Case 2: With Receiving end short–circuited

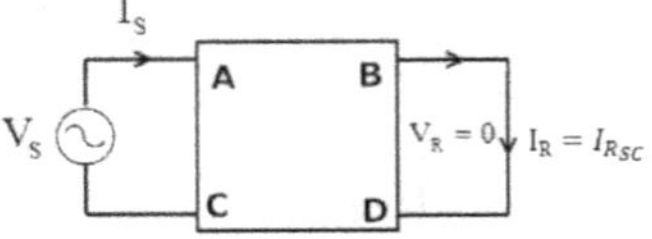

Figure 1.19 With receiving end short–circuited

As receiving end is short circuited, $V_R = 0$ and $I_R = I_{R_{SC}}$

$$V_S = AV_R + BI_R$$

$$V_S = B\ I_{R_{SC}} \qquad\qquad \dots..(1.8)$$

$$B = \frac{V_S}{I_{R_{SC}}}$$

$$I_S = CV_R + DI_R$$

$$I_S = D.I_{R_{SC}} \qquad\qquad .....(1.9)$$

$$D = \frac{I_S}{I_{R_{SC}}}$$

$$\frac{B}{D} = \frac{V_S}{I_S} \qquad\qquad .....(1.10)$$

Multiply (1.7) & (1.10)

$$\frac{V_S}{I_S} \times \frac{V_S}{I_S} = \frac{A}{C} \times \frac{B}{D}$$

$$\left(\frac{V_S}{I_S}\right)^2 = \frac{AB}{CD}$$

Assuming symmetrical line i.e., $A = D$

$$\left(\frac{V_S}{I_S}\right)^2 = \frac{B}{C}$$

$$\left(\frac{V_S}{I_S}\right) = \sqrt{\frac{B}{C}}$$

$$\left(\frac{V_S}{I_S}\right) = Z_C\sqrt{\frac{B}{C}} = \sqrt{\frac{Z_{SC}}{Y_{OC}}}$$

$$\frac{V_S}{I_S} = Z_C\sqrt{\frac{B}{C}} = \sqrt{\frac{Z_{SC}}{Y_{OC}}}$$

$$Z_C = \frac{V_S}{I_S} = \sqrt{Z_{sc}Y_{oc}}$$

Characteristic Impedance, Z_c: The Impedance of a transmission line with losses is known as Characteristic Impedance, Z_c

    * $Z_C = 400\ \Omega$ for overhead transmission lines

    * $Z_C = 40\ \Omega$ for underground cables

Surge impedance Z_S: The impedance of a transmission line without losses is known as Surge impedance Z_S.

    i.e., the impedance of an ideal transmission line.

1.8   CONDITIONS FOR ZERO VOLTAGE REGULATION AND MAXIMUM VOLTAGE REGULATION OF A TRANSMISSION LINE

* Consider a short transmission line, by taking I_R as reference.

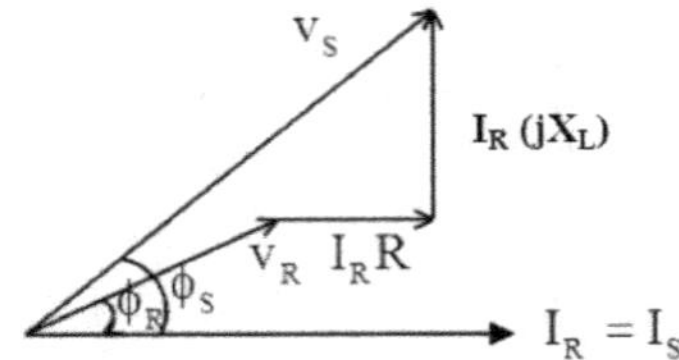

Figure 1.20 Vector diagram of short transmission line

From the above vector diagram,

$$V_S \cong V_R + I_R\,R\,\cos\phi_R + I_R\,X_L\,\sin\phi_R$$

$$V_S - V_R \cong I_R R \cos\phi_R + I_R X_L \sin\phi_R$$

$$\%\,\epsilon = \frac{V_S - V_R}{V_R} \times 100$$

$$= \frac{I_R R}{V_R}\cos\phi_R \times 100 + \frac{I_R X_L}{V_R}\sin\phi_R \times 100$$

$$= \left(\frac{I_R R}{V_R}\right)\cos\phi_R \times 100 + \left(\frac{I_R X_L}{V_R}\right)\sin\phi_R \times 100$$

$$= v_{rpu}\cos\phi_R \times 100 + v_{xpu}\sin\phi_R \times 100$$

$$= \%\,V_R \cos\phi_R + \%\,V_X \sin\phi_R \qquad\qquad .....(1.11)$$

In general, % Regulation

$$\%\,\epsilon = \%\,V_R \cos\phi_R + \%\,V_X \sin\phi_R$$

In general the sign is positive for lagging power factor load and negative for lead power factor load.

Case 1: Condition for maximum voltage regulation

Maximum %voltage regulation occurs at lagging power factor load

$$\%\,\epsilon = \%\,V_R \cos\phi_R + \%\,V_X \sin\phi_R$$

$$\frac{d}{d\phi_R}(\%\,\epsilon) = 0$$

$$\%\,V_R\,(-\sin\phi_R) + \%\,V_X\,(\cos\phi_R) = 0$$

$$\%\,V_R \sin\phi_R = \%\,V_X \cos\phi_R$$

$$\text{Tan }\phi_R = \frac{\%\,V_X}{\%\,V_R}$$

$$= \frac{\dfrac{I_R X_L \times 100}{V_R}}{\dfrac{I_R R \times 100}{V_R}}$$

$$\text{Tan } \phi_R = \frac{X_L}{R}$$

$$\phi_R = \text{Tan}^{-1}\left(\frac{X_L}{R}\right)$$

Impedance of the short line: $Z = R + j X_L$

Impedance Triangle:

Where θ = Impedance phase angle

From the Figure 1.21,

$$\text{Tan } \phi_R = \frac{X_L}{R}$$

$$\theta = \phi_R$$

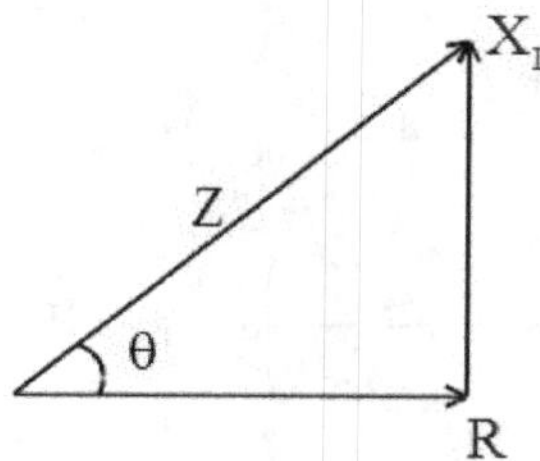

Figure 1.21 Impedance triangle of short line

Case 2: Condition for zero voltage regulation

* Zero voltage regulation occurs at leading p.f load st

$$\% \in = \% V_R \cos \phi_R - \% V_X \sin \phi_R$$

$$\% \in = 0 \Rightarrow \% V_R \cos \phi_R - \% V_X \sin \phi_R = 0$$

$$\% V_R \cos \phi_R - \% V_X \sin \phi_R$$

$$\text{Tan } \phi_R = \frac{\% V_R}{\% V_X} = \frac{\dfrac{I_R R}{V_R} \times 100}{\dfrac{I_R X_L}{R} \times 100} = \frac{R}{X_L}$$

$$\theta = \phi_R$$

$$\phi_R = \text{Tan}^{-1}\left(\frac{R}{X_L}\right) \qquad \qquad \dots\dots(1.12)$$

From Impedance triangle,

$$\text{Tan } \phi_R = \frac{R}{X_L} = \cot\theta = \tan\left(\frac{\pi}{2} - \theta\right)$$

$$\phi_R = \frac{\pi}{2} - \theta$$

1.9   RELATION BETWEEN ACTUAL LOADING OF TRANSMISSION LINE AND SURGE IMPEDANCE LOADING

(A)   **Characteristic Impedance Loading (CIL):** Consider a transmission line with losses connected to a load as in the Figure 1.22.

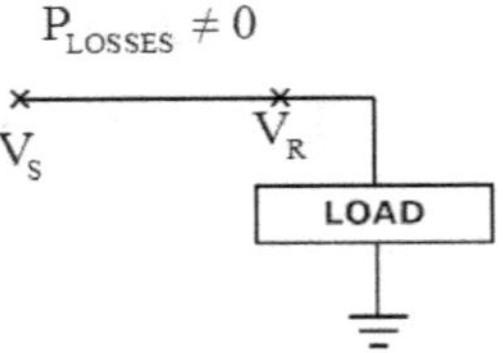

Figure 1.22 A practical transmission line

$$CIL = \left| \frac{|V_S| \, |V_R|}{Z_C} \right| \, W \text{ (or) kW (or) MW}$$

"The maximum active power transmitted through a line with losses and further through the load at unity power factor is known as Characteristic Impedance Loading" (CIL).

(B)   **Surge Impedance Loading (SIL):** Consider a transmission line without losses connected to a load as in the figure 1.23.

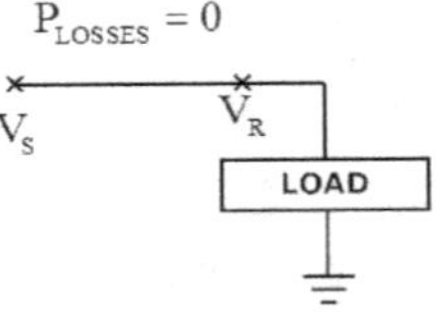

Figure 1.23 An ideal transmission line

$$SIL = \left| \frac{|V_S| \, |V_R|}{Z_S} \right| = \frac{|V|^2}{Z_S} \, W \text{ or kW or MW}$$

$$\because \qquad |V_S| = |V_R| = |V|$$

"The maximum active power transmitted through a loss less line and further to the Load at unity p.f. is known as Surge Impedance Loading"

Condition 1 – Actual Loading of a transmission line is greater than surge impedance Loading of transmission line

1. As power carrying capacity increases, current flowing through the transmission line increases, electromagnetic energy stored by inductor in the magnetic field $= \dfrac{1}{2} LI^2$ increases

   $\therefore$ Inductor is dominant, p.f is Lagging, $|V_R| < |V_S|$ and Ferranti effect do not occur.

Condition 2 – Actual Loading Actual Loading of a transmission line is lesser than surge impedance Loading of transmission line

1. As power carrying capacity decreases, current flowing through the transmission line decreases & Electromagnetic energy stored by inductor in the magnetic field $= \dfrac{1}{2}LI^2$ decreases

   $\therefore$ Capacitor is dominant, p.f is Leading, $|V_R| > |V_S|$ and Ferranti effect occurs.

   $\therefore$ Ferranti effect can be eliminating by loading the transmission line beyond their surge impedance loading capacity.

Q1. The sending end voltage & receiving end voltage of a transmission line are 220kV, 200kV respectively. Determine the Characteristic Impedance Loading of transmission line.

Solution:

$$CIL = \frac{|V_S||V_R|}{Z_C}$$

$$CIL = \frac{(220 \times 10^3)(200 \times 10^3)}{400}$$

$$CIL = 110MW$$

Q2. Determine the surge impedance loading of an underground cable operating at 400 kV.

Solution:

$$SIL = \frac{|V|^2}{Z_S} = \frac{(400 \times 10^3)}{40}$$

$$SIL = 4000MW$$

Q3. The open circuit & short circuit impedance of a transmission line are $16 \times 10^4\ \Omega$, & $1\ \Omega$ respectively. Determine the characteristic impedance of the transmission line.

Solution:

$$Z_C = \sqrt{Z_{OC}Z_{SC}} = \sqrt{16 \times 10^4 \times 1}$$

$$Z_C = 400\ \Omega$$

Q4. The ABCD constants of a 220kV transmission line are $A = D = 0.94\underline{|1^0}\ \underline{|1^0}$, $B = 130\underline{|73^0}$ (Ω) and $c = 0.0001\underline{|90^0}\ (\Omega)$. If the sending voltage of the Line for a given Load delivered at nominal voltage is 240kV then %voltage regulation of the line is

Solution:

For a No – Load condition $I_R = 0$

$$V_S = AV_R + BI_R = A\ V_{R_o}$$

$$V_{R_o} = \frac{V_S}{A}$$

$$\% \in = \frac{|V_{R_0}| - |V_R|}{|V_R|} \times 100$$

$$= \frac{\left|\dfrac{V_S}{A}\right| - |V_R|}{|V_R|} \times 100$$

$$= \frac{\left|\dfrac{240}{0.94}\right| - |220|}{|220|} \times 100$$

$$\% \in = 16.05\%$$

Q5. The ABCD–constants of a three phase transmission line are $A = D = 0.8\angle 1^0$, $B = 170\angle 85^0$ Ω, $C = 2 \times 10^{-3} \angle 90.4^0$ S. The sending end voltage is $V_S = 400$kV. Determine the receiving end voltage under no–load condition.

Solution:

No – Load receiving end voltage,

$$V_{R_0} = \frac{V_S}{A} \quad (\because \text{No–Load condition } I_R = 0)$$

$$= \frac{400}{0.8} \quad V_S = AV_R + BI_R$$

$$V_{R_0} = 500 \text{ kV}$$

Q6. A 220kV transmission line is represented by nominal π–parameters $A = 0.9\angle 5^0$, $B = 80\angle 65^0$ Ω, the sending end voltage is maintained at 220kV. Calculate the rise in voltage

Solution:

The no–load receiving end voltage is

$$V_{R_0} = \frac{V_S}{A}$$

$$= \frac{200}{0.9}$$

$$V_{R_0} = 244.4 \text{ kV}$$

The no–load rise in the voltage is

$$V_{R_0} - V_S = 244.4 - 220 = 24.4 \text{kV}$$

Q7. For a 500Hz frequency excitation, how can a 50km long power transmission line be modeled ?

Solution:

As frequency increases by 10 times, physical length decreases by 10 times,

Short line: $0 - 8$ km

Medium line: $8 - 16$ km

Long line: greater than 16 km

$\therefore$ 50 km line can be modeled as a long line.

Q8. In the matrix form, equations of a 4– terminal network representing a transmission line is given by

$$\begin{bmatrix} V_S \\ I_S \end{bmatrix} = \begin{bmatrix} A & B \\ C & D \end{bmatrix} \begin{bmatrix} V_R \\ I_R \end{bmatrix}$$

The two –networks considered are

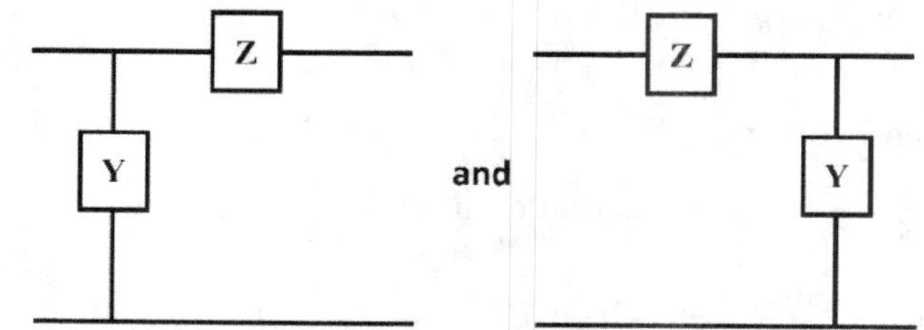

Determine the matrices for the networks A & B

Solution:

$$\begin{bmatrix} A & B \\ C & D \end{bmatrix} = \begin{bmatrix} 1 & 0 \\ Y & 1 \end{bmatrix}\begin{bmatrix} 1 & Z \\ 0 & 1 \end{bmatrix} = \begin{bmatrix} 1 & Z \\ Y & 1+YZ \end{bmatrix}$$

$$\begin{bmatrix} A & B \\ C & D \end{bmatrix} = \begin{bmatrix} 1 & Z \\ 0 & 1 \end{bmatrix}\begin{bmatrix} 1 & 0 \\ Y & 1 \end{bmatrix} = \begin{bmatrix} 1+YZ & Z \\ Y & 1 \end{bmatrix}$$

Q9. Two networks are connected as shown in the figure, the equivalent ABCD–constants are further obtained, given that $Z_1 = 10\lfloor 30^0$ Ω, C= $0.025\lfloor 45^0$, find the value of Z_2?

$$\begin{bmatrix} A & B \\ C & D \end{bmatrix} = \begin{bmatrix} 1 & Z_1 \\ 0 & 1 \end{bmatrix}\begin{bmatrix} 1 & 0 \\ \dfrac{1}{Z_2} & 1 \end{bmatrix}$$

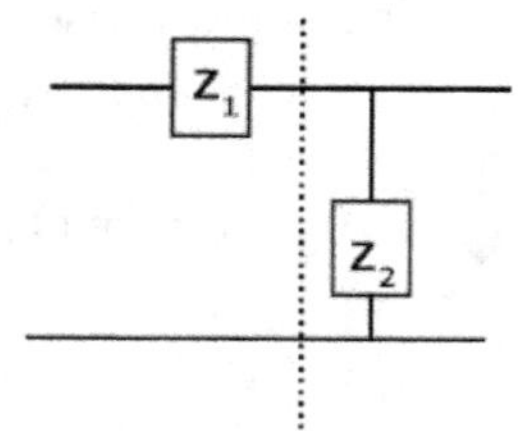

$$\begin{bmatrix} 1 + \dfrac{Z_1}{Z_2} & Z_1 \\ 1/Z_2 & 1 \end{bmatrix}$$

$$C = \frac{1}{Z_2} \Rightarrow Z_2 = \frac{1}{C} = \frac{1}{0.025\underline{|45^0}}$$

$$Z_2 = 40\underline{|30^0}\ \Omega$$

Q10. A 220kV, 20km Long three phase transmission line has the following ABCD constants. A $= D = 0.96\underline{|3^0}$, $B = 55\underline{|65^0}\ \Omega$, $C = 0.510^{-4}\underline{|80^0}$ S. Determine the charging current per phase.

Solution:

Definition: The source current with receiving end open–circuited in source condenser method is called "Charging Current".

$$I_R = 0$$

$$I_C = I_S = CV_R + DI_R$$

$$= CV_R$$

$$= (0.5 \times 10^{-4})\left(\frac{220\times 10^3}{\sqrt{3}}\right)$$

$$= \frac{11}{\sqrt{3}}\ A$$

Q11. A 50HZ, 3–ϕ transmission line of length 100km has a capacitance of $\dfrac{0.03}{\pi}\dfrac{\mu F}{km}$. It is represented by π– model. Determine the shunt admittance at each end of the transmission line?

Solution:

$$C = \frac{0.03}{\pi}\frac{\mu F}{km} \times 100km$$

$$C = \frac{3}{\pi}\mu F$$

The shunt admittance at the end of each transmission line is

$$\frac{Y}{2} = \frac{1}{2}(jB_C) = \frac{1}{2}(j\omega_C) = \frac{1}{2}(j2\pi \times 50 \times \frac{3}{\pi} \times 10^{-6} = j150 \times 10^{-6}\,S$$

Q12. The Generalized circuit constants of a 3–ϕ, 220kv rated voltage, medium length line are A $= D = 0.936\underline{|98^0}$, $B = 142\underline{|76.4^0}\ \Omega$, load at the receiving end is 50MW at 220kV with a p.f. of 0.9(lagging). Calculate the magnitude of Line–to–Line sending end voltage?

Solution:

$$V_R = 220kV$$

$$V_S = AV_R + BI_R$$

$$= (0.936\angle 98^0)\left(\frac{220}{\sqrt{3}}\right) + (142\underline{|76.4^0})I_R$$

$$P_R = \sqrt{3}\, V_R I_R \cos\phi_R$$

$$50 \times 10^6 = \sqrt{3} \times 220 \times 10^3 \times I_R \times 0.9$$

$$I_R = 145.79A = 145.79\underline{|{-}\cos^{-1}(0.9)} = 145.79\underline{|{-}25.84^0}\ \ A$$

$$V_{S_{L-N}} = A\, V_R + B\, I_R = (0.936\underline{|98^0})\left(\frac{220}{\sqrt{3}} \times 10^3\right) + (142\underline{|76.4^0}\ (145.79\underline{|{-}25.84^0}$$

$$V_{S_{LL}} = \sqrt{3}V_{S_{L-N}} = \sqrt{3} \times 133kV = 233kV$$

Q13. Calculate the % rise in voltage at the receiving end of transmission line of length 200km, operating at 50Hz?

Solution:

$$\% \text{ rise in voltage } = \frac{V_S - V_R}{V_R} \times 100 = \frac{\omega^2 l^2}{18} \times 10^{-8}$$

$$= \frac{(2\pi \times 50)^2 (200)^2}{18} \times 10^{-8}$$

$$= 2.1932\%$$

Q14. Calculate the time taken by voltage wave to travel 600km long overhead transmission line is.

Solution:

$$\text{Time, } t = \frac{\text{length, } l}{\text{velocity, } v} = \frac{600}{3 \times 10^5} = 2 \text{ msec}$$

Q15. A transmission line is having resistance 18 Ω & reactance 12 Ω and supplies a load of 5MW at a voltage 'v'. The supply voltage is "V_S". If $V = V_s$ then determine the p.f. of the load.

Solution:

$$V_R = V \text{ and } V_s = V$$

$$V_S = V_R + I_R R \cos\phi_R \pm IR\, X_L \sin\phi_R$$

for $V_S = V_R \Rightarrow I_R R \cos\phi_R - I_R X_L \sin\phi_R = 0$

$$\tan \phi_R = \frac{R}{X_L} = \frac{18}{12} = 1.5 \rightarrow \tan^{-1}(1.5) = \phi_R \rightarrow \phi_R = 56.3$$

$\cos \phi_R = 0.53$, Lead

1.10 INTERCONNECTION OF TRANSMISSION LINE

1.10.1   Transmission Lines Connected in Cascade

The cascade connection of transmission lines is shown in the figure 1.24.

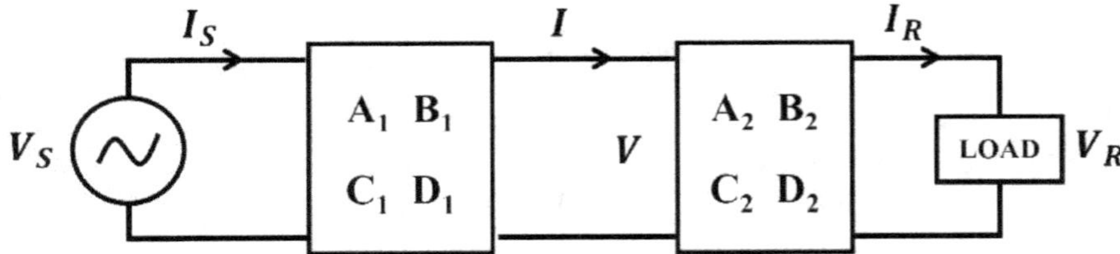

Figure 1.24 Cascade connection of transmission lines

$$\begin{bmatrix} V_S \\ I_S \end{bmatrix} = \begin{bmatrix} A_1 & B_1 \\ C_1 & D_1 \end{bmatrix}\begin{bmatrix} V \\ I \end{bmatrix}$$

$$\begin{bmatrix} V_S \\ I_S \end{bmatrix} = \begin{bmatrix} A_1 & B_1 \\ C_1 & D_1 \end{bmatrix}\begin{bmatrix} A_2 & B_2 \\ C_2 & D_2 \end{bmatrix}\begin{bmatrix} V_R \\ I_R \end{bmatrix}$$

$$\begin{bmatrix} V_S \\ I_S \end{bmatrix} = \begin{bmatrix} A_0 & B_0 \\ C_0 & D_0 \end{bmatrix}\begin{bmatrix} V_R \\ I_R \end{bmatrix}$$

$$\begin{bmatrix} A_0 & B_0 \\ C_0 & D_0 \end{bmatrix} = \begin{bmatrix} A_1 & B_1 \\ C_1 & D_1 \end{bmatrix}\begin{bmatrix} A_2 & B_2 \\ C_2 & D_2 \end{bmatrix}$$

Q16. Two transmission lines are connected in cascade, whose ABCD parameters are

$$\begin{bmatrix} A_1 & B_1 \\ C_1 & D_1 \end{bmatrix} = \begin{bmatrix} 1 & 10\underline{|30^\circ} \\ 0 & 1 \end{bmatrix} \text{ and } \begin{bmatrix} A_2 & B_2 \\ C_2 & D_2 \end{bmatrix} = \begin{bmatrix} 1 & 0 \\ 0.025\underline{|-30^\circ} & 1 \end{bmatrix}.$$

Determine the resultant ABCD Parameters.

Solution:

$$\begin{bmatrix} A_0 & B_0 \\ C_0 & D_0 \end{bmatrix} = \begin{bmatrix} A_1 & B_1 \\ C_1 & D_1 \end{bmatrix}\begin{bmatrix} A_2 & B_2 \\ C_2 & D_2 \end{bmatrix} = \begin{bmatrix} 1 & 10\underline{|-30^\circ} \\ 0 & 1 \end{bmatrix}\begin{bmatrix} 1 & 0 \\ 0.025\underline{|-30^\circ} & 1 \end{bmatrix}$$

1.10.2 Parallel Connection of Transmission Lines

The parallel connection of transmission lines is shown in the figure 1.25.

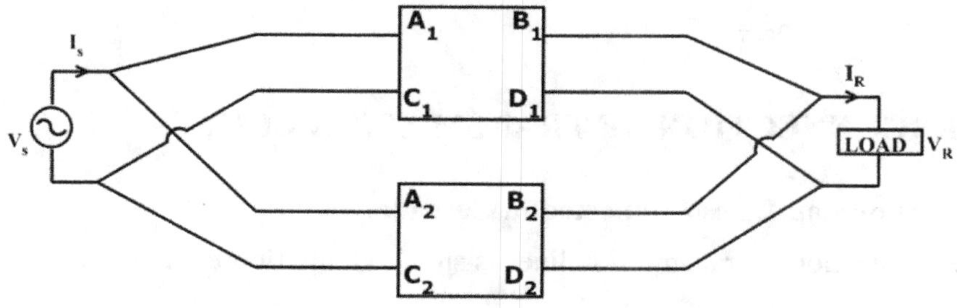

Figure 1.25 Parallel connection of transmission lines

$$A_0 = \frac{A_1 B_2 + A_2 B_1}{B_1 + B_2} \quad D_0 = \frac{D_1 B_2 + D_2 B_1}{B_1 + B_2}$$

$$B_0 = \frac{B_1 B_2}{B_1 + B_2} \quad C_0 = C_1 + C_2 + \frac{(A_1 - A_2)(D_2 - D_1)}{B_1 + B_2}$$

Case – With identical transmission lines connected in parallel

$$A_1 = A_2 = A \;;\; B_1 = B_2 = B \;;\; C_1 = C_2 = C;\; D_1 = D_2 = D$$

$$A_O = A$$

$$D_O = D$$

$$B_O = \frac{B}{2}$$

$$C_O = 2C$$

Q17. A medium line with parameters ABCD is extended by connecting a short line of impedance Z in series. Determine the overall ABCD–parameters of the series combination.

Solution:

Assume nominal–T line.

$$A_{new} = 1 + \frac{YZ}{2} = 1 + YZ_1 = A_{old}$$

$$C_{new} = Y = C_{old}$$

$$B_{old} = Z\left(1 + \frac{YZ}{4}\right) = Z + \frac{YZ^2}{4}$$

$$= \left(\frac{Z}{2} + \frac{Z}{2}\right) + Y\left(\frac{Z}{2} \cdot \frac{Z}{2}\right)$$

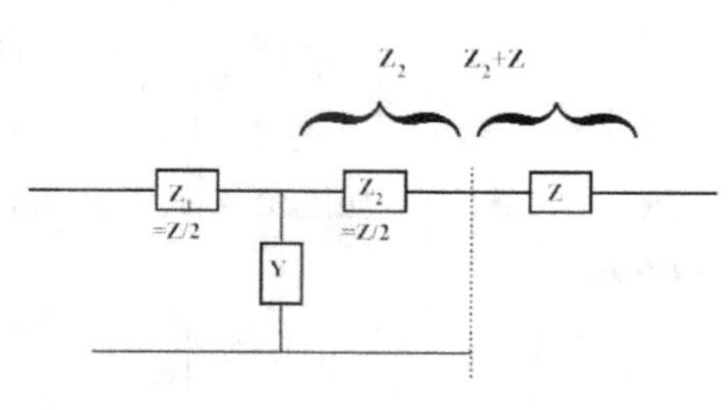

$$(Z_1 + Z_2) + Y(Z_1.Z_2)$$

$$B_{new} = \{Z_1 + (Z_2 + Z)\} + + YZ_1(Z_2 + Z)$$

$$= \{(Z_1 + Z_2 + YZ_1Z_2)\} + \{Z + YZ_1Z\}$$

$$B_{new} = B_{old} + ZA_{old}$$

$$D_{old} = 1 + YZ_2 = 1 + \frac{YZ}{2}$$

$$= 1 + Y(Z_2 + Z)$$

$$= 1 + YZ_2 + YZ$$

$$= D_{old} + YZ$$

$$D_{new} = D_{old} + Z\,C_{old}$$

OBJECTIVE QUESTIONS

1. A transmission line consists of R,L in ______________ and G,C in __________ .
2. Transmission lines are primarily classified based on ______________ of the travelling wave.
3. Transmission lines are secondarily classified based on ________________________________ .
4. Ferranti effect do not occur in ___________________ transmission line.
5. Ferranti effect occurs with receiving end ___________________________ .
6. Ferranti effect results in no–load receiving end voltage ____________than sending end voltage.
7. The impedance of a practical transmission line is known as ___________________________ .
8. The impedance of an ideal transmission line is known as ______________________________ .
9. The characteristic impedance of an overhead transmission line is ___________ohms.
10. The characteristic impedance of an underground cable is ______________ ohms.

11. The maximum active power transmitted through a practical transmission line is known as ________________________________ .

12. The maximum active power transmitted through an ideal transmission line is known as _____________________ .

13. ______________________________ varies inversely to the length of the line.
14. ______________________________ is independent of the length of the line.
15. The constants A,C can be determined with receiving ___________________ .
16. The constants B,D can be determined with receiving end ______________________________ .
17. Characteristic impedance is the geometric mean of ___________________ respectively.

2   Per Unit Quantities

2.1   INTRODUCTION

In chapter −1, the basics of transmission line has been discussed. In this chapter, the per unit quantities and their representation in power system network shall be discussed. In the earlier days, absolute system has been adopted to analyze the power system network. But due to increase in complexity of the power system network, absolute system of analysis has been replaced by per unit system of analysis.

2.2   PER UNIT REPRESENTATION OF POWER SYSTEM COMPONENTS

2.2.1 Absolute System of Analysis

In absolute system of anlaysis, number of equations required to analyze power system network equal no. of voltage levels.

Case 1: Consider an alternator connected directly to a load as in figure 2.1.

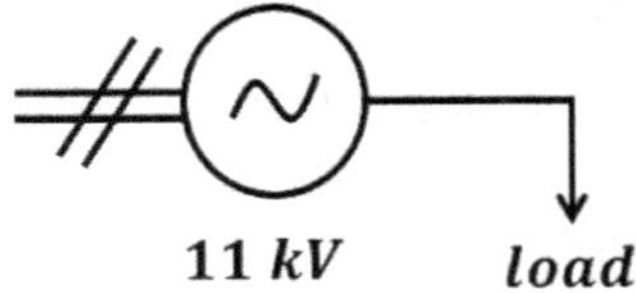

Figure 2.1 Alternator connected to load

As shown in the Figure 2.1., number of operating voltage levels = 1

No. of equations required to analyze the performance of the power system network = 1

Case 2: Consider an alternator connected to load through step up transformer, transmission line and load as in Figure 2.2.

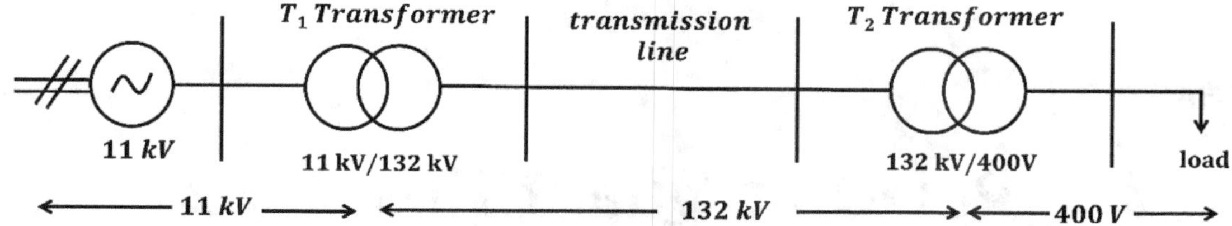

Figure 2.2 Alternator connected to load through step up transformer, transmission line and load

As shown in the Figure 2.2, number of operating voltage levels = 3

No. of equations required to analyze the performance of the power system network = 3

The modern power system network operates at multiple voltage levels therefore no. of equations increases and therefore no. voltages increases and evaluation time increases.

To overcome this limitation, absolute system of analysis is replaced with power system analysis.

Absolute system	Per unit system (p.u.)
All electrical quantities have specified units	All electrical quantities expressed in per unit
All power system components are analysed sequentially	All power system components are analysed simultaneously
Evaluation time is high	Evaluation time is low
No. of equations required = No. of voltage levels	No. of equations = 1 irrespective of number of voltage levels

$$\text{Per unit value} = \frac{\text{Actual value of an electrical quantity expressed in specified units}}{\text{Base value of an electrical quantity expressed in same unit}}$$

Selection of base value is optional. However, alternator ratings are selected as base value since alternator is the first component in the power system network.

2.3 EXPRESSING PER UNIT REACTANCE IN TERMS OF BASE VOLTAGE AND BASE CURRENT

Consider two machines.

Let machine – 1 ratings be V_1, I_1

Machine – 2 ratings be V_2, I_2

Let $\qquad V_2 = V_b$, $I_2 = I_b$

Per unit voltages:

$$V_{1\,p.u.} = \frac{V_1}{V_b} = \frac{V_1}{V_2}$$

$$V_{2\,p.u.} = \frac{V_2}{V_b} = \frac{V_2}{V_2} = 1\,\text{p.u.}$$

Per unit currents:

$$I_{1\ p.u.} = \frac{I_1}{I_b} = \frac{I_2}{I_2}$$

$$I_{2\ p.u.} = \frac{I_2}{I_b} = \frac{I_2}{I_2} = 1 p.u.$$

Base reactance (Ω)

$$X_b = \frac{V_b}{I_b}$$

Per unit reactances:

$$X_{1p.u.} = \frac{X_1}{X_b} = \frac{X_1}{\dfrac{V_b}{I_b}} = X_1\left(\frac{I_b}{V_b}\right)$$

$$X_{2p.u.} = \frac{X_2}{X_b} = \frac{X_2}{\dfrac{V_b}{I_b}} = X_2\left(\frac{I_b}{V_b}\right)$$

$$X_{p.u.} = X_{in}\left(\frac{I_b}{V_b}\right) \qquad\qquad .....(2.1)$$

2.4 EXPRESSING PER UNIT REACTANCE IN TERMS OF BASE VOLT AMPERE AND BASE VOLTAGE

From equation

$$X_{p.u.} = X\frac{I_b}{V_b}\frac{\dfrac{V_b}{10^6}}{\dfrac{V_b}{10^6}}$$

$$= X\frac{\dfrac{I_b V_b}{10^6}}{\dfrac{V_b^2}{10^6}} \qquad\qquad .....(2.2)$$

$$X_{p.u.} = X\frac{MVA}{(kV_b)^2} \qquad\qquad .....(2.3)$$

2.5 EXPRESSING BASE REACTANCE (Ω) IN TERMS OF BASE VA AND BASE CURRENT

From equation ....(2.4)

$$X_b = \frac{V_b}{I_b}$$

$$= \frac{V_b \times \dfrac{V_b}{10^6}}{\dfrac{I_b \times V_b}{10^6}}$$ ....(2.5)

$$= \frac{\left(\dfrac{V_b}{10^3}\right)^2}{\dfrac{I_b \times V_b}{10^6}}$$

$$X_b = \frac{(kV_b)^2}{MVA_b}$$ ....(2.6)

2.6 EXPRESSING $X_{P.U., NEW}$ IN TERMS OF $X_{P.U., OLD}$

From equation (2.5)

$$X_{p.u.(new)} = X_{in\ ohms}\left(\frac{MVA_{b(new)}}{kV^2_{b\ (new)}}\right)$$

$$X_{p.u.(old)} = X_{in\ ohms}\left(\frac{MVA_{b(old)}}{kV^2_{b\ (old)}}\right)$$

$$X_{p.u.(new)} = X_{p.u.\ old}\left[\frac{MVA_{b(new)}}{MVA_{b(old)}}\right]\left[\frac{kV^2_{b\ (old)}}{kV^2_{b\ (new)}}\right]$$ ....(2.7)

Q1. The base impedance & base voltage for a given power system are 5 Ω and 200 V. Calculate base KVA and base current?

Solution:

$$\text{Base current} = \frac{\text{Base voltage}}{\text{Base impedance}} = \frac{200}{5} = 40\ A$$

$$\text{Base Volt Amperes} = V_b * I_b = 200 * 40 = 8000\ VA = 8\ KVA$$

Q2. The base current and base voltage of 345 KV system are choosen to be 2000 A & 200 KV respectively. Determine per unit voltage and the base impedance for the system?

Solution:

$$\text{Base Impedance, } Z_b = \frac{V_b}{I_b} = \frac{200 \times 1000}{200} = 1000\,\Omega$$

$$\text{Per unit voltage} = \frac{\text{Actual voltage}}{\text{Base voltage}} = \frac{345\text{ KV}}{200\text{ KV}} = 1.725 \text{ p.u.}$$

Q3. If the rating of the system is 1380 MVA (1–Φ). Calculate the per unit current reduced to the base of question-2?

Solution:

$$\text{Per unit current} = \frac{\text{Actual current}}{\text{Base current}}$$

$$\text{Volt Ampere} = V_A I_A$$

$$1380 = 345 * I_A$$

$$I_A = 4000 \text{ A}$$

$$I \text{ p.u.} = \frac{4000}{2000} = 2 \text{ A}$$

Q4. Determine a 80 Ω impedance, 50 A current & 240 V voltage as per unit quantities referred to the base values of Q_1

Solution:

From Question 1: $Z_b = 5\ \Omega$, $V_b = 200$ V, $I_b = 40$ A

$$\text{p.u. impedance} = \frac{\text{Acutal impedance}}{\text{Base impedance}} = \frac{80}{5} = 16 \text{ p.u.}$$

$$\text{p.u. voltage} = \frac{\text{Acutal voltage}}{\text{Base voltage}} = \frac{240}{200} = 1.2 \text{ p.u.}$$

$$\text{p.u. current} = \frac{\text{Acutal current}}{\text{Base current}} = \frac{50}{40} = 1.25 \text{ p.u.}$$

Q5. A 1–Φ, 10 kVA, 200 V generator has an internal resistance of 1 Ω using the ratings of generator as base values. Determine the generated per unit voltage that is required to produced full load current under short circuit condition?

Solution:

$$\text{Base voltage} = 200 \text{ V} = 1 \text{ p.u.}$$

$$Z_G = 1\ \Omega$$

$$kVA_b = 10 \text{ KVA} = 1 \text{ p.u.}$$

$$I_b = \frac{10 \times 1000}{100} = 100 \text{ A} = 1 \text{ p.u.}$$

Generator voltage required to produce the rated current under short circuit $= I_G Z_G = 100 \times 1 = 100$ volt.

Q6. Consider at 10 KVA, 200/100 V transformer (1–ϕ) be approximately represented by 2Ω reactance required to lower voltage side considering the rated values as base quantities express the transfer reactance as a p.u. quantity?

Solution:

$$\text{Base volt amperes} = 10000 \text{ VA}$$

$$\text{Base voltage} = 100 \text{ V}$$

$$\text{Base current} = \frac{V_{Ab}}{V_b} = \frac{10000}{100} = 100 \text{ A}$$

$$Z_b = \frac{100}{100} = 1 \ \Omega$$

$$\text{Per unit referred to LV side, Z p.u. (LV)} = \frac{2}{1} = 2 \text{ p.u.}$$

Q7. Repeat Q6 expressing all quantities in terms of high voltage side?

Solution:

$$\text{Base voltage} = 200 \text{ V}$$

$$\text{Base current} = \frac{10000}{200} = 50 \text{ A}$$

$$\text{Base impedance} = \frac{\text{Base voltage}}{\text{Base current}} = \frac{200}{50} = 4 \text{ A}$$

$$Z \text{ actual HV side} = \left(\frac{\text{HV side}}{\text{LV side}}\right)^2 Z \text{ actual LV side}$$

$$= \left(\frac{200}{100}\right)^2 2 = 8 \ \Omega$$

$$\text{p.u. impedance referred to HV side, } Z_{\text{p.u. HV side}} = \frac{Z_{\text{acutal}} \text{ HV}}{Z_{\text{base}}} = \frac{8}{4} = 2 \text{ p.u.}$$

Note: The per unit impedance measured with respect to low voltage or high voltage side is same.

Q8. A 345 kV transmission line has a series impedance of $(4 + j60) \ \Omega$ and a shunt admittance of $j\,2*10^{-3}$ S using 50 MVA and line voltage has base values, calculate the per unit impedance & per unit admittance of the line?

Solution:

$$\text{Base MVA} = 50 \text{ MVA}$$

$$\text{Base KV} = 345 \text{ KV}$$

$$Z_{p.u.} = \frac{MVA_b}{(KV_b)^2} \times Z_b = \frac{50}{(345)^2} \times (4 + j60) = (1.65 + j25.2) \times 10^{-3} \, p.u.$$

$$Y_{p.u.} = Y \times \frac{(KV_b)^2}{100} = j2 \times 10^{-3}(3.45)^2 = j2.38 \, p.u,$$

Q9. A 3–Φ, Y connected system is rated at 40 MVA, 120 kV. Express 40 MVA of 3–Φ apparent power as a 3–Φ p.u. value referred to 3–Φ per kVA as base, per phase system kVA?

Solution:

(a) For 3 phase base

$$kVA_b = 40000 \ KVA = 1 \ p.u.$$

$$kV_b = 120 \ LV \ (L\text{–}L) = 1 \ p.u.$$

$$p.u. \ kVA = \frac{40000}{40000} = 1 \ p.u.$$

(b) For one phase base

$$kVA_b = \frac{40000 \ KVA}{3} = 13333 \ kVA = 1 \ p.u.$$

$$kV_b = \frac{120 \times 10^3}{\sqrt{3}} = 69.28 \ KV$$

$$p.u. \ kVA = \frac{13334}{13333} = 1 \ p.u.$$

For a three-phase star–Y connected system, p.u. values remain same evaluated with respect to three phase base or per phase base.

Q10. A 3 phase 6.5 KV transmission line delivers 8 MVA of load. The per phase impedance of the line is 0.01+j 0.05 p.u. referred to a 13 KV, 8 MVA base. What is the voltage drop across the line?

Solution:

$$\text{Base KVA} = 8000 \ KVA = 1 \ p.u.$$

$$\text{Base KV} = 13 \ KV = 1 \ p.u.$$

$$(VA)_b = \sqrt{3V_b I_b}$$

$$8000 \times 10^3 = \sqrt{3} \times 13 \times 10^3 \times I_b$$

$$I_b = 355.3 = 1 \ p.u.$$

$$\text{Base impedance, } Z_b = \frac{\text{Base voltage}}{I_b} = \frac{6500}{355.3} = 18.3 \ \Omega$$

$$Z \ p.u. = 0.01 + j0.05$$

Actual impedance in (Ω) = Z p.u. $\times$ Z_b = $(0.01 + j\,0.05) \times 18.3 = (0.183 + j0.915)\,\Omega$

Voltage drop = $Z_A \times I_b = (0.183 + j0.915) \times 355.3 = 65 + j325 = 331.55$ V

Q11. A 3–ϕ, Y connected 6.25 KVA, 220 V synchronous generator has a reactance of 8.4 Ω/ph. Using the rated KVA & voltage as base values, determine the p.u. reactance? Express this per unit values to a 750 KVA and 220 V base.

Solution:

(a)
$$kVA_b = 6.25 \text{ KVA} = 1 \text{ p.u.}$$
$$Z_b = 8.4 \text{ } \Omega/ph = 1 \text{ p.u.}$$
$$V_b = 220 \text{ V} = 1 \text{ p.u.}$$
$$I_b = \frac{KVA_b}{V_b} = \frac{6.25 \times 1000}{\sqrt{3} \times 220} = 16.4 \text{ A} = 1 \text{ p.u.}$$
$$Z_b = \frac{V_b}{I_b} = 13.41 \text{ } \Omega = 1 \text{ p.u.}$$
$$Z_{p.u.} = \frac{8.4}{13.41} = 0.626 \text{ p.u.}$$

(b)
$$X_{p.u.new} = X_{p.u.old} \times \left[\left(\frac{MVA_{b,new}}{MVA_{,old}} \right) \times \left(\frac{KV_{b,old}}{KV_{b,new}} \right)^2 \right]$$

$$= 0.027 \times \left[\left(\frac{7500}{6250} \right) \times \left(\frac{220}{220} \right)^2 \right] = 0.0324 \text{ p.u.}$$

Q12. A portion of a power system consists of 2 generators in parallel connected to a step–up transformer that links them with a 230 KV transmission line. The ratting of the components are,

G_1:   100 MVA, % X =12%

G_2:   5 MVA, % X = 8%

Transformer: 15 MVA, % X = 6%

Transmission line $(4 + j60)\,\Omega$, 230 KV

where the % reactances are computed on the basis of individual component rating.

Express the reactance and impedance in % with 15 MVA as base value.

Solution:

For G_1,

$$\% X = 100 X_{p.u.}$$
$$= X\,(\Omega) \times \frac{MVA_b}{(KV_b)^2}$$

$\%\ X \propto MVA_b$

$12\% \propto 10\ MVA_b$

$\%\ X_{G1} \propto 15\ MVA_b$

$$\frac{12}{X_{G1}} = \frac{10}{15}\ ,\ X_{G1} = 18\ \%$$

For G_2,

$\%X = 8\%$ ——————for 5 MVA

$? <$ —————— for 15 MVA

$X_{G2} = 24\%$

For transformer

$\%X = 6\%$————————→ 15 MVA

$? <$ ——————15 MVA

$\%X$ of transformer $= 6\%$

For Transmission line

$\%\ Z_{TL} = 100\ Z$ p.u.

$$= 100 \times Z\ \Omega \times \frac{MVA_b}{(kV_b)^2}$$

$$= 100 \times (4 + j\ 60) \times \frac{15 \times 10000000}{(230 \times 1000)^2} = (0.11 + j\ 0.17)\ \text{p.u.}$$

$\%X_{TL} = 17\%$

OBJECTIVE QUESTIONS

1. Define per unit value?

   Ans: The per unit value of any quantity is defined as the ratio of the actual value of the quantity to the base value expressed as a decimal.

2. What are the quantities whose base values are required to represent the power system by reactance diagram?

   Ans: The base values of voltage, current, power and impedance are required to represent the power system by reactance diagram. Usually the base values of voltage and power are chosen in KV and MVA. The base values of current and impedance are calculated using the chosen values.

3. What is the need for base values?

   Ans: The power system components may operate at different voltage levels and power levels. It will be convenient for analysis of power system if the voltage, power, current and impedance ratings of power system components are expressed with reference to a common value called base value.

4. What are the advantages of per unit computations?

   Ans:

   (a)   Manufacturers usually specify the impedance of a device or machine in per unit on the base of the name plate rating.

   (b)   The per unit values of widely different rating machines lie within a narrow range, even though the ohmic values has a very large range.

   (c)   c)The per unit impedance of a circuit element connected by transformers expressed on a proper base will be same referred to either side of the transformer.

   (d)   d) The per unit impedance of a three phase transformer is independent of the type of winding connection either star or delta.

5. All the electrical quantities have specified units in _______________ of analysis.

6. All the electrical quantities are expressed in per unit in _______________ of analysis.

7. The power system components are analyzed simultaneously in _______________ of analysis.

8. In absolute system of analysis, number of equations required to analyze the performance of power system network is equal to number of _______________.

9. In per unit system of analysis, number of equations required to analyze the performance of power system network is equal to _______________.

10. The evaluation time of power system network is less in _______________ of analysis.

11. Actual value of an electrical quantity is also known as _______________.

12. Base value of an electrical quantity is also known as _______________.

13. As _______________ is the first component in power system network, _______________ ratings are considered as base values in per unit system analysis.

14. _______________ value do not depend on the base value.

15. The per unit primary and secondary reactance of a transformer are _______________.

16. Base MVA refers to _______________ phase value in per unit system of analysis.

17. Base KV refers to _______________ voltage in per unit system of analysis.

18. Percentage reactance is directly proportional to _______________ and inversely proportional to the _______________.

3 Graph Theory

3.1 INTRODUCTION

The geometric structure of a network can be described by replacing the network components by single line segments irrespective of the characteristic of the components.

These line segments are known as "elements" and their terminals are known as "nodes".

Graph theory or network topology is the study of interconnection of network elements in the power system network.

Example: Consider the power system network as shown in the figure 3.1.

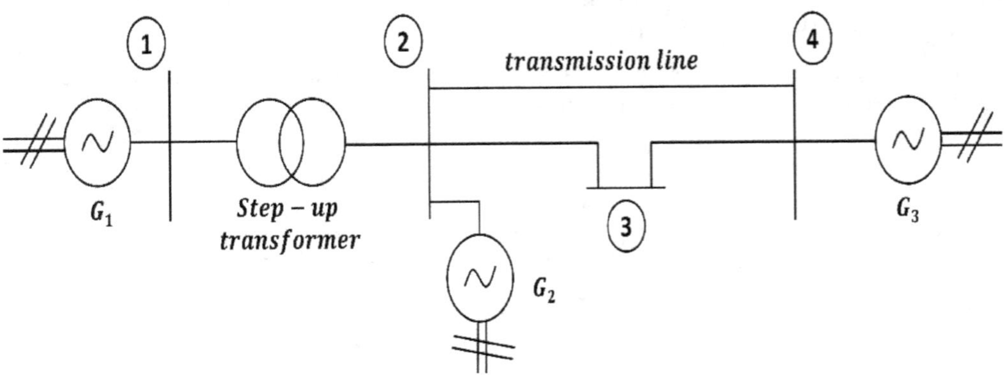

Figure 3.1 Power system network

For the above single line diagram of the power system network, graph can be represented as follows in the figure 3.2.

In order to describe the geometrical structure of a power system network, it is sufficient to replace the network components by single line segments irrespective of characteristics of the components.

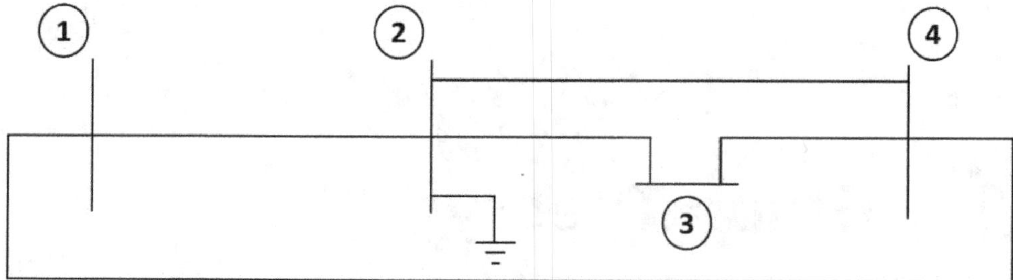

Figure 3.2 Graph for the power system network as in figure 3.1

These line segments are known as elements and their terminals are known as nodes.

A node and an element are incident if the node is terminal of the element.

Nodes can be incident to one or more elements.

3.2 DEFINITIONS

Graph: A graph represents the geometrical interconnection of the elements of a network.

Sub graph: A sub graph is a subset of elements of the graph.

Path: A path is a sub graph of connected elements with not more than two elements connected to any one node.

Connected graph: A graph is said to be connected if and only if there is a path between every pair of nodes.

Directed/Oriented graphs: A graph is said to be directed or oriented if each element of the connected graph is assigned a direction.

Analysis of graph theory: Consider a directed graph consisting of 7 branches (elements) and 5 nodes. The Directed graph for the graph in figure 3.2 is shown in the figure 3.3.

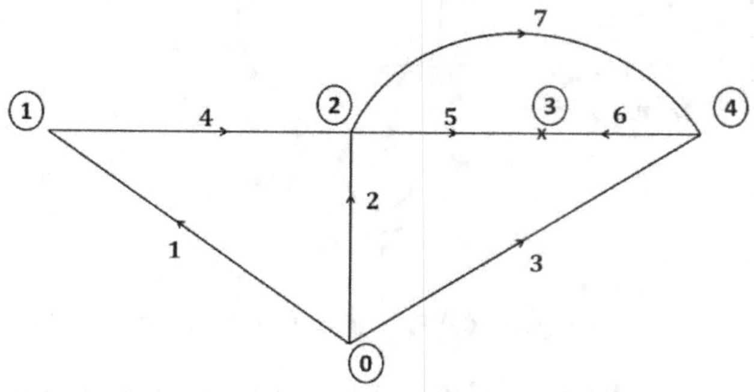

Figure 3.3 Directed graph for the graph in figure 3.2

Tree: A connected sub graph consisting of all the nodes of a graph without any closed path.

The no. of elements required to form a tree, b = n − 1, where n is the no. of nodes including reference node in graph.

The elements 1, 2, 3 & 6 are known as branches and form a subset of elements of the connected graph.

Links: The elements of the connected graph that are not included in the tree are known as link and forms a sub graph, not necessary connected.

No. of links of a connected graph with e, elements $l = e - b = e - n + 1$

3.3 INCIDENCE MATRICES

3.3.1 Element–Node Incident Matrix $[\overline{A}]$

The incidence of elements to nodes in a connected graph is represented by element–node incidence matrix.

The elements of the matrix are as follows

$$\overline{A} =$$

Elements	Nodes				
	0	1	2	3	4
1	+1	−1	0	0	0
2	+1	0	−1	0	0
3	+1	0	0	0	−1
4	0	+1	−1	0	0
5	0	0	+1	−1	0
6	0	0	0	−1	+1
7	0	0	+1	0	−1

$a_{ij} = 1$ (if the ith element is incident to & oriented away from the jth node)

$a_{ij} = -1$ (if the ith element is incident to & oriented toward the jth node)

$a_{ij} = 0$ (if the ith element is not incident to the jth node)

The order/dimension of the matrix is $e \times n = 7 \times 5$

3.3.2 Bus Incidence Matrix (A)

This matrix is obtained by deleting the row corresponding to reference node.

The selection of reference node in a connected graph is optional i.e. any node of connected graph can be selected as reference node.

The dimension of this matrix $= e \times (n - 1)$.

It is denoted by A.

$$
A = \begin{array}{c} \\ 1 \\ 2 \\ 3 \\ 4 \\ 5 \\ 6 \\ 7 \end{array}
\begin{array}{c} \begin{array}{cccc} 1 & 2 & 3 & 4 \end{array} \\
\left[\begin{array}{cccc}
-1 & 0 & 0 & 0 \\
0 & -1 & 0 & 0 \\
0 & 0 & 0 & -1 \\
+1 & -1 & 0 & 0 \\
0 & +1 & -1 & 0 \\
0 & 0 & -1 & +1 \\
0 & +1 & 0 & -1
\end{array} \right] \end{array}
$$

Number of trees for a given connected graph $= \Delta \left| AA^{T} \right|$.

Considering the bus incidence matrix, the tree and co-tree combination can be further represented as follows.

	Elements		Nodes				
			0	1	2	3	4
	Branches	1					
		2			A_b		
$A =$		3					
		4					
	Links	5			A_l		
		6					
		7					

Matrix A_b gives one to one correspondence between branches and nodes.

The size of the matrix $A_b = b \times (n - 1)$

The matrix A_l gives one to one correspondence between links and nodes.

The size of matrix $A_l = l \times (n - 1)$

3.3.3 Branch Path Incidence Matrix (K)

The incidence of branches to paths in a tree is shown by branch–path incidence matrix, where a path is oriented from a bus to the reference node.

The incidence of branches to paths in a tree is represented using this matrix, where path is oriented to the reference node from a bus.

Consider the connected graph for the graph in Figure 3.2. as in Figure 3.4.

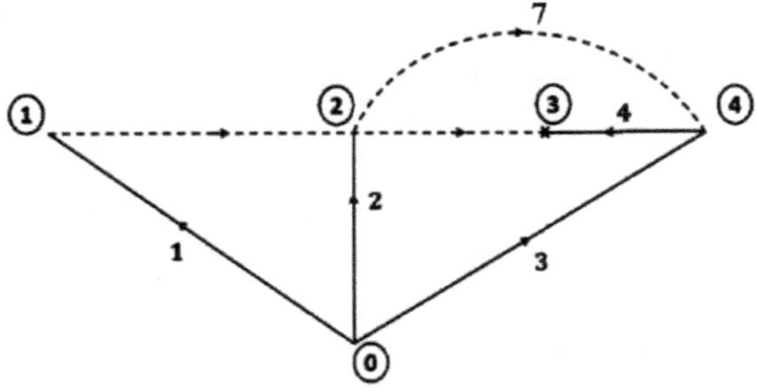

Figure 3.4: Connected graph

The elements of this matrix are

$K_{ij}= 1$ (if i^{th} branch is in the path from j^{th} bus to reference node and is oriented in same direction)

$K_{ij}=-1$ (if i^{th} branch is in the path from j^{th} bus to reference and oriented in opposite direction)

$K_{ij}= 0$ (if i^{th} branch is not connected in the path from j^{th} bus to reference)

$$[K] = \begin{array}{c|cccc} \text{Branch} & (1) & (2) & (3) & (4) \\ \hline 1 & -1 & 0 & 0 & 0 \\ 2 & 0 & -1 & 0 & 0 \\ 3 & 0 & 0 & -1 & -1 \\ 4 & 0 & 0 & -1 & 0 \end{array}$$

The size of matrix K is $= b \times (n-)$.

The branch path incidence matrix relates branches to path.

Sub matrix A_b relates branches to buses.

$\therefore$ One to one correspondence exists between paths and buses.

$A_b\,K^T = [I]$

$$[A_l] = \begin{array}{c} 5 \\ 6 \\ 7 \end{array} \begin{bmatrix} 0 & 1 & -1 & 0 \\ 1 & -1 & 0 & 0 \\ 0 & 1 & 0 & -1 \end{bmatrix} \begin{array}{c} \end{array}$$

with columns labelled $1\ 2\ 3\ 4$

Proof:

$$[K^T] = \begin{bmatrix} -1 & 0 & 0 & 0 \\ 0 & -1 & 0 & 0 \\ 0 & 0 & -1 & -1 \\ 0 & 0 & -1 & 0 \end{bmatrix}$$

$$[A_b] = \begin{bmatrix} -1 & 0 & 0 & 0 \\ 0 & -1 & 0 & 0 \\ 0 & 0 & 0 & -1 \\ 0 & 0 & -1 & 1 \end{bmatrix}$$

$$A_b K^T = \begin{bmatrix} -1 & 0 & 0 & 0 \\ 0 & -1 & 0 & 0 \\ 0 & 0 & 0 & -1 \\ 0 & 0 & -1 & 1 \end{bmatrix} \begin{bmatrix} -1 & 0 & 0 & 0 \\ 0 & -1 & 0 & 0 \\ 0 & 0 & -1 & -1 \\ 0 & 0 & -1 & 0 \end{bmatrix}$$

$$= \begin{bmatrix} 1 & 0 & 0 & 0 \\ 0 & 1 & 0 & 0 \\ 0 & 0 & 0 & 1 \\ 0 & 0 & 1 & 0 \end{bmatrix} = I$$

Therefore, the paths & buses have one to one correspondence between them.

3.3.4 Basic Cutset Incidence Matrix [B]

A cutset is defined as a set of elements, if disconnected divides the connected graph into two connected subgraphs.

The incidence of elements to basic cutsets of a connected graph is given by Basic cut–set incidence matrix, B.

A unique independent group of cutsets may be chosen if each cutset contains only one branch.

These independent cutsets are known as basic cutsets.

No. of cutsets = No. of branches and orientation of cutset is same as that of branch.

A cutset eliminates a non–reference node.

The elements of this matrix are:

$B_{ij}= 1$ (if the ith element is incident to and oriented in same direction as the jth cutset).

$B_{ij}=-1$ (if the ith element is incident to and oriented in a direction opposite to the jth cutset).

$B_{ij}= 0$ (if the ith element is not incident to the jth basic cutset).

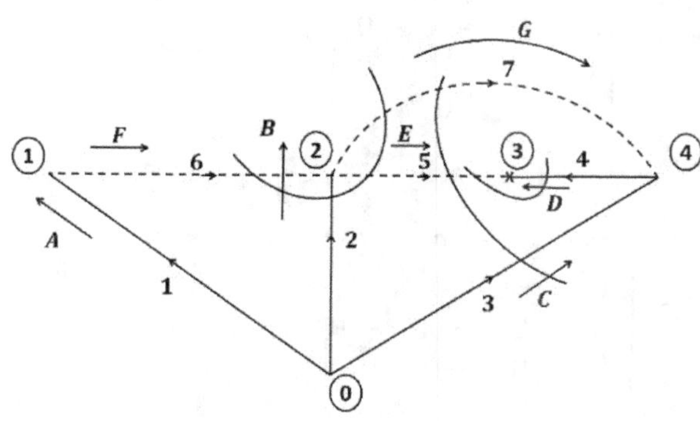

Figure 3.5 Connected graph with links

$$[B] = \begin{array}{c} \text{branches} \\ \\ \text{links} \end{array} \begin{array}{c|cccc} Elements & A & B & C & D \\ 1 & +1 & 0 & 0 & 0 \\ 2 & 0 & +1 & 0 & 0 \\ 3 & 0 & 0 & 1 & 0 \\ 4 & 0 & 0 & 0 & 1 \\ \hline 5 & 0 & -1 & 1 & 1 \\ 6 & -1 & 1 & 0 & 0 \\ 7 & 0 & -1 & 1 & 0 \end{array}$$

The incidence of elements to a basic cutset of a connected graph is given by basic cutset incidence matrix.

The matric B can be partitioned into submatrices U_b & B_l

where the rows of U_b corresponds to branches and rows of B_l correspondence to links.

U_b shows the one-one correspondence of the branches and basic cutsets.

The submatrix B_l can be obtained from bus incidence matrix.

The incidence of links to buses is shown by the sub matrix Al and the incidence of branches to buses is shown by matrix A_b.

Since there is one to one correspondence of branches and basic cutsets.

$$Bl \, A_b = Al$$

$$Bl \, A_b = \begin{bmatrix} 0 & -1 & +1 & +1 \\ -1 & +1 & 0 & 0 \\ 0 & -1 & +1 & 0 \end{bmatrix} \begin{bmatrix} -1 & 0 & 0 & 0 \\ 0 & -1 & 0 & 0 \\ 0 & 0 & 0 & -1 \\ 0 & 0 & -1 & +1 \end{bmatrix}$$

$$\underbrace{\qquad}_{A_b} \quad \underbrace{\qquad}_{B_l}$$

$$= \begin{bmatrix} 0 & 1 & -1 & 0 \\ +1 & -1 & 0 & 0 \\ 0 & 1 & 0 & -1 \end{bmatrix} = Al$$

$B_l \, A_b$ shows the incidence of links to buses

$$B_l \, A_b = A_l \, A_b^{-1}$$

$$B_l = A_l K^T$$

3.3.5 Augmented Cut-Set Incidence Matrix $[\bar{B} \ or \ \hat{B}]$

Imaginary Cutsets known as Tie cutsets that are introduced such that no. of cutsets equal to number of elements.

An augmented cut–set incidence matrix is formed by adjoining to the basic cutset incidence matrix, additional columns corresponding to tie–cutsets.

Each Tie Cutset contains only one link of the connected graph.

The Tie Cutset is oriented in the same direction as the associate link.

An Augmented Cutset incidence matrix is formed by adjoining to the basic cutset incidence matrix additional columns corresponding to cutset.

	Elements	Basic Cutsets				Tie Sets		
		A	B	C	D	E	F	G
$[\bar{B}] =$	1	1	0	0	0	0	0	0
	2	0	+1	0	0	0	0	0
	3	0	0	1	0	0	0	0
	4	0	0	0	1	0	0	0
	5	0	−1	1	1	1	0	0
	6	−1	+1	0	0	0	1	0
	7	0	−1	1	0	0	0	1

Elements	Basic Cutsets	Tiesets
Branches	U_{bbxb}	O_{bxl}
links	B_{lixb}	U_{lixl}

3.3.6 Basic Loop Incidence Matrix (C)

The incidence of elements to the basic loops of a connected graph is given by basic loop incidence matrix.

A path between adjacent nodes connected by a graph is known as <u>open–loop.</u>

No. of basic loops = No. of links

The loops which contain only one link and are independent and are known as basic loop.

Orientation of a basic loop is same as that of link,

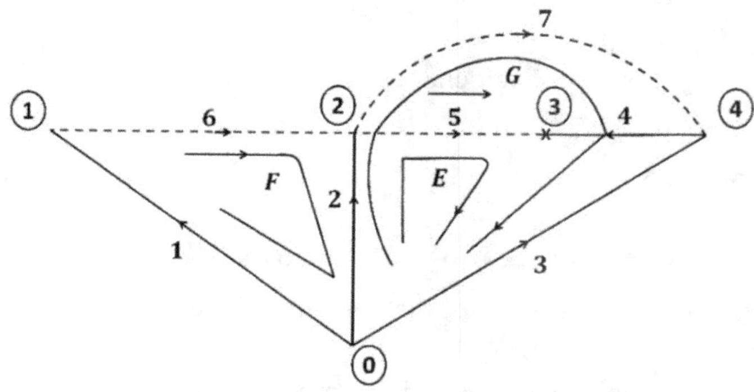

Figure 3.6 Connected graph for augmented loop incidence matrix

Elements		Basic loop	
	E	F	G
1	0	1	1
2	1	−1	0
3	−1	0	−1
4	−1	0	0
5	1	0	0
6	0	1	0
7	0	0	1

$$[C] =$$

Elements	Basic loops
Branches	$C_{b\,b\times l}$
links	$C_{l\,l\times l}$

3.3.7 Augmented Loop Incidence Matrix ($\bar{C}$)

An augmented loop incidence matrix is formed by adjoining to the basic loop incidence matrix the columns showing the incidence of elements to open loops.

Augmented loop incidence matrix is introduced in order that no. of loops equals the no. of elements.

No. of open loops = no. of branches.

An open loop is defined as a path between adjacent nodes connected by a branch.

The orientation of open loop is same as associate branch.

$$[\bar{C}] =$$

Elements	Open loops				Basic loops		
	A	B	C	D	E	F	G
1	1	0	0	0	0	1	1
2	0	1	0	0	1	−1	0
3	0	0	1	0	−1	0	−1
4	0	0	0	1	−1	0	0
5	0	0	0	0	1	0	0
6	0	0	0	0	0	1	0
7	0	0	0	0	0	0	1

Elements	Open loops	Basic loops
Branches	$U_{b\,b\times b}$	$C_{b\,b\times l}$
links	$O_{l\times b}$	$U_{l\,l\times l}$

Q1. For the power system network shown in figure 3.9, draw the connected graph and determine all the incidence matrices.

1. Element – node incident matrix $(\overline{A})$
2. Bus incidence matrix (A)
3. Branch path incidence matrix (K)
4. Basic cutset incidence matrix (B).
5. Augmented Cutset Incidence Matrix $(\overline{B})$

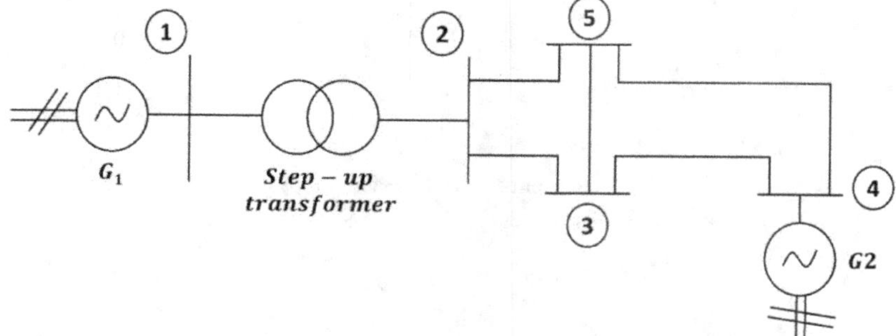

Assuming solid neutral connection for alternators

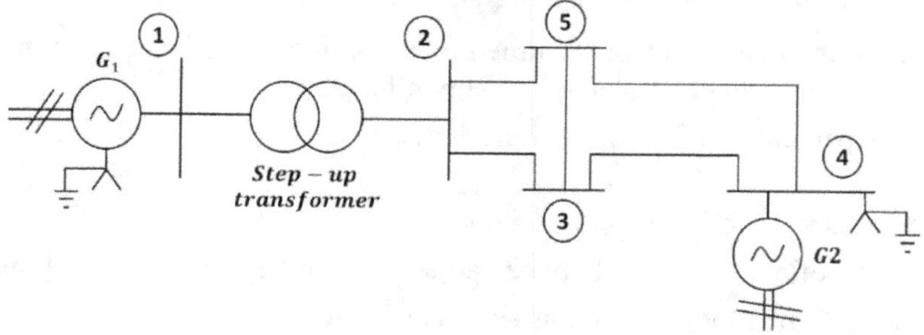

Figure 3.7 Power system network

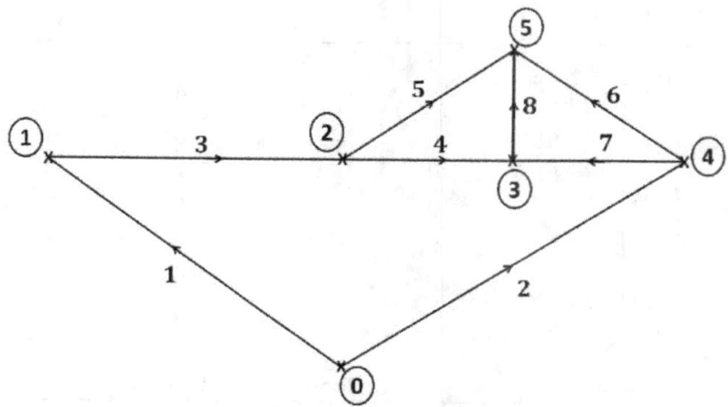

Figure 3.8 Connected graph for the power system network

1. **Element–node incident matrix ($\overline{A}$):** The incidence of elements to nodes in a connected graph is shown by element–node incidence matrix.

$[\overline{A}] =$

Elements	Nodes					
	0	1	2	3	4	5
1	+1	−1	0	0	0	0
2	+1	0	0	0	−1	0
3	0	+1	−1	0	0	0
4	0	0	1	−1	0	0
5	0	0	1	0	0	−1
6	0	0	0	0	1	−1
7	0	0	0	−1	1	0
8	0	0	0	1	0	−1

Order of $\overline{A} = 8 \times 6$

Verification: Summation of each row element $= 0$

2. **Bus incidence matrix (A):** The matrix obtained by deleting the column corresponding to the reference node is known as Bus incidence matrix, A.

   Delete the reference corresponding node (0)

$$[A] = \begin{bmatrix} -1 & 0 & 0 & 0 & 0 \\ 0 & 0 & 0 & -1 & 0 \\ 1 & -1 & 0 & 0 & 0 \\ 0 & 1 & -1 & 0 & 0 \\ 0 & 1 & 0 & 0 & -1 \\ 0 & 0 & 0 & 1 & -1 \\ 0 & 0 & -1 & 1 & 0 \\ 0 & 0 & 1 & 0 & -1 \end{bmatrix} \quad \text{Order} = e \times (n-1) = 8 \times 5$$

3. **Branch path incidence matrix (K):** The incidence of branches to paths in a tree is shown by branch–path incidence matrix, where a path is oriented from a bus to the reference node.

$[K] =$

Branch Elements	Paths				
	1	2	3	4	5
1	−1	−1	−1	0	−1
2	0	0	0	−1	0
3	0	−1	−1	0	−1
4	0	0	−1	0	0
5	0	0	0	0	−1

Tree and co-tree representation is shown it the figure 3.10

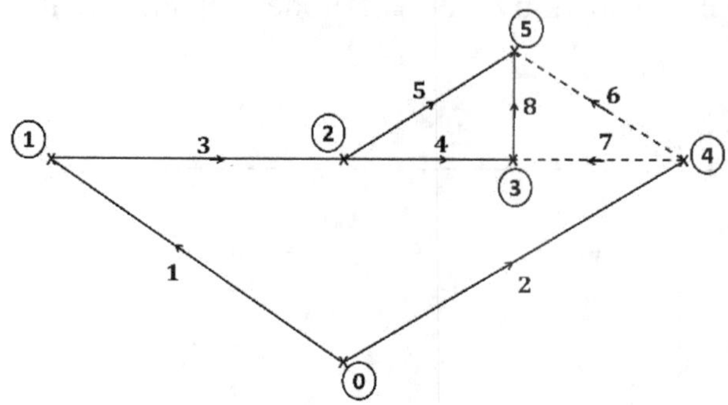

Figure 3.9 Tree and cotree of the directed graph for the power system network

4. **Basic cutset incidence matrix (B):** The incidence of elements to basic cutsets of a connected graph is given by Basic cut–set incidence matrix, B.

$$[B] =$$

Elements	Basic Cutsets				
	A	B	C	D	E
1	1	0	0	0	0
2	0	1	0	0	0
3	0	0	1	0	0
4	0	0	0	1	0
5	0	0	0	0	1
6	1	−1	−1	0	1
7	1	−1	1	1	0
8	0	0	0	−1	1

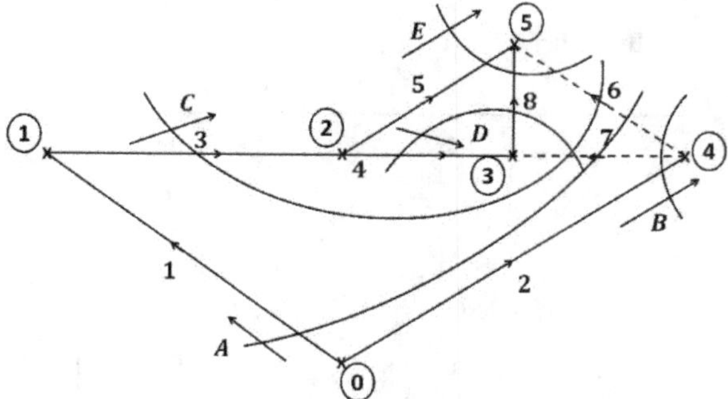

Figure 3.10 Twigs for the directed graph for the power system network

5. **Augmented Cutset Incidence Matrix ($\overline{B}$):** An augmented cut–set incidence matrix is formed by adjoining to the basic cutset incidence matrix, additional columns corresponding to tie–cutsets.

$[\overline{B}] =$

Elements	Cutsets					Tie Sets		
	A	B	C	D	E	F	G	H
1	1	0	0	0	0	0	0	0
2	0	1	0	0	0	0	0	0
3	0	0	1	0	0	0	0	0
4	0	0	0	1	0	0	0	0
5	0	0	0	0	1	0	0	0
6	1	−1	−1	0	1	1	0	0
7	1	−1	1	1	0	0	1	0
8	0	0	0	−1	1	0	0	1

6. **Augmented loop incidence matrix ($\overline{C}$):** The incidence of elements to basic loops of a connected graph is given by Basic loop incidence matrix, C.

A path between adjacent nodes connected by a branch is known as open–loop.

An augmented loop incidence matrix is formed by adjoining to the basic loop incidence matrix the columns showing the incidence of elements to open loops.

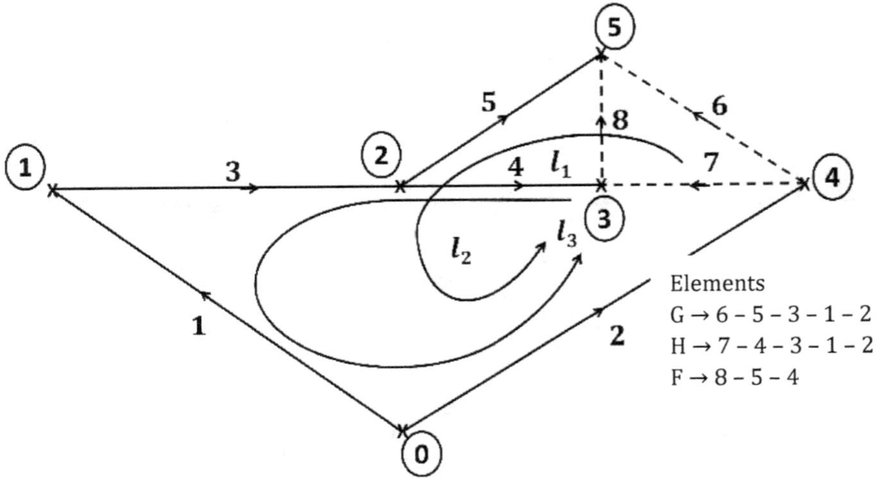

Figure 3.11 Links for the directed graph for the power system network

$[\bar{C}] =$

Elements	Open loops					Basic Loops		
	A	B	C	D	E	F	G	H
1	1	0	0	0	0	0	−1	−1
2	0	1	0	0	0	0	1	1
3	0	0	1	0	0	0	−1	−1
4	0	0	0	1	0	−1	0	−1
5	0	0	0	0	1	1	−1	0
6	0	0	0	0	0	0	1	0
7	0	0	0	0	0	0	0	1
8	0	0	0	0	0	−1	0	0

Q2. Considering bus1 as the reference bus for the given four bus system as in figure 3.13 with elements 1,2 & 5 as branches. Prove the following relations

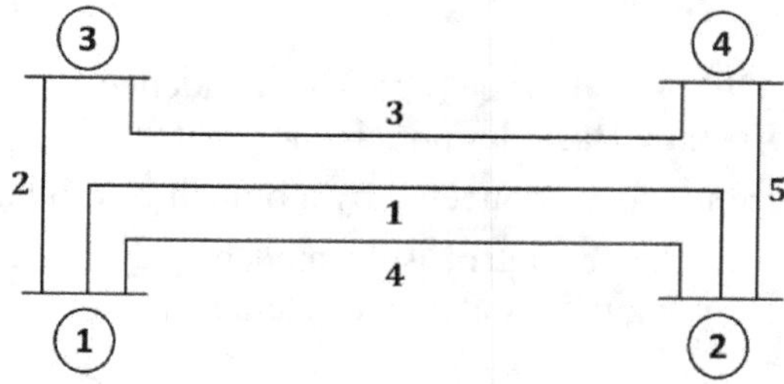

Figure 3.12 Power system network

$$1.\ A_b\,K^T = U \qquad 2.\ A l\,K^T = B l \qquad 3.\ -B_l^T = C_b$$

Solution:

The Connected graph for given single line diagram is shown in figure 3.13.

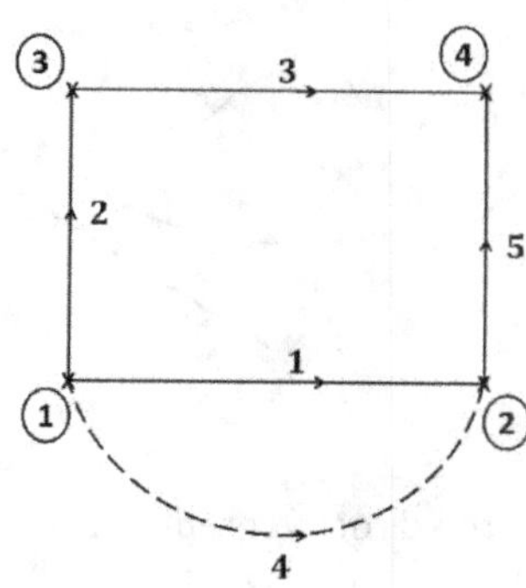

Figure 3.13 Connected graph for the power system network as in figure 3.14

1. **Element–node incident matrix ($\overline{A}$):** The incidence of elements to nodes in a connected graph is shown by element–node incidence matrix.

$$[\overline{A}] = \begin{array}{c|cccc} & 1 & 2 & 3 & 4 \\ \hline 1 & 1 & -1 & 0 & 0 \\ 2 & 1 & 0 & -1 & 0 \\ 3 & 0 & 0 & 1 & -1 \\ 4 & 1 & -1 & 0 & 0 \\ 5 & 0 & 1 & 0 & -1 \end{array}$$

Verification: Summation of each row element $= 0$

2. **Bus incidence matrix (A):** The matrix obtained by deleting the column corresponding to the reference node is known as <u>Bus incidence matrix, A.</u>

Delete the reference corresponding node (0)

$$[A] = \begin{array}{c|ccc} & 2 & 3 & 4 \\ 1 & -1 & 0 & 0 \\ 2 & 0 & -1 & 0 \\ 3 & 0 & 1 & -1 \\ 4 & -1 & 0 & 0 \\ 5 & 1 & 0 & -1 \end{array} \qquad \text{Deleting ref node 1}$$

Given 1,2,5 are branches then [A] can be rewritten as

$$[A] = \begin{array}{cc|ccc} & & 2 & 3 & 4 \\ \hline & 1 & -1 & 0 & 0 \\ b & 2 & 0 & -1 & 0 \\ & 5 & 1 & 0 & -1 \\ \hline l & 3 & 0 & 1 & -1 \\ & 4 & -1 & 0 & 0 \end{array} \qquad \begin{array}{c} A_b \\ \\ A_l \end{array}$$

$$[A_b] = \begin{bmatrix} -1 & 0 & 0 \\ 0 & -1 & 0 \\ 1 & 0 & -1 \end{bmatrix} \quad ; [A_l] = \begin{bmatrix} 0 & 1 & -1 \\ -1 & 0 & 0 \end{bmatrix}$$

3. **Branch path incidence matrix (K):** The incidence of branches to paths in a tree is shown by branch–path incidence matrix, where a path is oriented from a bus to the reference node.

$$[K] = \begin{array}{c|ccc} \text{Branch} & & \text{Paths} & \\ \text{Elements} & 2 & 3 & 4 \\ \hline 1 & -1 & 0 & -1 \\ 2 & 0 & -1 & 0 \\ 5 & 0 & 0 & -1 \end{array}$$

4. **Basic cutset incidence matrix (B):** The incidence of elements to basic cutsets of a connected graph is given by Basic cut–set incidence matrix, B.

Elements	Basic Cutsets			
	A	B	C	
1	1	0	0	
2	0	1	0	$B_b = U_b$
5	0	0	1	
3	1	−1	1	B_l
4	1	0	0	

$$[B] =$$

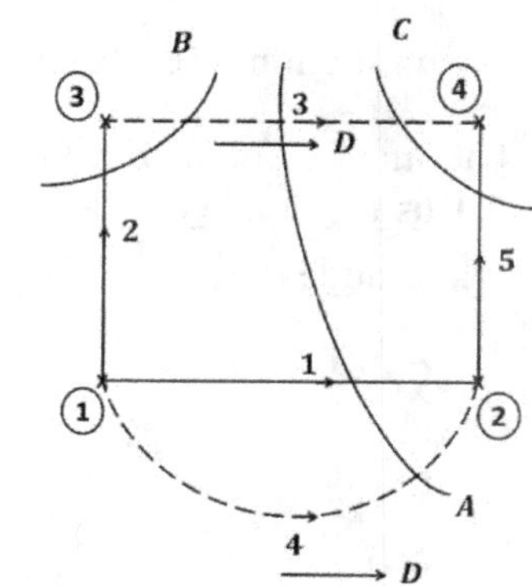

Figure 3.14 Cutsets for the Connected graph as in figure 3.14

$$[B_l] = \begin{bmatrix} 1 & -1 & 1 \\ 1 & 0 & 0 \end{bmatrix}$$

$$[A_b] = \begin{bmatrix} -1 & 0 & 0 \\ 0 & -1 & 0 \\ 1 & 0 & -1 \end{bmatrix}$$

1. $A_b\, K^T = U$

$$\begin{bmatrix} -1 & 0 & 0 \\ 0 & -1 & 0 \\ 1 & 0 & -1 \end{bmatrix} \begin{bmatrix} -1 & 0 & 0 \\ 0 & -1 & 0 \\ -1 & 0 & -1 \end{bmatrix} = \begin{bmatrix} 1 & 0 & 0 \\ 0 & 1 & 0 \\ 0 & 0 & 1 \end{bmatrix}$$

5. **Augmented Cut-set Incidence Matrix ($\bar{B}$):** An augmented cut–set incidence matrix is formed by adjoining to the basic cutset incidence matrix, additional columns corresponding to tie–cutsets.

Cutsets			Tie Sets	
A	B	C	D	E
1	0	0	0	0
0	1	0	0	0
0	0	1	0	0
1	−1	1	1	0
1	0	0	0	1

$$[\bar{B}] =$$

6. Basic loop incidence matrix, C:

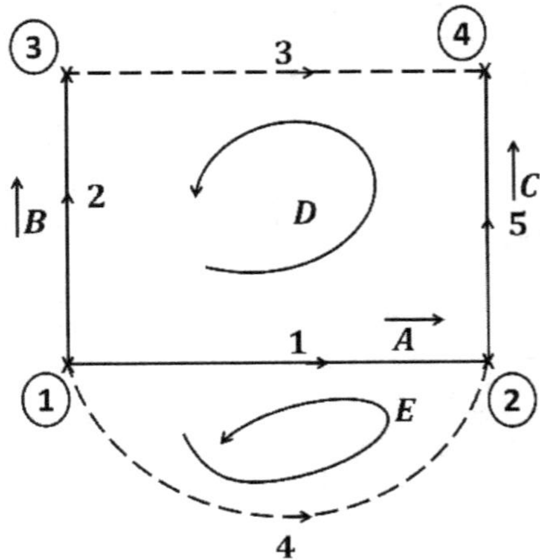

Figure 3.15 Links for the directed graph for the power system network

The incidence of elements to basic loops of a connected graph is given by Basic loop incidence matrix, C.

$$
C = \quad
\begin{array}{c|cc}
\text{Elements} & \multicolumn{2}{c}{\text{Basic Loops}} \\
 & D & E \\
\hline
1 & -1 & -1 \\
2 & +1 & 0 \\
5 & -1 & 0 \\
3 & +1 & 0 \\
4 & 0 & 1 \\
\end{array}
$$

7. Augmented loop incidence matrix ($\bar{C}$):

A path between adjacent nodes connected by a branch is known as open–loop.

An augmented loop incidence matrix is formed by adjoining to the basic loop incidence matrix, the columns showing the incidence of elements to open loops.

$$
[\bar{C}] =
$$

Elements	Open loops			Basic loops	
	A	B	C	D	E
1	1	0	0	-1	-1
2	0	1	0	+1	0
5	0	0	1	-1	0
3	0	0	0	1	0
4	0	0	0	0	1

$$[C_b] = \begin{bmatrix} -1 & -1 \\ 1 & 0 \\ -1 & 0 \end{bmatrix}$$

1. $\qquad A_l K^T = B_l$

2. $\qquad \begin{bmatrix} 0 & 1 & -1 \\ -1 & 0 & 0 \end{bmatrix} \begin{bmatrix} -1 & 0 & 0 \\ 0 & -1 & 0 \\ -1 & 0 & -1 \end{bmatrix} = \begin{bmatrix} 1 & -1 & 1 \\ 1 & 0 & 0 \end{bmatrix}$

3. $\qquad -B_l^T = C_b$

$$[B_l]^T = \begin{bmatrix} 1 & 1 \\ -1 & 0 \\ 1 & 0 \end{bmatrix}$$

$$[B_l]^T = \begin{bmatrix} -1 & -1 \\ 1 & 0 \\ -1 & 0 \end{bmatrix}$$

$$-B_l^T = C_b$$

Q3. The given connected graph consists of four buses prove the relations

$$A_b K^T = U$$
$$A_l K^T = B_l$$

Elements 1, 2, 4 are branches.

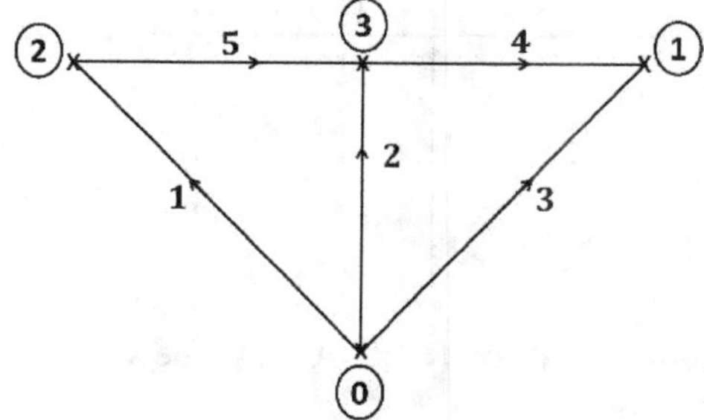

Figure 3.16 Connected graph

1. Element–node incident matrix $(\overline{A})$
2. Bus incidence matrix (A)
3. Branch path incidence matrix (K)
4. Basic cutset incidence matrix (B).
5. Augmented Cutset Incidence Matrix $(\overline{B})$
6. Augmented loop incidence matrix $(\overline{C})$

1. **Element–node incident matrix $(\overline{A})$:** The incidence of elements to nodes in a connected graph is shown by element–node incidence matrix.

$$[\bar{A}] =$$

Elements \ Nodes	0	1	2	3
1	1	0	−1	0
2	1	0	0	−1
3	1	−1	0	0
4	0	−1	0	+1
5	0	0	1	−1

2. Bus incidence matrix (A): The matrix obtained by deleting the column corresponding to the reference node is known as Bus incidence matrix, A.

Delete the reference corresponding node (0)

$$[A] =$$

Elements \ Nodes	1	2	3
1	0	−1	0
2	0	0	−1
4	−1	0	+1
3	−1	0	0
5	0	1	−1

$$[A] =$$

Elements		Nodes 1	2	3	
b	1	0	−1	0	
	2	0	0	−1	A_b
	4	−1	0	+1	
l	3	−1	0	0	A_l
	4	0	1	−1	

$$[A_b] = \begin{bmatrix} 0 & -1 & 0 \\ 0 & 0 & -1 \\ -1 & 0 & 1 \end{bmatrix} \; ; [A_l] = \begin{bmatrix} -1 & 0 & 0 \\ 0 & 1 & -1 \end{bmatrix}$$

3. Branch path incidence matrix (K): The incidence of branches to paths in a tree is shown by branch–path incidence matrix, where a path is oriented from a bus to the reference node.

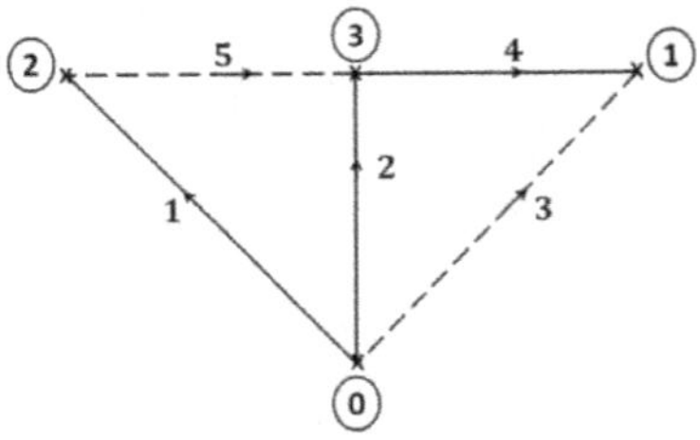

Figure 3.17 Links for the oriented graph as in figure 3.18

$$[K] =$$

	Branch	Paths		
	Elements	1	2	3
	1	0	−1	0
	2	−1	0	−1
	4	−1	0	0

4. **Basic cutset incidence matrix (B):** The incidence of elements to basic cutsets of a connected graph is given by Basic cut–set incidence matrix, B.

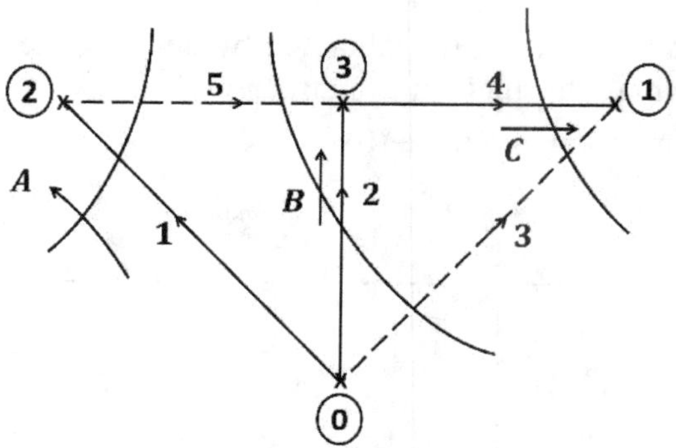

Figure 3.18 Basic cutsets for the oriented graph as in figure 3.18

$$[B] =$$

Elements	Basic Cutsets			
	A	B	C	
1	1	0	0	$B_b=U_b$
2	0	1	0	
4	0	0	1	
3	0	1	1	B_l
5	−1	1	0	

5. **Augmented Cutset Incidence Matrix ($\bar{B}$):** An augmented cut–set incidence matrix is formed by adjoining to the basic cutset incidence matrix, additional columns corresponding to tie–cutsets.

$$[\bar{B}] =$$

Cutsets			Tie Sets	
A	B	C	D	E
1	0	0	0	0
0	1	0	0	0
0	0	1	0	0
0	1	1	1	0
−1	1	0	0	1

$$[B_l] = \begin{bmatrix} 0 & 1 & 1 \\ -1 & 1 & 0 \end{bmatrix}$$

Now

1. $A_b K^T = U$

$$\begin{bmatrix} 0 & -1 & 0 \\ 0 & 0 & -1 \\ -1 & 0 & 1 \end{bmatrix} \begin{bmatrix} 0 & -1 & -1 \\ -1 & 0 & 0 \\ 0 & -1 & 0 \end{bmatrix} = \begin{bmatrix} 1 & 0 & 0 \\ 0 & 1 & 0 \\ 0 & 0 & 1 \end{bmatrix} = U$$

2. $A_1 K^T = B_1$

$$\begin{bmatrix} -1 & 0 & 0 \\ 0 & 1 & -1 \end{bmatrix} \begin{bmatrix} 0 & -1 & -1 \\ -1 & 0 & 0 \\ 0 & -1 & 0 \end{bmatrix} = \begin{bmatrix} 0 & 1 & 1 \\ -1 & 1 & 0 \end{bmatrix} = [B_l]$$

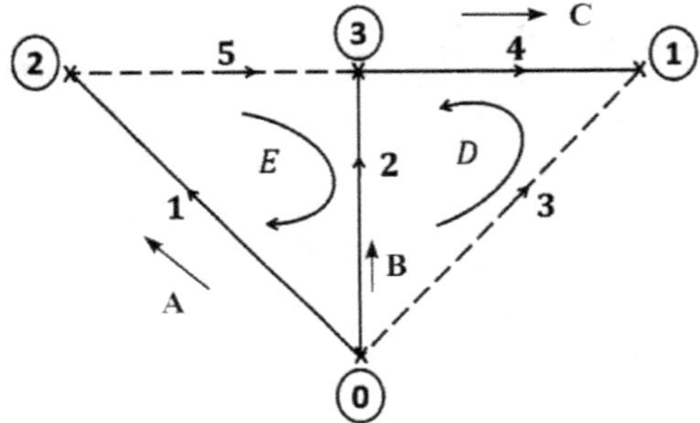

Figure 3.19 Links for the oriented graph as in figure 3.18

6. **Augmented loop incidence matrix ($\overline{C}$):** The incidence of elements to basic loops of a connected graph is given by Basic loop incidence matrix, C.

A path between adjacent nodes connected by a graph is known as open–loop.

An augmented loop incidence matrix is formed by adjoining to the basic loop incidence matrix the columns showing the incidence of elements to open loops.

$[\overline{C}] =$

Elements	Open loops			Basic loops	
	A	B	C	D	E
1	1	0	0	0	1
2	0	1	0	-1	-1
4	0	0	1	-1	0
3	0	0	0	1	0
5	0	0	0	0	1

Elements	Open loops	Basic loops
b	$U_{b b X b}$	$C_{b b X l}$
l	$O_{l X b}$	$U_{l X l}$

Q4.   A four–bus power system network is shown in Figure 3.20. Formulate the incidence matrices and prove the following

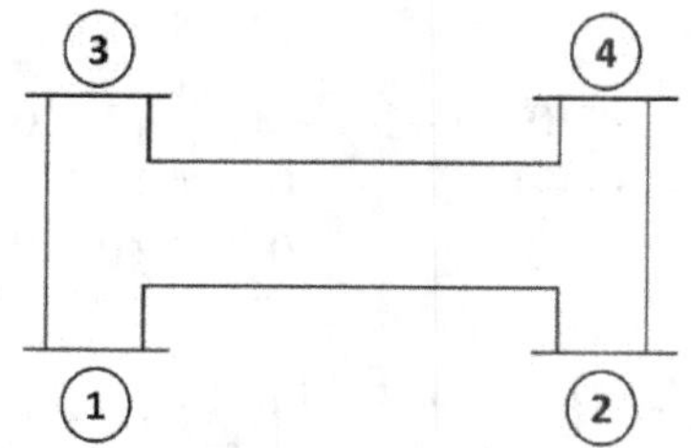

Figure 3.20 Power system network

1. $B_l^T = U_b$ 2. $A_b\,K^T = U$ 3. $A_l\,K^T = B_l$ 4. $[\bar{C}][\bar{B}]^T = U$

Solution:

The connected graph of given line diagram.

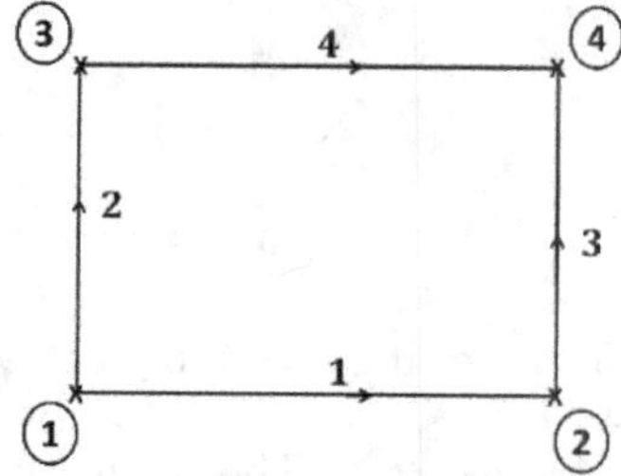

Figure 3.21 Oriented graph for the power system network in figure 3.22

Let node–1 be the reference node.

1. **Element–node incident matrix ($\bar{A}$):** The incidence of elements to nodes in a connected graph is shown by element–node incidence matrix.

	Elements	Nodes 1	2	3	4
$[\bar{A}] =$	1	1	−1	0	0
	2	1	0	−1	0
	3	0	1	0	−1
	4	0	0	1	−1

2. **Bus incidence matrix (A):** The matrix obtained by deleting the column corresponding to the reference node is known as Bus incidence matrix, A.

	Elements	Nodes 2	3	4
$[\bar{A}] =$	1	−1	0	0
	2	0	−1	0
	3	1	0	−1
	4	0	1	−1

Representation of tree and cotree:

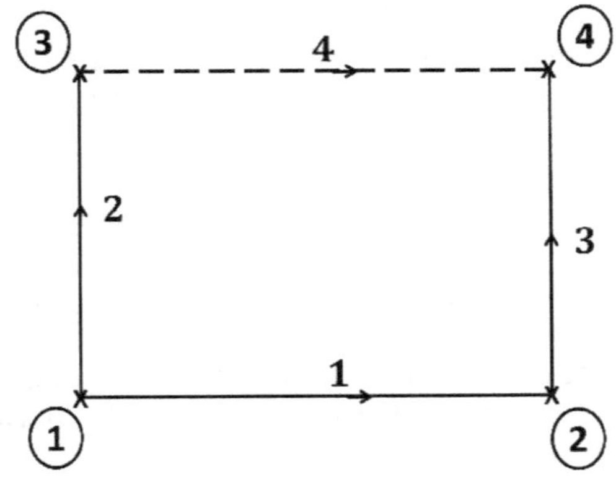

Figure 3.22: Representation of tree and co-tree of Oriented graph

Elements	Open loops	Basic loops
b	$A_{b_{bX(n-1)}}$	$A_{b_{bX3}}$
l	$A_{l_{lX(n-1)}}$	$A_{l_{1X3}}$

$$[A_b] = \begin{bmatrix} -1 & 0 & 0 \\ 0 & -1 & 0 \\ 1 & 0 & -1 \end{bmatrix} \; ; [A_l] = \begin{bmatrix} 0 & 1 & -1 \end{bmatrix}$$

3. **Branch path incidence matrix (K):** The incidence of branches to paths in a tree is shown by branch–path incidence matrix, where a path is oriented from a bus to the reference node.

Branch Elements	Paths 2	3	4
1	−1	0	−1
2	0	−1	0
3	0	0	−1

$[K] =$ (table above)

4. **Basic cutset incidence matrix (B):** The incidence of elements to basic cutsets of a connected graph is given by Basic cut–set incidence matrix, B.

Elements	Basic Cutsets A	B	C
1	1	0	0
2	0	1	0
3	0	0	1
4	1	−1	1

$[B] =$ (table above)

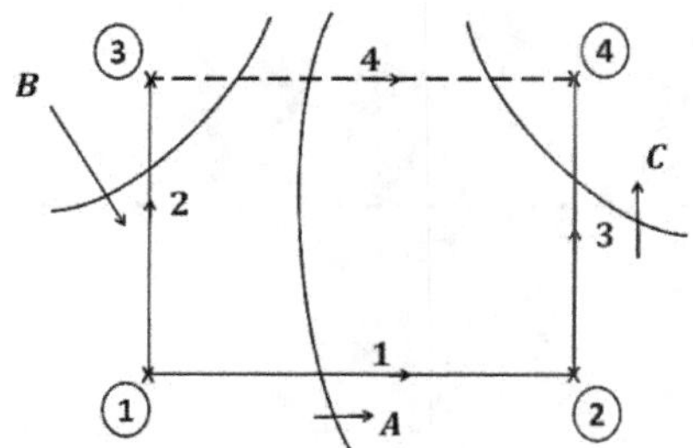

Figure 3.23 Basic cutsets of Oriented graph

5. **Augmented Cutset Incidence Matrix ($\overline{B}$):** An augmented cut–set incidence matrix is formed by adjoining to the basic cutset incidence matrix, additional columns corresponding to tie–cutsets.

$$\overline{B} = \begin{array}{c|ccc|c} & \multicolumn{3}{c|}{\textbf{Cutsets}} & \textbf{Tie Sets} \\ & A & B & C & D \\ \hline & 1 & 0 & 0 & 0 \\ & 0 & 1 & 0 & 0 \\ & 0 & 0 & 1 & 0 \\ & 1 & -1 & 1 & 1 \end{array}$$

$$[B_l] = \begin{bmatrix} 1 & -1 & 1 \end{bmatrix}$$

$$-B_l^T = \begin{bmatrix} -1 \\ 1 \\ -1 \end{bmatrix}$$

6. **Basic Loop Incidence matrix, C:** The incidence of elements to basic loops of a connected graph is given by <u>Basic loop incidence matrix, C.</u>

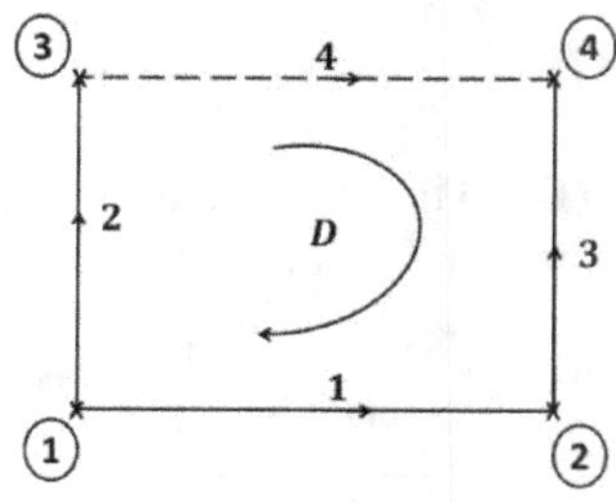

Figure 3.24 Basic loops of Oriented graph

$$[C] = \begin{bmatrix} -1 \\ 1 \\ -1 \\ 1 \end{bmatrix}$$

7. **Augmented loop incidence matrix ($\bar{C}$):** A path between adjacent nodes connected by a graph is known as open–loop.

An augmented loop incidence matrix is formed by adjoining to the basic loop incidence matrix the columns showing the incidence of elements to open loops.

	Open loops			Basic loops
	A	B	C	D
	1	0	0	−1
$[\bar{C}] =$	0	1	0	1
	0	0	1	−1
	0	0	0	1

Elements	Open loops	Basic loops
6	U_b	C_b
l	O	U_l

$$C_b = \begin{bmatrix} -1 \\ 1 \\ -1 \end{bmatrix}$$

1. $\quad -B_l^T = C_b$

2. $\quad A_b K^T = \begin{bmatrix} -1 & 0 & 0 \\ 0 & -1 & 0 \\ 1 & 0 & -1 \end{bmatrix} \begin{bmatrix} -1 & 0 & 0 \\ 0 & -1 & 0 \\ -1 & 0 & -1 \end{bmatrix} = \begin{bmatrix} 1 & 0 & 0 \\ 0 & 1 & 0 \\ 0 & 0 & 1 \end{bmatrix} = I$

3. $\quad A_l K^T = B_l$

$$\begin{bmatrix} 0 & 1 & -1 \end{bmatrix} \begin{bmatrix} -1 & 0 & 0 \\ 0 & -1 & 0 \\ -1 & 0 & -1 \end{bmatrix} = \begin{bmatrix} 1 & -1 & 1 \end{bmatrix} = [B_l]$$

4. $\quad [\bar{C}][\bar{B}]^T = U$

Q5. The power system network consisting of 5 buses and 7 elements are given in the figure 3.25. Formulate the incidence matrix and prove the relations.

1. $A_l K^T = B_l$ 2. $-B_l^T = C_b$ 3. $A_l A_b^{-1} = B_l$

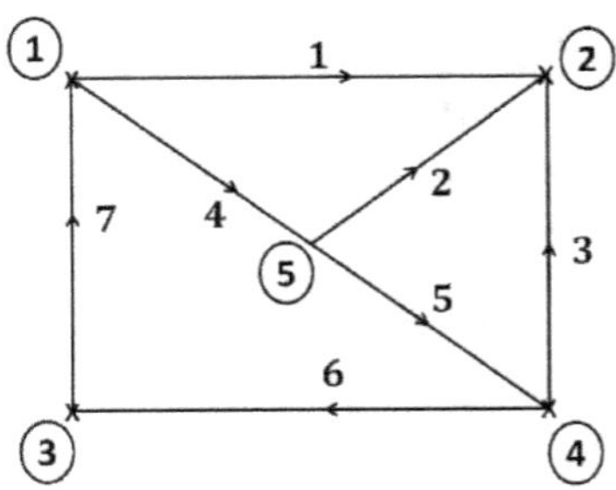

Figure 3.25 Oriented graph

Elements   2,4,5,6 forms a tree

1. **Element–node incident matrix ($\overline{A}$):** The incidence of elements to nodes in a connected graph is shown by element–node incidence matrix.

$[\overline{A}] =$

Elements	Nodes				
	1	2	3	4	5
1	1	−1	0	0	0
2	0	−1	0	0	1
3	0	−1	0	1	0
4	1	0	0	0	−1
5	0	0	0	−1	1
6	0	0	−1	1	0
7	−1	0	1	0	0

2. **Bus incidence matrix (A):** The matrix obtained by deleting the column corresponding to the reference node is known as <u>Bus incidence matrix, A.</u>

$[A] =$

Elements	Nodes			
	2	3	4	5
2	−1	0	0	1
4	0	0	0	−1
5	0	0	−1	1
6	0	−1	1	0
1	−1	0	0	0
3	−1	0	1	0
7	0	1	0	0

$$[A_b] = \begin{bmatrix} -1 & 0 & 0 & 1 \\ 0 & 0 & 0 & -1 \\ 0 & 0 & -1 & 1 \\ 0 & -1 & 1 & 0 \end{bmatrix} \; ; [A_l] = \begin{bmatrix} -1 & 0 & 0 & 0 \\ -1 & 0 & 1 & 0 \\ 0 & 1 & 0 & 0 \end{bmatrix}$$

3. **Branch path matrix (K):** The incidence of branches to paths in a tree is shown by <u>branch–path incidence</u> matrix, where path is oriented from a bus to the reference node.

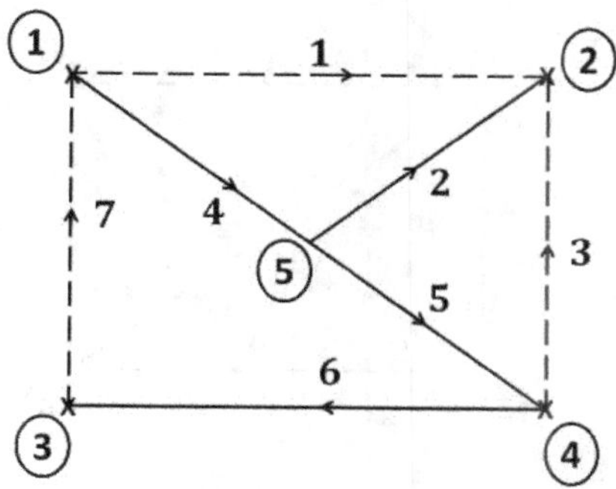

Figure 3.26 Twigs and links of the oriented graph

<table>
<tr><td rowspan="2">$[K] =$</td><td>**Branch**</td><td colspan="4">**Paths**</td></tr>
</table>

Branch Elements	2	3	4	5
2	−1	0	0	0
4	−1	−1	−1	−1
5	0	−1	−1	0
6	0	−1	0	0

Tree and cotree representation

4. **Basic cutset matrix (B):** The incidence of elements to basic cutsets of a connected graph is given by Basic cut–set incidence matrix, B.

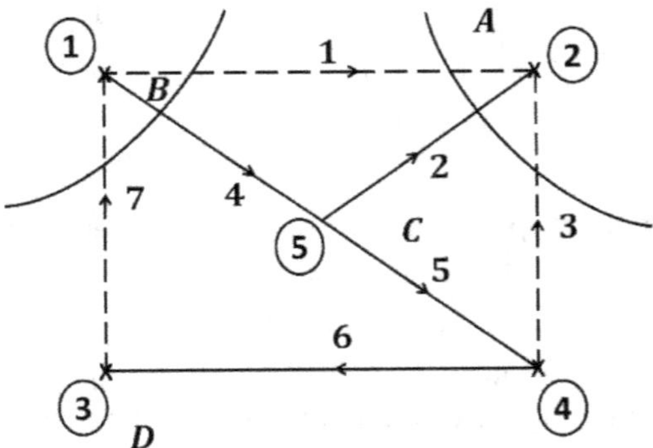

Figure 3.27 Basic cutsets of the oriented graph

Elements	A	B	C	D
1	1	1	0	0
2	1	0	0	0
3	1	0	−1	0
4	0	1	0	0
5	0	0	1	0
6	0	0	0	1
7	0	1	−1	−1

where the first column of the above table is headed "**Basic Cutsets**" and $[B] =$ labels the matrix.

Elements	Basic cutsets					
2	1	0	0	0		
4	0	1	0	0		
5	0	0	1	0	b	$B_b = U_b$
6	0	0	0	1		
1	1	1	0	0		
3	1	0	−1	0	l	B_l
7	0	1	−1	−1		

5. **Augmented Cutset Incidence Matrix ($\bar{B}$):** An augmented cut–set incidence matrix is formed by adjoining to the basic cutset incidence matrix, additional columns corresponding to tie–cutsets.

$$[\bar{B}] = \begin{array}{ccccccc} A & B & C & D & E & F & G \\ 1 & 0 & 0 & 0 & 0 & 0 & 0 \\ 0 & 1 & 0 & 0 & 0 & 0 & 0 \\ 0 & 0 & 1 & 0 & 0 & 0 & 0 \\ 0 & 0 & 0 & 1 & 0 & 0 & 0 \\ 1 & 1 & 0 & 0 & 1 & 0 & 0 \\ 1 & 0 & -1 & 0 & 0 & 1 & 0 \\ 0 & 1 & -1 & -1 & 0 & 0 & 1 \end{array}$$

6. **Basic loop incidence matrix [C]:** The incidence of elements to basic loops of a connected graph is given by Basic loop incidence matrix, C.

   A path between adjacent nodes connected by a graph is known as open–loop.

   An augmented loop incidence matrix is formed by adjoining to the basic loop incidence matrix the columns showing the incidence of elements to open loops.

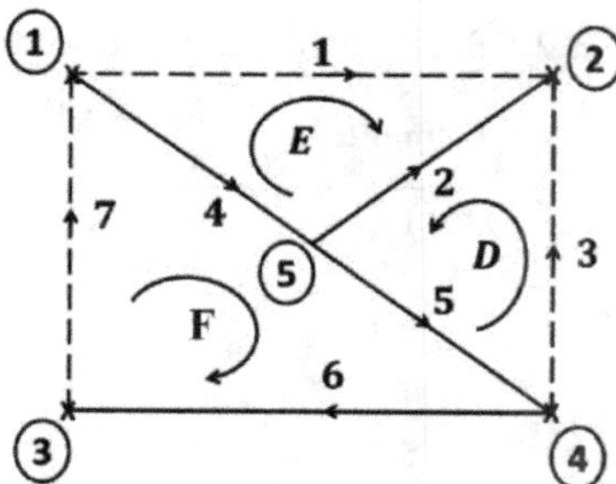

Figure 3.28 Basic loops of the oriented graph

$$[C] = \begin{array}{c|ccc} \text{Elements} & \multicolumn{3}{c}{\text{Basic Loops}} \\ 2 & -1 & -1 & 0 \\ 4 & 0 & -1 & 1 \\ 5 & 1 & 0 & 1 \\ 6 & 0 & 0 & 1 \\ 1 & 0 & 1 & 0 \\ 3 & 1 & 0 & 0 \\ 7 & 0 & 0 & 1 \end{array}$$

$$[\bar{C}] =$$

Elements	Open loops				Basic loops		
	A	B	C	D	D	E	F
2	1	0	0	0	−1	−1	0
4	0	1	0	0	0	−1	1
5	0	0	1	0	1	0	1
6	0	0	0	1	0	0	1
1	0	0	0	0	0	1	0
3	0	0	0	0	1	0	0
7	0	0	0	0	0	0	1

1. $A_1 K^T = B_1$

$$A_1 = \begin{bmatrix} -1 & 0 & 0 & 0 \\ -1 & 0 & 1 & 0 \\ 0 & 1 & 0 & 0 \end{bmatrix} \begin{bmatrix} -1 & -1 & 0 & 0 \\ 0 & -1 & -1 & -1 \\ 0 & -1 & -1 & 0 \\ 0 & -1 & 0 & 0 \end{bmatrix} = \begin{bmatrix} 1 & 1 & 0 & 0 \\ 1 & 0 & -1 & 0 \\ 0 & -1 & -1 & -1 \end{bmatrix}$$

2. $-B_l^T = C_b$

$$-B_l^T = -\begin{bmatrix} 1 & 1 & 0 \\ 1 & 0 & -1 \\ 0 & -1 & -1 \\ 0 & 0 & -1 \end{bmatrix} = \begin{bmatrix} -1 & -1 & 0 \\ -1 & 0 & 1 \\ 0 & 1 & 1 \\ 0 & 0 & 1 \end{bmatrix}$$

3. $A_1 A_b^{-1} = B_1$

$$A_b^{-1} = K^T$$

OBJECTIVE QUESTIONS

1. Define elements and nodes?

   Ans: The line segments replacing the power system network components irrespective of their characteristics and describing the geometrical structure of a network are known as elements and the terminals are known as nodes.

2. Define graph?

   Ans: The geometrical interconnection of the elements of a network is known as graph.

3. Define Sub graph?

   Ans: Any subset of elements of the graph is known as sub–graph.

4. Define Path?

   Ans: A subgraph of connected elements with no more than two elements connected to any node is known as a path.

5. Define Connected graph?

   Ans: A graph with a path between every pair of nodes is known as connected graph.

6. Define Oriented graph or directed graph?

Ans: A connected graph with each element assigned a direction is known as oriented or directed graph.

7. Define tree?

   Ans: A connected subgraph containing all nodes of a graph but no closed path is known as tree.

8. Define Co-tree?

   Ans: The connected graph that is not included in the tree is known as co–tree.

9. Define Branches?

   Ans: The elements of a tree are known as branches and number of branches, $b = n - 1$ where n is the number of nodes in the graph.

10. Define Links?

    Ans: The elements of the connected graph that are not included in the tree are known as links and number of links $= e - b$ where e is the number of elements of the graph.

11. Define Loop?

    Ans: The closed path with a link added to the tree in a graph is known as loop.

12. Define Basic Loops?

    Ans: Loops containing only one link are known as basic loops and number of basic loops= number of links.

13. Define Cutset?

    Ans: A cut–set is a set of elements that, if removed, divides a connected graph into two connected subgraphs.

14. Define Basic Cutset?

    Ans: Cutsets containing only one branch are known as basic cutsets and number of basic cutsets = number of branches.

15. The incidence of elements to nodes in a connected graph is shown by ______________.

16. The matrix obtained by deleting the column corresponding to the reference node is known as

    ______________.

17. The incidence of branches to paths in a tree is shown by ______________ matrix, where a path is oriented from a bus to the reference node.

18. The incidence of elements to basic cutsets of a connected graph is given by ______________.

19. An ______________ matrix is formed by adjoining to the basic cutset incidence matrix, additional columns corresponding to tie–cutsets.

20. The incidence of elements to basic loops of a connected graph is given by______________.

21. A path between adjacent nodes connected by a graph is known as ______________.

22. An ______________ matrix is formed by adjoining to the basic loop incidence matrix the columns showing the incidence of elements to open loops.

23. Define bus in a power system network?

    Ans: The meeting point of various components in a power system is called as bus. The bus is a conductor made of copper or aluminium having negligible resistance. Buses are considered as points of constant voltage in a power system.

24. What is the difference between an ordinary loop and basic loop?

    Ans: An ordinary loop is a closed path containing one or more links. A basic loop is a closed path having only one link.

25. Why is the sub matrix U_b of the basic cutest matrix, B is a unity matrix and what is its dimension?

    Ans: The sub matrix U_b describes the incidence of branches to basic cutsets. Since each cutset contains only one branch, this sub matrix U_b is a unity matrix. The number of basic cutsets being equal to the number of branches, U_b is a square matrix.

26. What does the sub matrix, C_b of the incidence matrix C represent?

    Ans: The sub matrix C_b describes the incidence of branches to basic loops.

4   Admittance Bus Matrix

4.1   INTRODUCTION

Admittance matrix is used for load flow analysis.

The incidence matrices are

1. Element–node incident matrix ($\overline{A}$)
2. Bus incidence matrix (A)
3. Branch path incidence matrix (K)
4. Basic cutset incidence matrix (B).
5. Augmented Cutset Incidence Matrix ($\overline{B}$)
6. Basic loop incidence matrix (C)
7. Augmented loop incidence matrix ($\overline{C}$)

The matrices of incidence network give the information about network connectivity, orientation cutsets and loops.

The matrices of incident network do not give the information about the nature of network elements forming the interconnected network.

The complete behaviour of network can be obtained from the knowledge of behaviour of element forming an interconnected network.

An element in an electrical network is characterised by relationship between current flowing through it and voltage across it.

Primitive element: A fundamental element which is not connected to any other element is known as primitive element.

A set of such unconnected elements is known as a primitive network.

Consider two unconnected coils as shown in the figure 4.1.

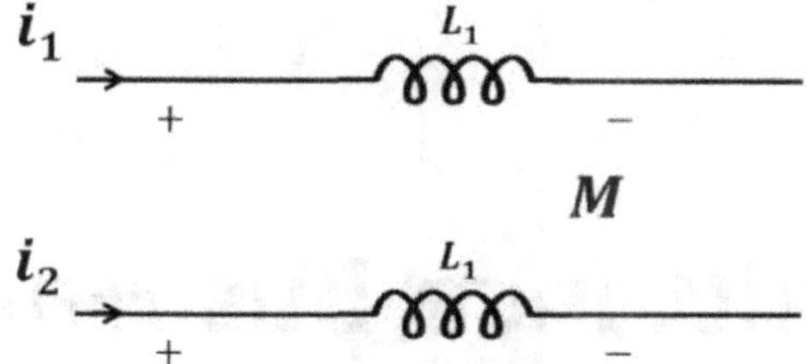

Figure 4.1 Two unconnected coils

The equations describing the elements are

$$\left.\begin{aligned} V_1 &= L_1\frac{di_1}{dt} + M\frac{di_2}{dt} \\ V_2 &= L_2\frac{di_2}{dt} + M\frac{di_1}{dt} \end{aligned}\right\} \qquad \qquad .....(4.1)$$

In impedance form,

$$\left.\begin{aligned} V_1 &= Z_{11}I_1 + Z_{12}I_2 \\ V_2 &= Z_{21}I_1 + Z_{22}I_2 \end{aligned}\right\} \qquad \qquad .....(4.2)$$

Equation (4.2) does not depend on connectivity of two coils but depends on individual currents and voltages and they are primitive elements and together form a primitive network.

4.2 REPRESENTATION OF NETWORK COMPONENTS

Performance of network elements can be expressed either in impedance or admittance form.

The parameters and variables are:

Let voltage across the element p-q be V_{pq}

Current flowing through the element p-q be I_{pq}

i.e. each element has two variables V_{pq} & I_{pq}

(a)   Impedance form

$$V_{pq} = E_p - E_q$$

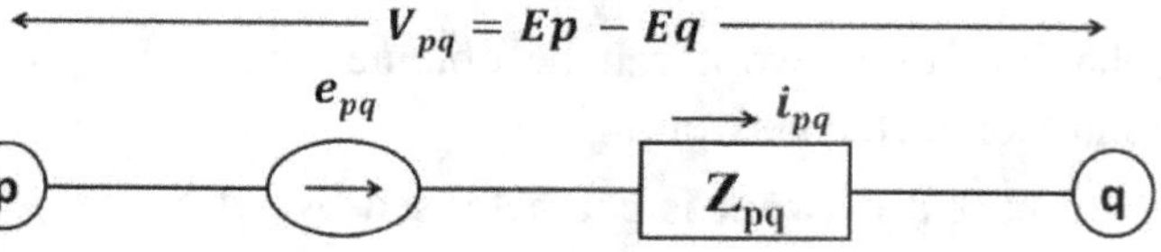

Figure 4.2 Impedance form representation

E_{pq} is source voltage connected in series with the element pq.

Z_{pq} is self–impedance of the element pq.

apply KVL, $+ E_{pq} - I_{pq}Z_{pq} + V_{pq} = 0$

$$E_{pq} + V_{pq} = I_{pq}Z_{pq} \qquad \qquad .....(4.3)$$

(b)   Admittance form

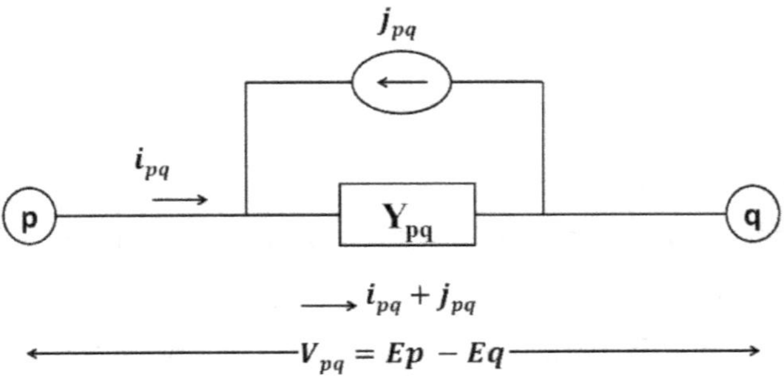

Figure 4.3 Admittance form representation

$$V_{pq} = E_p - E_q$$

j_{pq} is source current connected with element pq

Y_{pq} is self–admittance of the element pq

$$i_{pq} + j_{pq} = Y_{pq}V_{pq} \qquad\qquad .....(4.4)$$

The equations (4.3) and (4.4) represent performance equations of an element in impedance and admittance form.

In steady state, parameters and variables of Z_{pq} and Y_{pq} are real numbers for DC and complex numbers for AC.

In matrix form, equations (4.3) and (4.4) are

$$\bar{V} + \bar{E} = [Z]\bar{I}$$
$$\bar{I} + \bar{J} = [Y]\bar{V}$$

Where [Z] is primitive impedance and [Y] is primitive admittance matrix.

A diagonal element of the matrix $[\bar{Z}]$ or $[\bar{Y}]$ is the self impedance, $Z_{pq}\text{–}Z_{pq}$ or self-admittance, $Y_{pq}\text{–}Y_{pq}$

An off-diagonal element of the matrix $[\bar{Z}]$ or $[\bar{Y}]$ is the mutual impedance

$Z_{pq\text{–}rs}$ or mutual admittance $Y_{pq\text{–}rs}$ between the elements pq & rs

[Z] and [Y] are inverse matrices

$$[Z] = [Y]^{-1}$$

If there is no mutual coupling between the elements pq and rs then [Z] & [Y] do not contain off diagonal element.

4.3   FORMATION OF NETWORK MATRICES BY SINGULAR TRANSFORMATION

4.3.1 Network Performance Equations

A network is an interconnected set of elements.

In bus frame of reference, the performance of interconnected network is described by (n–1) nodal equations.

In matrix form

$$\overline{E}_{BUS} = \overline{Z}_{BUS}\,\overline{I}_{BUS}$$
$$\overline{I}_{BUS} = \overline{Y}_{BUS}\,\overline{E}_{BUS}$$

$\overline{E}_{BUS}$ is vector of bus voltages measured with respect to reference bus.

$\overline{I}_{BUS}$ is vector of impressed bus current.

$\overline{Z}_{BUS}$ bus impedance matrix with open circuited driving point and transfer impedance as the elements.

$\overline{Y}_{BUS}$ bus admittance matrix with short circuited driving point and transfer admittances as the elements.

4.3.2 Bus Admittance Matrix

Bus admittance matrix can be obtained by using bus incidence matrix to relate variables and parameters of primitive network quantities of an interconnected network.

The performance equation of a primitive network in admittance form.

$$\bar{i} + \bar{j} = \left[\overline{y}\right]\overline{v} \qquad\qquad .....(4.5)$$

$$A^t\bar{i} + A^t\bar{j} = \left[\overline{y}\right]A^t\overline{v} \qquad\qquad .....(4.6)$$

Matrix [A] is incidence of elements to buses.

$A^t\bar{i}$ is a vector in which each element in it is algebraic sum of currents flowing through it and terminating at a bus.

In KCL, algebraic sum of currents terminals at bus is zero.

$$A^t\bar{i} = 0 \qquad\qquad .....(4.7)$$

$A^t\bar{j}$ is a vector in which each element is the algebraic sum of source currents is equal to impressed bus current.

$$\overline{I}_{BUS} = A^t\bar{j} \qquad\qquad .....(4.8)$$

$$(\overline{I^*_{BUS}})^t = [j^*]^t A^*$$

As matrix A is real matrix $A^* = A$

$$(\overline{I^*_{BUS}})^t = [j^*]^t A) \qquad\qquad .....(4.9)$$

Power into the interconnected network $= (\overline{I^*_{BUS}})^t \overline{E}_{BUS}$

Power into the primitive network $= [j^*]^t \bar{v}$

As transformation of variables must be power invariant, power into the interconnected network = power into the primitive network.

$$(\overline{I^*_{BUS}})^t \overline{E_{BUS}} = [j*]^t \, \bar{v}$$

$$\overline{j}^{*t} \, A \, \overline{E_{BUS}} = [j*]^t \, \bar{v}$$

$$A \, \overline{E_{BUS}} = \bar{v} \qquad\qquad .....(4.10)$$

Substitute (4.7) and (4.8) in (4.6)

We get $\qquad \overline{I_{BUS}} = A^t[\bar{y}] \, \bar{v} \qquad\qquad .....(4.11)$

Substitute (4.10) in (4.11)

$$\overline{I_{BUS}} = A^t[\bar{y}] \, A \, \overline{E_{BUS}} \qquad\qquad .....(4.12)$$

$$Y_{BUS} \, E_{BUS} = A^t[\bar{y}] \, A\overline{E_{BUS}}$$

$$(\because \overline{I_{BUS}} = \overline{Y_{BUS}E_{BUS}})$$

$$Y_{BUS} = A^t[\bar{y}] \, A \qquad\qquad .....(4.13)$$

As $\overline{[A]}$ is singular matrix $A^t[\bar{y}] \, A$ is singular transformation of Y.

$$Z_{BUS} = Y_{BUS} = \{A^t[\bar{y}] \, A\}^{-1} \qquad\qquad .....(4.14)$$

4.4 FORMULATION OF [Y$_{BUS}$] FOR AN INTERCONNECTED NETWORK

Consider an interconnected power system network as shown in the figure 4.4

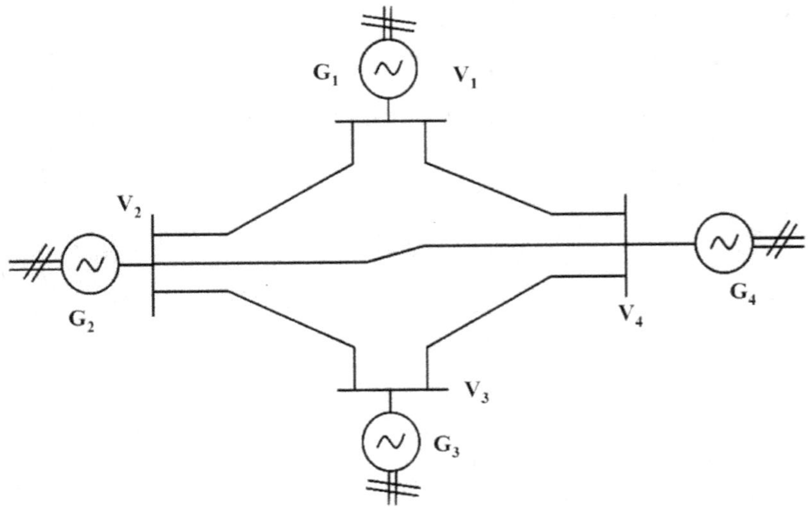

Figure 4.4 Interconnected power system network

As the nominal-π method is the accurate method to represent a transmission line, the nominal–π representation of transmission line is shown in figure 4.5.

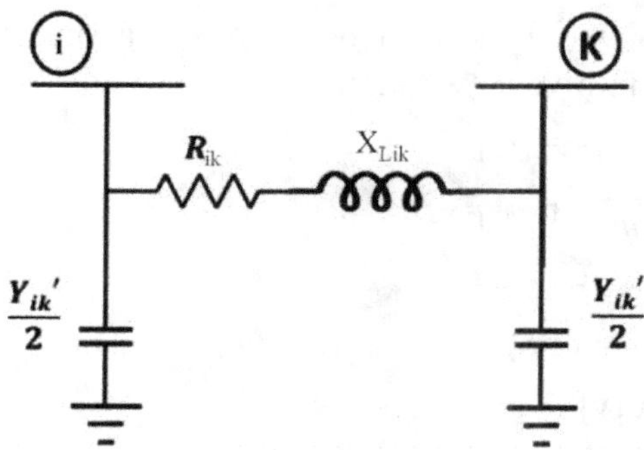

Figure 4.5 Interconnected power system network

General representation

Consider a transmission line connected between buses

1. Series impedance of transmission line

$$z_{ik} = R_{ik} + jX_{Lik} = R_{ik} + j\,\omega\,L_{ik} \qquad\qquad .....(4.15)$$

2. Series admittance of transmission line

$$y_{ik} = \frac{1}{Z_{ik}} = \frac{1}{R_{ik} + jX_{Lik}} \times \frac{R_{ik} - jX_{Lik}}{R_{ik} - jX_{Lik}}$$

$$y_{ik} = \frac{R}{R_{ik}^2 + X_{Lik}^2} - j\,\frac{X_{Lik}}{R_{ik}^2 + X_{Lik}^2}$$

$$y_{ik} = G_{ik} - j\,B_{ik} \qquad\qquad .....(4.16)$$

Half-line charging admittance

$$\frac{y_{ik}^|}{2} = \frac{1}{2}\left[y_{ik}^|\right] = \frac{1}{2}\left[j\,B_{cik}\right] = \frac{1}{2}\left[j\omega_{cik}\right] \qquad\qquad .....(4.17)$$

Representing each transmission line by nominal– π method as in figure 4.6

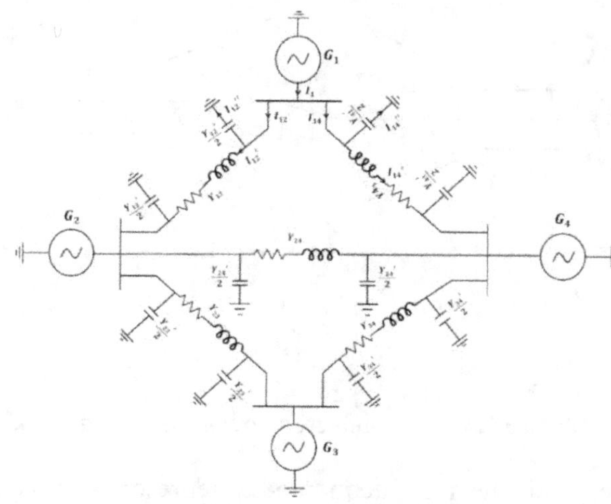

Figure 4.6 Nominal- π representation of power system network for the network shown in the figure 4.4.

Apply KCL at bus (1)

$$I_1 = I_{12} + I_{14}$$

$$= \{ I_{12}^{|} + I_{12}^{||} \} + \{ I_{14}^{|} + I_{14}^{||} \} \qquad \ldots..(4.18)$$

$$= \left[\left\{ y_{12}(V_1 - V_2) + + \frac{y_{i2}^{|}}{2}\,(V_1) \right\} + \left\{ y_{14}(V_1 - V_4) + \frac{y_{i4}^{|}}{2}\,(V_1) \right\} \right]$$

$$= \left(y_{12} + y_{14} + \frac{y_{i2}^{|}}{2} + \frac{y_{i4}^{|}}{2} \right) V_1 + y_{12}(-V_2) + 0.\,V_3 + y_{14}(-V_4)$$

$$\ldots..(4.19)$$

In standard form,

$$I = Y_{11}V_1 + Y_{12}V_2 + Y_{13}V_3 + Y_{14}V_4$$

Current injected into i^{th} bus is,

$$I_i = Y_{11}V_1 + Y_{12}V_2 + Y_{13}V_3 + \ldots\ldots\ldots\ldots\ldots = \sum_{k=1}^{n} Y_{ik}V_k \qquad \ldots..(4.20)$$

The admittance bus matrix for four bus network is as shown below.

$$\begin{bmatrix} I_1 \\ I_2 \\ I_3 \\ I_4 \end{bmatrix} = \begin{bmatrix} Y_{11} & Y_{12} & Y_{13} & Y_{14} \\ Y_{21} & Y_{22} & Y_{23} & Y_{24} \\ Y_{31} & Y_{32} & Y_{33} & Y_{34} \\ Y_{41} & Y_{42} & Y_{43} & Y_{44} \end{bmatrix} \begin{bmatrix} V_1 \\ V_2 \\ V_3 \\ V_4 \end{bmatrix}$$

$$[I_{BUS}] = [Y_{BUS}][V_{BUS}]$$

4.5 PROPERTIES OF $[Y_{BUS}]$

1. The diagonal elements consists of series admittance connected between buses and half line charging admittance connected to a bus.
2. The off-diagonal elements consist of only the series admittances connected between the buses.
3. $[Y_{BUS}] = [Y_{BUS}]^{T}$ therefore it is a symmetric matrix.
4. $[Y_{BUS}]$ is a square matrix i.e. most of the elements of $[Y_{BUS}]$ elements are zeros.

Q1. Determine primitive admittance matrix for a given power system network for the figure 4.7.

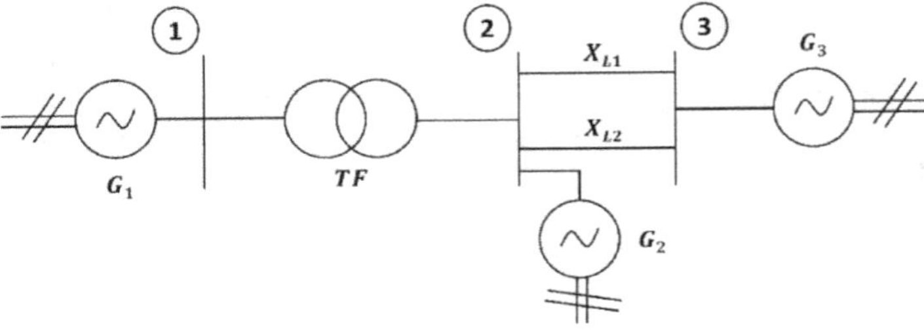

Figure 4.7 Power system network

Solution:

Represent each power system component by diagonal element.

$[y]=$

Elements	Elements					
	G_1	G_2	G_1	TF	T_{L1}	T_{L2}
G_1	Y_{G1}	0	0	0	0	0
G_2	0	Y_{G2}	0	0	0	0
G_3	0	0	Y_{G3}	0	0	0
TF	0	0	0	Y_{TF}	0	0
T_{L1}	0	0	0	0	Y_{XL1}	0
T_{L2}	0	0	0	0	0	Y_{XL2}

Q2. Formulate primitive admittance matrix with reactance of each element to be j0.1 for the figure 4.2

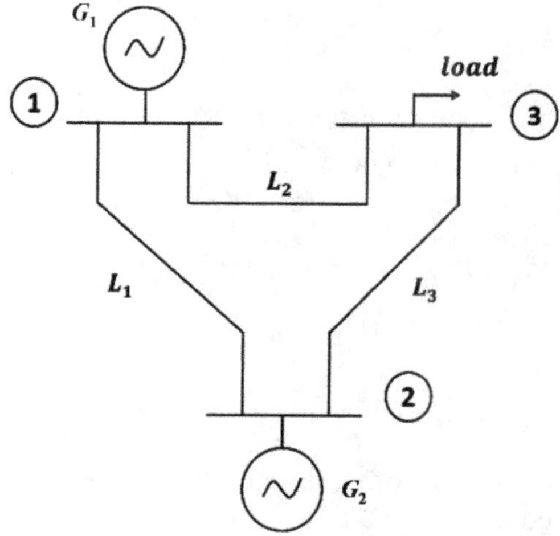

Figure 4.8 Power system network

Solution:

Represent each power system component by diagonal element.

Primitive impedance Matrix $[Z] =$ j

Elements	G_1	G_2	L_1	L_2	L_3
G_1	0.1	0	0	0	0
G_2	0	0.1	0	0	0
L_1	0	0	0.1	0	0
L_2	0	0	0	0.1	0
L_3	0	0	0	0	0.1

Primitive admittance matrix:

$$[\mathbf{y}] = \begin{array}{c} G_1 \\ G_2 \\ L_1 \\ L_2 \\ L_3 \end{array} \begin{bmatrix} 10 & 0 & 0 & 0 & 0 \\ 0 & 10 & 0 & 0 & 0 \\ 0 & 0 & 10 & 0 & 0 \\ 0 & 0 & 0 & 10 & 0 \\ 0 & 0 & 0 & 0 & 10 \end{bmatrix}$$

Q3. A power system network is shown in figure 4.9. The reactance of generator is $j0.1$ per unit and reactance for each line is $j0.3$ per unit. The two lines have mutual reactance of $j0.01$ per unit. Formulate primitive impedance and primitive admittance matrix.

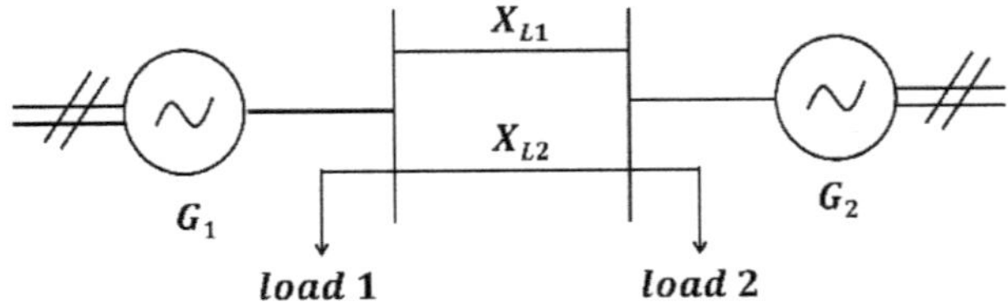

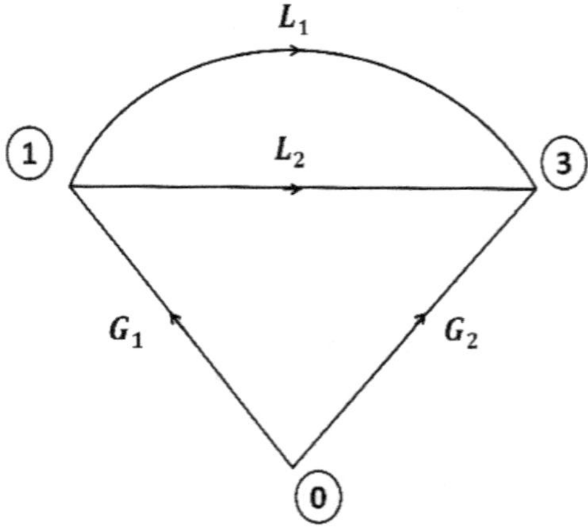

Figure 4.9 Power system network

Solution:

Represent each power system component by diagonal element and mutual coupling in the corresponding off-diagonal elements.

$$[Z] = j \begin{array}{c} G_1 \\ G_2 \\ XL_1 \\ XL_2 \end{array} \begin{array}{cccc} G_1 & G_2 & X_{L1} & X_{L2} \end{array} \begin{bmatrix} 0.1 & 0 & 0 & 0 \\ 0 & 0.1 & 0 & 0 \\ 0 & 0 & 0.3 & 0.01 \\ 0 & 0 & 0.01 & 0.3 \end{bmatrix}$$

$$[Y] = j \begin{array}{c} \\ G_1 \\ G_2 \\ X_{L1} \\ X_{L2} \end{array} \begin{array}{cccc} G_1 & G_2 & X_{L1} & X_{L2} \\ \begin{bmatrix} 10 & 0 & 0 & 0 \\ 0 & 10 & 0 & 0 \\ 0 & 0 & 3.33 & 0.01 \\ 0 & 0 & 0.01 & 3.33 \end{bmatrix} \end{array}$$

Q4. Form the Y bus by using singular transformation for the network shown in figure 4.4

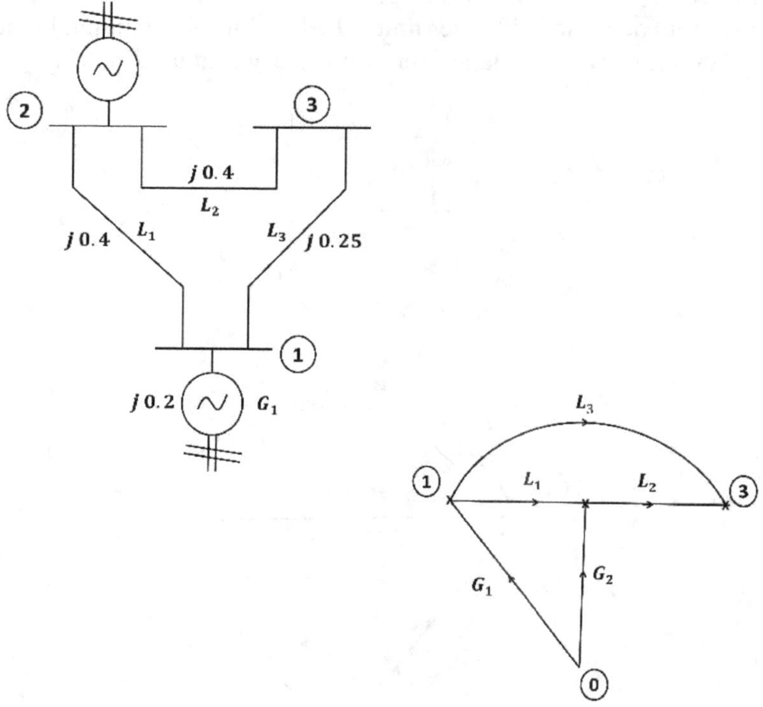

Figure 4.10 Power system network

Solution:

Method I: Singular Transformation Method

Bus incidence matrix is obtained as follows.

	Elements	Nodes		
		1	2	3
[A]=	G_1	−1	0	0
	G_2	0	−1	0
	L_1	1	−1	0
	L_2	0	1	−1
	L_2	1	0	−1

Primitive admittance matrix,

$$[\mathbf{y}] = [\mathbf{z}]^{-1} = \begin{matrix} G_1 \\ G_2 \\ L_1 \\ L_2 \\ L_3 \end{matrix}\begin{bmatrix} 0.2 & 0 & 0 & 0 & 0 \\ 0 & 0.1 & 0 & 0 & 0 \\ 0 & 0 & 0.4 & 0 & 0 \\ 0 & 0 & 0 & 0.4 & 0 \\ 0 & 0 & 0 & 0 & 0.25 \end{bmatrix}^{-1}$$

$$= +j \begin{bmatrix} -5 & 0 & 0 & 0 & 0 \\ 0 & -1 & 0 & 0 & 0 \\ 0 & 0 & 2.5 & 0 & 0 \\ 0 & 0 & 0 & 2.5 & 0 \\ 0 & 0 & 0 & 0 & 4 \end{bmatrix}$$

Now as per singular transformation,

$$Y_{BUS} = [A^t][\overline{Y}]\,[A]$$

$$= j \begin{bmatrix} -1 & 0 & 1 & 0 & 1 \\ 0 & -1 & -1 & 1 & 0 \\ 0 & 0 & 0 & -1 & -1 \end{bmatrix} \begin{bmatrix} -5 & 0 & 0 & 0 & 0 \\ 0 & -1 & 0 & 0 & 0 \\ 0 & 0 & 2.5 & 0 & 0 \\ 0 & 0 & 0 & 2.5 & 0 \\ 0 & 0 & 0 & 0 & 4 \end{bmatrix} \begin{bmatrix} -1 & 0 & 0 \\ 0 & -1 & 0 \\ 0 & -1 & 0 \\ 0 & 1 & -1 \\ 1 & 0 & -1 \end{bmatrix}$$

$$= j \begin{bmatrix} 5 & 0 & -2.5 & 0 & -4 \\ 0 & 10 & 2.5 & 2.5 & 0 \\ 0 & 0 & 0 & 2.5 & 4 \end{bmatrix} \begin{bmatrix} -1 & 0 & 0 \\ 0 & -1 & 0 \\ 0 & -1 & 0 \\ 0 & 1 & -1 \\ 1 & 0 & -1 \end{bmatrix}$$

$$= j \begin{bmatrix} -11.5 & 2.5 & 4 \\ 2.5 & 0 & 2.5 \\ 4 & 2.5 & -6.5 \end{bmatrix}$$

Method II: Direct Inspection Method

The diagonal element is the sum of the reciprocals of all the impedances connected to a bus.

The off-diagonal element is the negative of the reciprocal of the impedances connected between the buses.

$$Y_{bus} =$$

Bus	Node 1	Node 2	Node 3
1	$\dfrac{1}{j0.2} + \dfrac{1}{j0.4} + \dfrac{1}{j0.25}$	$\dfrac{-1}{j0.4}$	$\dfrac{-1}{j0.25}$
2	$\dfrac{-1}{j0.4}$	$\dfrac{1}{j0.1} + \dfrac{1}{j0.4} + \dfrac{1}{j0.4}$	$\dfrac{-1}{j0.4}$
3	$\dfrac{-1}{j0.25}$	$\dfrac{-1}{j0.4}$	$\dfrac{1}{j0.4} + \dfrac{1}{j0.25}$

$$Y_{bus} =$$

Bus	1	2	3
1	$-j11.5$	$j\,2.5$	$j4$
2	$j\,2.5$	$-j15$	$j\,2.5$
3	$j4$	$j\,2.5$	$-j6.5$

Q5. For the power system network in figure 4.11, obtain primitive admittance matrix by singular transformation method. Reactance of each line is j0.1 p.u.

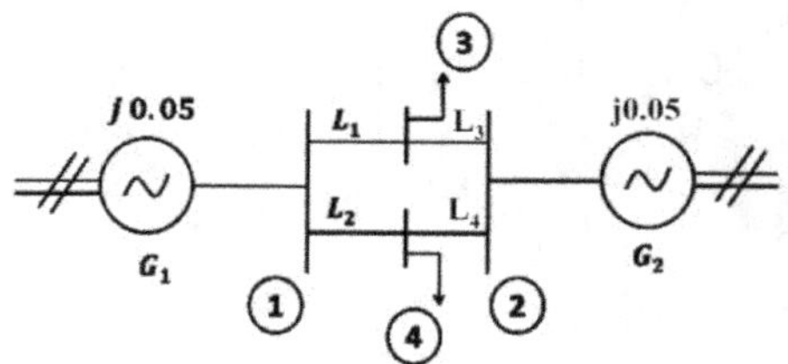

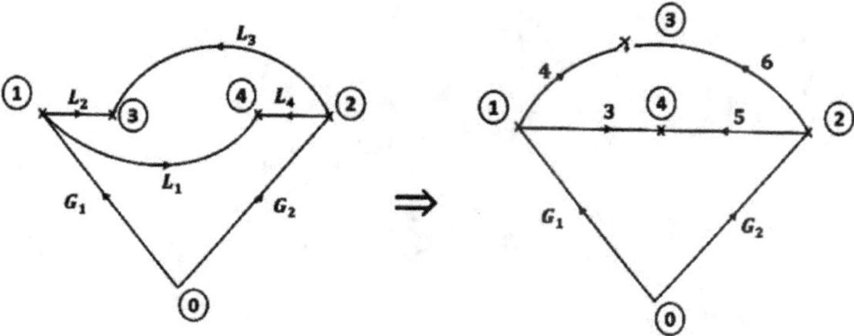

Figure 4.11 Power system network

Solution:

Represent each power system component by diagonal element and mutual coupling in the corresponding off-diagonal elements.

$$[A] = \begin{array}{c} \\ G_1 \\ G_2 \\ L_1 \\ L_2 \\ L_3 \\ L_4 \end{array} \begin{array}{cccc} 1 & 2 & 3 & 4 \\ \left[\begin{array}{cccc} -1 & 0 & 0 & 0 \\ 0 & -1 & 0 & 0 \\ 1 & 0 & 0 & -1 \\ 1 & 0 & -1 & 0 \\ 0 & 1 & 0 & -1 \\ 0 & 1 & -1 & 0 \end{array}\right] \end{array}$$

$$[y] = [z]^{-1} = j \begin{array}{c} 1 \\ 2 \\ 3 \\ 4 \\ 5 \\ 6 \end{array} \left[\begin{array}{cccccc} 0.05 & 0 & 0 & 0 & 0 & 0 \\ 0 & 0.05 & 0 & 0 & 0 & 0 \\ 0 & 0 & 0.1 & 0 & 0 & 0 \\ 1 & 0 & 0 & 0.1 & 0 & 0 \\ 0 & 1 & 0 & 0 & 0.1 & 0 \\ 0 & 1 & 0 & 0 & 0 & 0.1 \end{array}\right]$$

Q6. Consider the linear graph of a power system as shown in figure 4.12. Each element has series impedance of $(0.01 + j0.06)\,\Omega$. Compute $[Y_{BUS}]$ by singular transformation and verify.

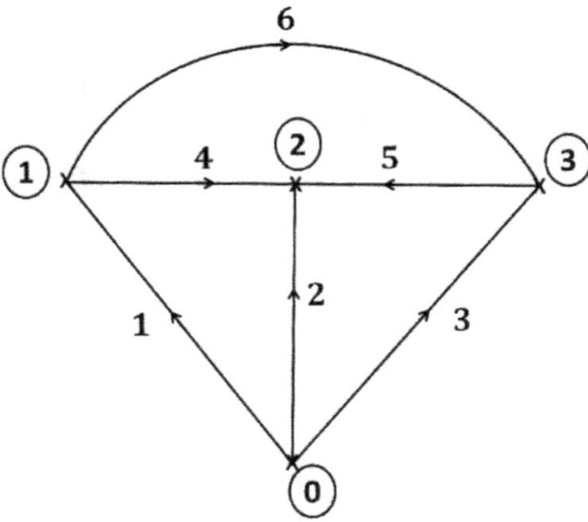

Figure 4.12 Power system network

Solution:

Represent each power system component by diagonal element.

Bus incidence matrix is,

$$[A] = \begin{array}{c} 1 \\ 2 \\ 3 \\ 4 \\ 5 \\ 6 \end{array}\begin{bmatrix} -1 & 0 & 0 \\ 0 & -1 & 0 \\ 0 & 0 & -1 \\ 1 & -1 & 0 \\ 0 & -1 & 1 \\ 1 & 0 & -1 \end{bmatrix}_{6 \times 3}$$

Method I: Singular Transformation Method

Priminitive admittance matrix,

$$[y]=[z]^{-1}$$

$$= j\begin{array}{c} 1 \\ 2 \\ 3 \\ 4 \\ 5 \\ 6 \end{array}\begin{bmatrix} (0.01+j0.06) & 0 & 0 & 0 & 0 & 0 \\ 0 & (0.01+j0.06) & 0 & 0 & 0 & 0 \\ 0 & 0 & (0.01+j0.06) & 0 & 0 & 0 \\ 0 & 0 & 0 & (0.01+j0.06) & 0 & 0 \\ 0 & 0 & 0 & 0 & (0.01+j0.06) & 0 \\ 0 & 0 & 0 & 0 & 0 & (0.01+j0.06) \end{bmatrix}^{-1}$$

$$= \begin{bmatrix} 2.7-j16.2 & 0 & 0 & 0 & 0 & 0 \\ 0 & 2.7-j16.2 & 0 & 0 & 0 & 0 \\ 0 & 0 & 2.7-j16.2 & 0 & 0 & 0 \\ 0 & 0 & 0 & 2.7-j16.2 & 0 & 0 \\ 0 & 1 & 0 & 0 & 2.7-j16.2 & 0 \\ 0 & 1 & 0 & 0 & 0 & 2.7-j16.2 \end{bmatrix}$$

$$[A]^T[Y] = \begin{bmatrix} -1 & 0 & 0 & 1 & 0 & 1 \\ 0 & -1 & 0 & -1 & -1 & 0 \\ 0 & 0 & -1 & 0 & 1 & -1 \end{bmatrix} \begin{bmatrix} 1 & 0 & 0 & 0 & 0 & 0 \\ 0 & 1 & 0 & 0 & 0 & 0 \\ 0 & 0 & 1 & 0 & 0 & 0 \\ 0 & 0 & 0 & 1 & 0 & 0 \\ 0 & 0 & 0 & 0 & 1 & 0 \\ 0 & 0 & 0 & 0 & 0 & 1 \end{bmatrix} [2.7 - j16.2]$$

$$= \begin{bmatrix} -2.7 + j16.2 & 0 & 0 & 2.7 - j16.2 & 0 & 2.7 - j16.2 \\ 0 & -2.7 + j16.2 & 0 & -2.7 + j16.2 & -2.7 + j16.2 & 0 \\ 0 & 0 & -2.7 + j16.2 & 0 & 2.7 - j16.2 & -2.7 + j16.2 \end{bmatrix}$$

$$[Y_{BUS}]\,[A]^T\,[Y]\,[A]$$

$$= \begin{bmatrix} -2.7 + j16.2 & 0 & 0 & 2.7 - j16.2 & 0 & 2.7 - j16.2 \\ 0 & -2.7 + j16.2 & 0 & -2.7 + j16.2 & -2.7 + j16.2 & 0 \\ 0 & 0 & -2.7 + j16.2 & 0 & 2.7 - j16.2 & -2.7 + j16.2 \end{bmatrix}$$

$$\times \begin{bmatrix} -1 & 0 & 0 \\ 0 & -1 & 0 \\ 0 & 0 & -1 \\ 1 & -1 & 0 \\ 0 & -1 & 1 \\ 1 & 0 & -1 \end{bmatrix}$$

$$= \begin{bmatrix} 8.1 - j48.6 & -2.7 + j16.2 & -2.7 + j16.2 \\ -2.7 + j16.2 & 8.1 - j48.6 & -2.7 + j16.2 \\ -2.7 + j16.2 & -2.7 + j16.2 & 8.1 - j48.6 \end{bmatrix}$$

Method II: Direct Inspection Method

The diagonal element is the sum of the reciprocals of all the impedances connected to a bus.

The off-diagonal element is the negative of the reciprocal of the impedances connected between the buses.

$$Y_{bus} = \begin{array}{c} \\ 1 \\ 2 \\ 3 \end{array} \begin{array}{ccc} 1 & 2 & 3 \\ \begin{bmatrix} 8.1 - j48.6 & -2.7 + j16.2 & -2.7 + j16.2 \\ -2.7 + j16.2 & 8.1 - j48.6 & -2.7 + j16.2 \\ -2.7 + j16.2 & -2.7 + j16.2 & 8.1 - j48.6 \end{bmatrix} \end{array}$$

Q7. For the power system transmission network shown in the figure below, formulate Y_{BUS} by singular transformation method where bus (1) is the reference.

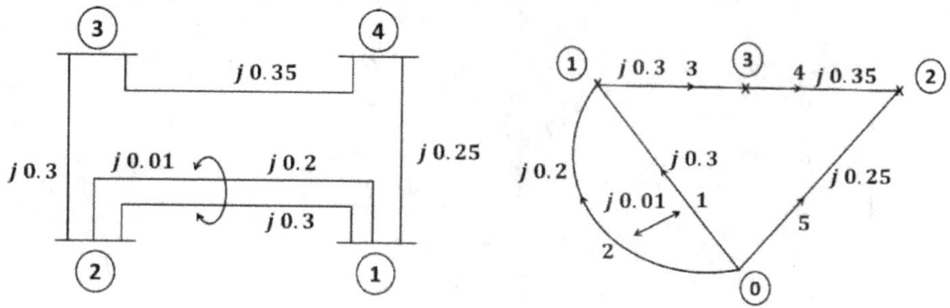

Figure 4.13 Power system network

Solution:

The bus incidence matrix is

$$
[A] = \begin{array}{c} \\ 1 \\ 2 \\ 3 \\ 4 \\ 5 \end{array}
\begin{array}{ccc} 2 & 3 & 4 \\ \end{array}
\begin{bmatrix} -1 & 0 & 0 \\ -1 & 0 & 0 \\ 1 & -1 & 0 \\ 0 & 1 & -1 \\ 0 & 0 & -1 \end{bmatrix}
$$

$$
[y] = [z]^{-1} = \begin{array}{c} \\ 1 \\ 2 \\ 3 \\ 4 \\ 5 \end{array}
\begin{array}{ccccc} 1 & 2 & 3 & 4 & 5 \\ \end{array}
\begin{bmatrix} j0.3 & j0.01 & 0 & 0 & 0 \\ j0.01 & j0.2 & 0 & 0 & 0 \\ 0 & 0 & j0.3 & 0 & 0 \\ 0 & 0 & 0 & j0.35 & 0 \\ 0 & 0 & 0 & 0 & j0.25 \end{bmatrix}^{-1}
$$

$[Y_{bus}] = [A^t][\overline{Y}]\,[A]$

$$
= j \begin{bmatrix} -1 & -1 & 1 & 0 & 0 \\ 0 & 0 & -1 & 1 & 0 \\ 0 & 0 & 0 & -1 & -1 \end{bmatrix}
\begin{bmatrix} 0.3 & 0.01 & 0 & 0 & 0 \\ 0.01 & 0.2 & 0 & 0 & 0 \\ 0 & 0 & 0.3 & 0 & 0 \\ 0 & 0 & 0 & 0.35 & 0 \\ 0 & 0 & 0 & 0 & 0.25 \end{bmatrix}^{-1}
\begin{bmatrix} -1 & 0 & 0 \\ -1 & 0 & 0 \\ 1 & -1 & 0 \\ 0 & 1 & -1 \\ 0 & 0 & -1 \end{bmatrix}
$$

$$
= j \begin{bmatrix} -1 & -1 & 1 & 0 & 0 \\ 0 & 0 & -1 & 1 & 0 \\ 0 & 0 & 0 & -1 & -1 \end{bmatrix}
\begin{bmatrix} -3.33 & -100 & 0 & 0 & 0 \\ -100 & -5 & 0 & 0 & 0 \\ 0 & 0 & -3.33 & 0 & 0 \\ 0 & 0 & 0 & -2.8 & 0 \\ 0 & 0 & 0 & 0 & -4 \end{bmatrix}
\begin{bmatrix} -1 & 0 & 0 \\ -1 & 0 & 0 \\ 1 & -1 & 0 \\ 0 & 1 & -1 \\ 0 & 0 & -1 \end{bmatrix}
$$

$$
= -j \begin{bmatrix} -103.33 & -105 & 3.33 & 0 & 0 \\ 0 & 0 & -3.33 & 2.8 & 0 \\ 0 & 0 & 0 & -2.8 & -4 \end{bmatrix}
\begin{bmatrix} -1 & 0 & 0 \\ -1 & 0 & 0 \\ 1 & -1 & 0 \\ 0 & 1 & -1 \\ 0 & 0 & -1 \end{bmatrix}
$$

$$
= -j \begin{bmatrix} 211.66 & -3.33 & 0 \\ -3.33 & 6.13 & 2.8 \\ 0 & 0 & 6.8 \end{bmatrix}
$$

$$
= j \begin{bmatrix} -211.66 & -3.33 & 0 \\ 3.33 & -6.13 & -2.8 \\ 0 & 0 & -6.8 \end{bmatrix}_{3 \times 3}
$$

Q8. A three-bus network is shown in figure 4.14, indicating the per unit impedance of each element

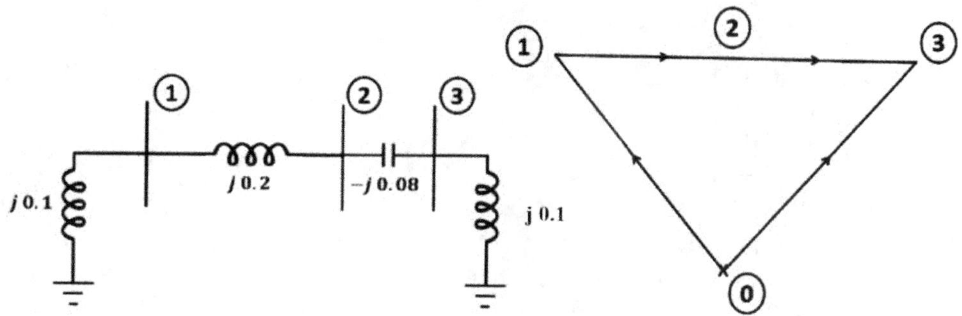

Figure 4.14 Power system network

Solution:

The diagonal element is the sum of the reciprocals of all the impedances connected to a bus.

The off-diagonal element is the negative of the reciprocal of the impedances connected between the buses.

The bus admittance matrix is

$$= j \begin{bmatrix} -15 & 5 & 0 \\ 5 & -7.5 & -12.5 \\ 0 & -12.5 & 2.5 \end{bmatrix}_{3 \times 3}$$

Q9. The reactances of the network elements are shown. Determine the Y_{bus} matrix?

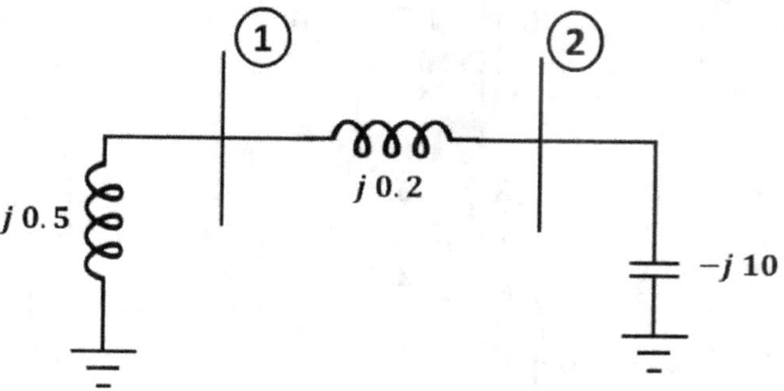

Figure 4.15 Power system network

Solution:

The diagonal element is the sum of the reciprocals of all the impedances connected to a bus.

The off-diagonal element is the negative of the reciprocal of the impedances connected between the buses.

$$
Y_{bus} = \begin{bmatrix} \dfrac{1}{j0.5} + \dfrac{1}{j0.2} & \dfrac{-1}{+j0.2} \\[2mm] \dfrac{-1}{+j0.2} & \dfrac{+1}{-j10} + \dfrac{1}{j0.2} \end{bmatrix}
$$

$$
Y_{bus} = j \begin{array}{cc} & \begin{array}{cc} 1 & \ 2 \end{array} \\ \begin{array}{c} 1 \\ 2 \end{array} & \begin{bmatrix} -7 & 5 \\ 5 & -4.9 \end{bmatrix} \end{array}
$$

Q10. Determine the $[Y_{BUS}]$ for the given network in figure 4.10?

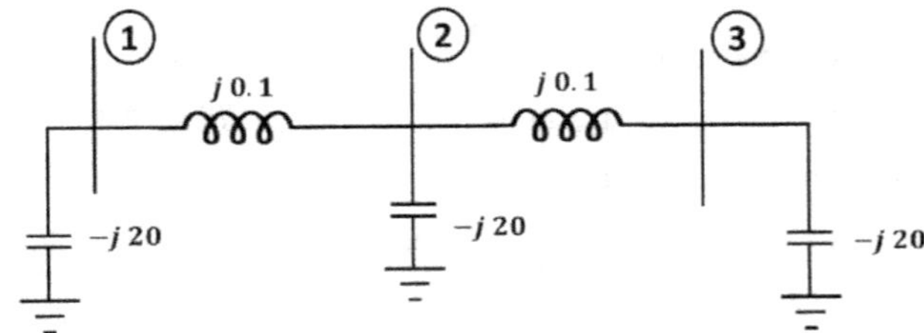

Figure 4.16 Power system network

Solution:

The diagonal element is the sum of the reciprocals of all the impedances connected to a bus.

The off-diagonal element is the negative of the reciprocal of the impedances connected between the buses.

$$
[Y_{BUS}] = \begin{array}{c} \\ 1 \\ 2 \\ 3 \end{array} \begin{array}{ccc} 1 & \quad 2 & \quad 3 \end{array} \begin{bmatrix} -9.95 & 10 & 0 \\ 10 & -19.95 & 10 \\ 0 & 10 & -9.95 \end{bmatrix}
$$

Q11. The power system network is having 20 buses, the $[Y_{BUS}]$ matrix of the power system network is 80% sparse. Determine the minimum no. of the transmission line required in the interconnected power system network.

Solution:

No. of elements in $[Y_{BUS}]$ matrix $= 20 \times 20 = 400$

80% are of zero elements

$$
400 \times \frac{80}{100} = 320
$$

No. of non–zero elements $= 80$

No. of non–zero diagonal elements $=$ no. of buses $= 20$

No. of non–zero off diagonal elements $= 80{-}20{=}60$

$$
\text{Therefore no. of transmission lines} = \frac{\text{no. of non - zero off diagnol elements}}{2} = \frac{60}{2} = 30
$$

Shortcut formula:

Formula $= \dfrac{n^2(1-x)-n}{2}$, n = bus number, x = fraction of sparsity

$$= \frac{20^2(1-0.8)-20}{2} = \frac{400\,(0.2)-20}{2} = \frac{80-20}{2} = 30$$

Q12. The Y_{BUS} matrix of a three-bus power system network is given

$$\begin{bmatrix} -j30 & j20 & j10 \\ j20 & -j20 & j5 \\ j10 & j5 & -j20 \end{bmatrix} \text{ p.u.}$$

Determine

(a) The impedance of transmission line between 1 and 2 buses

(b) The impedance of transmission line between 2 and 3 buses

(c) The impedance of transmission line between 3 and 1 buses

(d) Which among the 3 buses are having the shunt elements

Solution:

(a) The impedance of transmission line between buses 1 and 2 buses

$$z_{12} = \frac{1}{y_{12}} = \frac{1}{-Y_{12}} = \frac{1}{-j20} = +j0.05 \text{ p.u.}$$

(b) The impedance of transmission line between buses 2 and 3 buses

$$z_{23} = \frac{1}{y_{23}} = \frac{1}{-Y_{23}} = \frac{1}{-j5} = +j0.2 \text{ p.u.}$$

(c) Impedance of the transmission line between buses 3 and 1

$$z_{31} = \frac{1}{y_{31}} = \frac{1}{-Y_{31}} = \frac{1}{-j10} = +j0.1 \text{ p.u.}$$

(d) Which among the 3 buses are having the shunt elements

Bus 1:

$$Y_{11} = y_{12} + y_{13} + y_{10}$$
$$-j30 = -j20 - j10 + y_0$$
$$y_{10} = 0$$

Therefore shunt element is not connected to bus-1

Bus 2:

$$Y_{22} = y_{21} + y_{23} + y_{20}$$
$$-j20 = -j20 - j5 + y_{20}$$
$$j5 = y_{20}$$

A shunt capacitor is connected at bus2.

Impedance of shunt capacitor is,

$$z_{20} = \frac{1}{y_{20}} = \frac{1}{j5} = -j0.2$$

Bus 3:

$$Y_{33} = y_{31} + y_{32} + y_{30}$$
$$-j20 = -y_{31} - y_{32} + y_{30}$$
$$-j20 = -j10 - j5 + y_{30}$$
$$-j5 = y_{30}$$

A shunt inductor is connected at bus2.

Impedance of shunt inductor is,

$$z_{30} = \frac{1}{Y_{30}} = \frac{1}{-j5} = j0.2 \text{ p.u.}$$

Q13. The $[Y_{BUS}]$ matrix of 3 bus power system network is

$$\begin{bmatrix} -j20 & j10 & j10 \\ j10 & -j20 & j5 \\ j10 & j5 & -j15 \end{bmatrix} \text{p.u.}$$

Determine the following:

(a) A shunt reactor having reactance j0.2 is added to bus 3
     Determine the value of $Y_{33,}$new?
(b) A shunt reactor having susceptance $(-j10)$ is added to bus 2.
     Determine the value of $Y_{22,new}$.
(c) A shunt capacitance having a susceptance of $+0.05j$ is added to the bus 1.
     Determine the value of $Y_{11,new}$.
(d) A transmission line between buses 1 and 2 is disconnected. Determine the modified $[Y_{BUS}]$ matrix?
(e) A transmission line between buses 2 and 3 is disconnected. Determine the modified $[Y_{BUS}]$ matrix?
(f) A transmission line between buses 3 and 1 is disconnected. Determine the modified $[Y_{BUS}]$ matrix?

Solution:

(a) A shunt reactor having reactance j0.2 is added to bus 3

     Shunt admittance, $y_{30} = \frac{1}{j0.2} = -j5 \text{ p.u.}$

$$Y_{33,new} = Y_{33,old} + y_{30} = -j15 - j5 = -j20 \text{ p.u.}$$
$$Y_{33,new} = -j20 \text{ p.u.}$$

(b) A shunt reactor having susceptance $-j10$ is added to bus 2

     Shunt admittance, $y_{20} = -j10 \text{ p.u.}$

$$Y_{22,new} = Y_{22,old} + y_{20} = -j20 - j10 = -j30 \text{ p.u.}$$

(c) A shunt capacitance having a susceptance of $+j\,0.05$ is added to the bus 1.

     Shunt admittance, $y_{10} = +j0.05$

$$Y_{11,new} = Y_{11,old} + y_{10} = -j20 + j0.05 = -j19.95 \text{ p.u.}$$

(d) A transmission line between buses 1 and 2 is disconnected.

Hence, $Y_{12} = 0$, $Y_{21} = 0$

At bus 1,

$$Y_{11} = Y_{12} + Y_{13} + y_{10}$$
$$= (-y_{12}) + (-y_{13}) + y_{10}$$
$$= 0 + (-j10) + 0$$
$$Y_{11} = -j10$$

At bus 2,

$$Y_{22} = Y_{21} + Y_{23} + y_{20}$$
$$= (-y_{21}) + (-y_{23}) + y_{20}$$
$$= 0 + (-j5) + (-j5)$$
$$Y_{22} = -j10$$

$$[Y_{BUS}] = \begin{bmatrix} -j10 & 0 & j10 \\ 0 & -j10 & j5 \\ j10 & j5 & -j15 \end{bmatrix}_{3 \times 3}$$

(e) Transmission line between buses 2 and 3 is disconnected.

Hence, $Y_{23} = 0$, $Y_{32} = 0$

At bus 2,

$$Y_{22} = Y_{21} + Y_{23} + y_{20}$$
$$= (-y_{21}) + (-y_{23}) + y_{20}$$
$$= (-j10) + 0 + (-j5)$$
$$Y_{22} = -j15$$

At bus3,

$$Y_{33} = Y_{31} + Y_{32} + y_{30}$$
$$= (-y_{31}) + (-y_{32}) + y_{30}$$
$$= (-j10) + 0 + 0$$
$$Y_{33} = -j10$$

$$[Y_{BUS}] = \begin{bmatrix} -j20 & j10 & j10 \\ j10 & -j15 & 0 \\ j10 & 0 & -j10 \end{bmatrix}_{3 \times 3}$$

(f) A transmission line between buses 3 and 1 is disconnected

Hence, $Y_{31} = 0$, $Y_{13} = 0$

Bus 1,

$$Y_{11} = Y_{12} + Y_{13} + Y_{10}$$
$$= (-y_{12}) + (-y_{13}) + y_{10}$$
$$= (-j10) + 0 + 0$$
$$Y_{11} = -j10$$

Bus 3,

$$Y_{33} = Y_{31} + Y_{32} + Y_{30}$$
$$= (-y_{31}) + (-y_{32}) + y_{30}$$
$$= 0 + (-j5) + 0$$
$$= -j5$$

$$[Y_{BUS}] = \begin{bmatrix} -j10 & j10 & 0 \\ j10 & -j20 & j5 \\ 0 & j5 & -j5 \end{bmatrix}_{3 \times 3}$$

OBJECTIVE QUESTIONS

1. What does the diagonal elements of Y_{bus} consists of?

    Ans: The diagonal elements of aY_{bus} matrix consists of series admittance of the elements between the buses and shunt elements connected to a bus.

2. What does the diagonal elements of Y_{bus} consists of?

    Ans: The off–diagonal elements of aY_{bus} matrix consists of only series admittances of the elements between the buses.

3. Where the series admittances between transmission lines are reflected in Y_{bus} matrix?

    Ans: The series admittances of elements between the buses are reflected in both diagonal and off–diagonal elements in Y_{bus} matrix.

4. Where the shunt admittances are reflected in Y_{bus} matrix?

    Ans: The shunt elements connected to a bus are reflected only in diagonal elements.

5. What are the properties of Y_{bus} matrix?

    Ans: (a) Y_{bus} is a square matrix (b) Y_{bus} is a sparse matrix (c) Y_{bus} is a symmetric matrix.

6. What are the methods to determine Y_{bus} matrix?

    Ans: Y_{bus} matrix of a power system network can be obtained by (a) Direct inspection method (b) Step by step method (c) Singular transformation method.

7. Is the reference node considered in Y_{bus} matrix formulation?

    Ans: Reference node is always not considered in the formation of Y_{bus} by any method.

8. What is the equation of current at i^{th} bus in Y_{bus} analysis?

    Ans: The current at an ith bus is $I_i =$

9. Y_{bus} by singular transformation matrix is $Y_{bus} = [A]^T [y] [A]$ where 'y' is primitive admittance matrix.

10. What is a primitive admittance matrix?

    Ans: A primitive admittance matrix is a matrix which is an inverse of primitive impedance matrix, the diagonal elements of the primitive impedance matrix being the self–impedances of the network elements and the off–diagonal elements being the mutual impedances between the network elements.

11. What is the nature of primitive impedance matrix without mutual coupling between the network elements?

    Ans: Primitive impedance matrix is a diagonal matrix.

12. How Y_{bus} is determined with the network elements mutually coupled?

    Ans: Y_{bus} is obtained by singular transformation.

13. Why the determination of Y_{bus} matrix using an incidence matrix is called singular transformation?

    Ans: The incidence matrix used in the determination of an interconnected network matrix is called a transformation matrix and is a singular matrix. Hence this transformation is called as singular transformation.

5   Impedance Bus Matrix

5.1   INTRODUCTION

Impedance bus matrix is used in short circuit analysis.

As the shunt elements are neglected and resistance of the transmission line is ignored,

$$Z_{BUS} \neq Y_{BUS}^{-1}$$

There are many method to determine impedance bus matrix.

5.2   CALCULATION OF IMPEDANCE MATRIX WITHOUT MUTUAL COUPLING BETWEEN THE TRANSMISSION LINES

Assumptions

1. The shunt elements connected to a bus are neglected.
2. Resistance of line is neglected.

$$Z_{BUS} \neq Y_{BUS}^{-1}$$

Consider a linear and passive network as shown in figure 5.1.

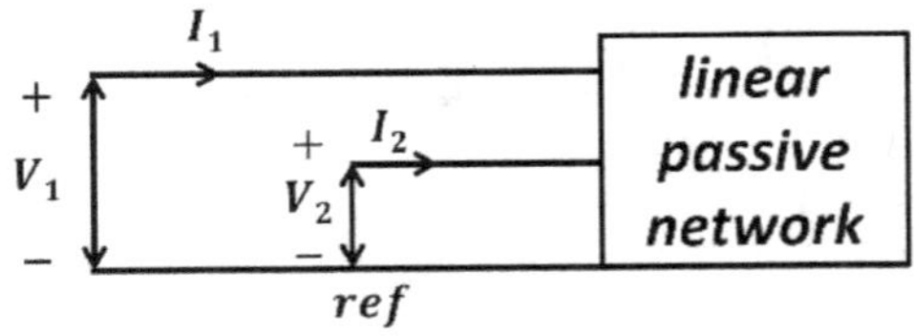

Figure 5.1 Linear and passive network

The basic two port network equations are

$$V_1 = Z_{11}\, I_1 + Z_{12}\, I_2$$
$$V_2 = Z_{21}\, I_1 + Z_{22}\, I_2$$

In the matrix form,

$$\begin{bmatrix} V_1 \\ V_2 \end{bmatrix} = \begin{bmatrix} Z_{11} & Z_{12} \\ Z_{21} & Z_{22} \end{bmatrix} \begin{bmatrix} I_1 \\ I_2 \end{bmatrix}$$

Type – 1 Modification - Adding an element between new bus and reference bus

Consider a linear and passive network as in figure 5.2.

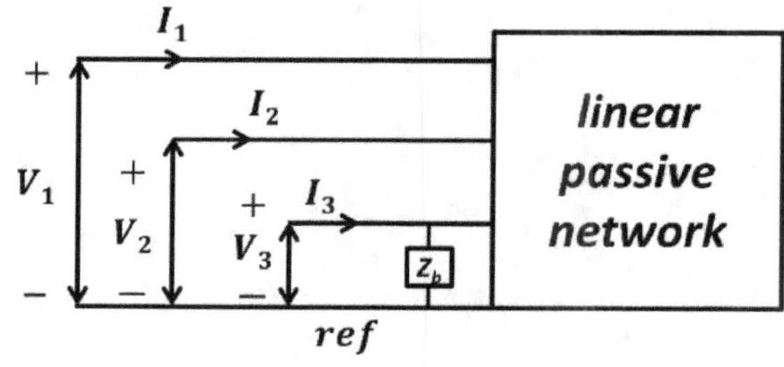

Figure 5.2 Linear and passive network

An element with impedance 'Z_b' is added to the reference bus. Therefore a new bus is created and size of Z_{bus} matrix increases by one and the added element is called branch.

The basic two port network equations in the matrix form are

$$\begin{bmatrix} V_1 \\ V_2 \\ V_3 \end{bmatrix} = \begin{bmatrix} Z_{11} & Z_{12} & Z_{13} \\ Z_{21} & Z_{22} & Z_{23} \\ Z_{31} & Z_{32} & Z_{33} \end{bmatrix} \begin{bmatrix} I_1 \\ I_2 \\ I_3 \end{bmatrix}$$

From the network $V_3 = Z_b I_3 = 0 . I_1 + 0 . I_2 + Z_b I_3$ $\qquad\qquad$ .....(5.1)

From the matrix $V_3 = Z_{31} I_1 + Z_{32} I_2 + Z_{33} I_3$ $\qquad\qquad$ .....(5.2)

Calculation of off–diagonal elements

From (5.1) and (5.2)

$$Z_{31} = 0, \; Z_{32} = 0$$

Let k = new bus created

∴ $\qquad\qquad Z_{ki} = 0$

Where i = 1,2,........n; n is size of previous Z-bus matrix

(No. of buses initially)

$$Z_{13} = Z_{31}$$
$$Z_{23} = Z_{32}$$
$$Z_{ik} = Z_{ki} = 0$$
$$i = 1,2,........n$$

(No. of buses initially)

i.e. size of previous Z_{BUS} matrix

Calculation of diagonal elements

From (5.1) and (5.2)

$$Z_{33} = Z_b \text{ in standard form } Z_{kk} = Z_b$$

Type–2 Modification

Adding an element from a new bus to old bus.

Consider a linear and passive network as in figure 5.3.

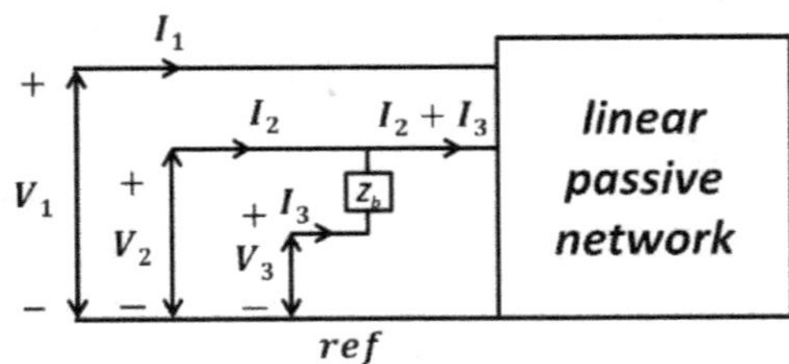

Figure 5.3 Linear and passive network

Consider an element added from an old bus to a new bus.

Therefore, size of $[Z_{BUS}]$ increases by one and the added element is known as branch.

$$V_1 = Z_{11} I_1 + Z_{12} (I_2 + I_3) \qquad \qquad .....(5.3)$$
$$= Z_{11} I_1 + Z_{12}I_2 + Z_{12}I_3$$
$$V_2 = Z_{21} I_1 + Z_{22} (I_2 + I_3) \qquad \qquad .....(5.4)$$
$$= Z_{21}I_1 + Z_{22}I_2 + Z_{22}I_3$$

Apply KVL to the Loop,

$$V_2 + V_b I_3 - V_3 = 0 \qquad \qquad .....(5.5)$$
$$V_3 = Z_b I_3 + Z_{21} I_1 + Z_{22}I_2 + Z_{22} I_3$$
$$V_3 = Z_{21} I_1 + Z_{22} I_2 + (Z_{22} + Z_b) I_3 \qquad \qquad .....(5.6)$$

Arranging the equation on matrix form

$$\begin{bmatrix} V_1 \\ V_2 \\ V_3 \end{bmatrix} = \begin{bmatrix} Z_{11} & Z_{12} & Z_{13} \\ Z_{21} & Z_{22} & Z_{23} \\ Z_{31} & Z_{32} & Z_{33} + Z_b \end{bmatrix} \begin{bmatrix} I_1 \\ I_2 \\ I_3 \end{bmatrix} \qquad \qquad .....(5.7)$$

The standard form is

$$\begin{bmatrix} V_1 \\ V_2 \\ V_3 \end{bmatrix} = \begin{bmatrix} Z_{11} & Z_{12} & Z_{13} \\ Z_{21} & Z_{22} & Z_{23} \\ Z_{31} & Z_{32} & Z_{33} \end{bmatrix} \begin{bmatrix} I_1 \\ I_2 \\ I_3 \end{bmatrix} \qquad \qquad .....(5.8)$$

Calculation of off diagonal elements

Comparing (5.7) and (5.8),

$$Z_{23} = Z_{22} \; ; \; Z_{31} = Z_{21} \; ; \; Z_{32} = Z_{22} \; ; Z_{13} = Z_{12}$$

Say new bus = k

Old bus = j

$$Z_{ki} = Z_{ji}$$

$i = 1,2,3\ldots\ldots.n$

n = size of previous Z_{BUS} matrix

Calculation of diagonal elements

$$Z_{33} = Z_{22} + Z_b$$

in standard form $Z_{kk} = Z_{jj} + Z_b$

Type – 3 Modification - Addition of element between reference bus and old bus,

Consider a linear and passive network as in figure 5.4.

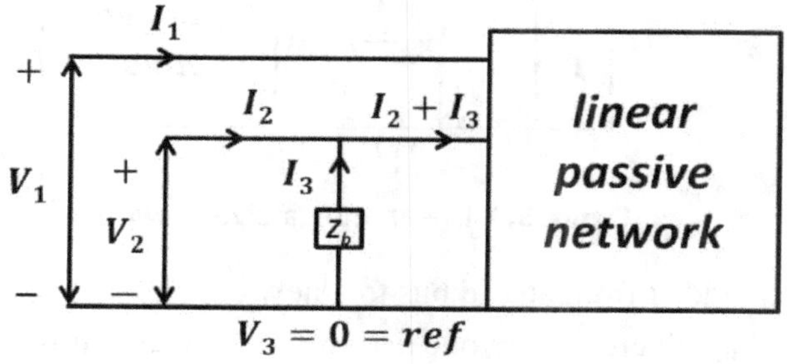

Figure 5.4 Linear and passive network

Connect bus 3 to the reference bus

As voltage of the reference bus is zero $V_3 = 0$

$$V_1 = Z_{11} I_1 + Z_{12} (I_2 + I_3)$$

$$V_1 = Z_{11} I_1 + Z_{12} I_2 + Z_{12}I_3 \qquad \ldots(5.9)$$

$$V_2 = Z_{21} I_1 + Z_{22} (I_2 + I_3)$$

$$V_2 = Z_{21}I_1 + Z_{22}I_2 + Z_{22}I_3 \qquad \ldots(5.10)$$

$$V_3 = V_2 + I_3 Z_b \qquad \ldots(5.11)$$

Substituting (5.10) in (5.11)

$$Z_{21} I_1 + Z_{22} I_2 + Z_{22} I_3 + Z_b I_3 = 0$$

$$Z_{21} I_1 + Z_{22} I_2 + (Z_{22} + Z_b) I_3 = 0$$

$$I_3 = \frac{-(Z_{21} I_1 + Z_{22} I_2)}{(Z_{22} + Z_b)} \qquad \ldots(5.12)$$

Substituting (5.12) in (5.9) and (5.10)

$$V_2 = Z_{21}I_1 + Z_{22} I_2 + Z_{22} \frac{(-Z_{21} I_1 - Z_{22} I_{22})}{(Z_{22} + Z_b)}$$

$$V_2 = \frac{1}{(Z_{22} + Z_b)}[(Z_{21} (Z_{22} + Z_b) - Z_{11}Z_{22})I_1 + (Z_{22}(Z_{22} + Z_b) - Z_{22}^2)I_2]$$

As new bus not creating the size of $[Z_{BUS}]$ do not increase by 1. Therefore, added element is known as link.

$$V_2 = \left[\left(Z_{21} - \frac{Z_{11}Z_{22}}{Z_{22} + Z_b}\right)I_1 + \left(Z_{22} - \frac{Z_{22}^2}{Z_{22} + Z_b}\right)I_2\right]$$

$$V_1 = \left[\left(Z_{11} - \frac{Z_{12}Z_{21}}{Z_{22}+ Z_b}\right)I_1 + \left(Z_{12} - \frac{Z_{12}Z_{22}}{Z_{22}+ Z_b}\right)I_2\right] \qquad .....(5.13)$$

$$\begin{bmatrix} V_1 \\ V_2 \end{bmatrix} = \begin{bmatrix} Z_{11} & Z_{12} \\ Z_{21} & Z_{22} \end{bmatrix}\begin{bmatrix} I_1 \\ I_2 \end{bmatrix} - \frac{1}{Z_{22} + Z_b}\begin{bmatrix} Z_{11} & Z_{12} \\ Z_{21} & Z_{22} \end{bmatrix}\begin{bmatrix} I_1 \\ I_2 \end{bmatrix}$$

$$= \begin{bmatrix} Z_{11} & Z_{12} \\ Z_{21} & Z_{22} \end{bmatrix}\begin{bmatrix} I_1 \\ I_2 \end{bmatrix} - \frac{1}{Z_{22}+ Z_b}\begin{bmatrix} Z_{12} \\ Z_{22} \end{bmatrix}\begin{bmatrix} Z_{21} & Z_{22} \end{bmatrix}\begin{bmatrix} I_1 \\ I_2 \end{bmatrix} \qquad .....(5.14)$$

From Equation (5.14),

$$Z_{BUS\,(new)} = Z_{BUS\,(old)} - \frac{\begin{bmatrix} Z_{12} \\ Z_{22} \end{bmatrix}\begin{bmatrix} Z_{21} & Z_{22} \end{bmatrix}}{Z_{22} + Z_b}$$

$$Z_{BUS\,(new)} = Z_{BUS\,(old)} - \frac{\begin{bmatrix} Z_{1k} \\ Z_{2k} \\ . \\ . \\ Z_{nk} \end{bmatrix}\begin{bmatrix} Z_{21} & Z_{22} & . & . & Z_{nk} \end{bmatrix}}{Z_{kk} + Z_b}$$

Corresponding to an element

$$Z_{BUS\,(new)} = Z_{BUS\,(old)} - \frac{[Z_{ik}][Z_{kj}]^T}{Z_{kk} + Z_b}$$

i = 1,2,3....n

j = 1,2,3....n

n = size of Z bus matrix

Type – 4 Modification- Addition of an element between old buses.

Consider a linear and passive network as in figure 5.5.

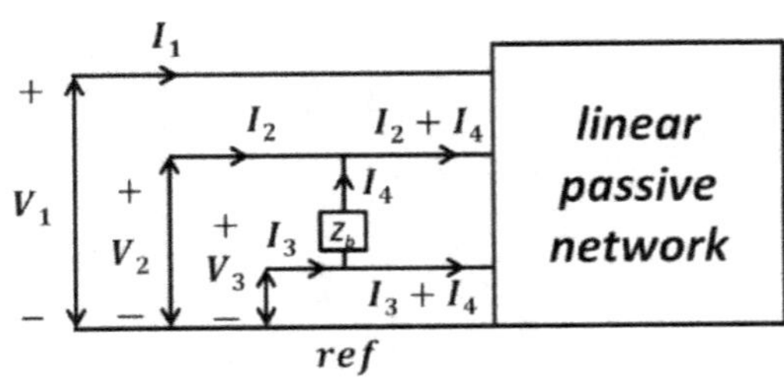

Figure 5.5 Linear and passive network

An element with impedance 'Z_b' is added between old buses 2 and 3. The size of $[Z_{BUS}]$ do not increase by one and the added element is called link.

$$V_1 = Z_{11}\,I_1 + Z_{12}\,(I_2 + I_4) + Z_{13}\,(I_3 - I_4)$$

$$V_1 = Z_{11} I_1 + Z_{12}I_2 + Z_{13}I_3 + (Z_{12} - Z_{13}) I_4$$

$$V_2 = Z_{12} I_1 + Z_{22} (I_2 + I_4) + Z_{23} (I_3 - I_4) \qquad .....(5.15)$$

$$V_2 = Z_{21} I_1 + Z_{22}I_2 + Z_{23}I_3 + (Z_{22} - Z_{23}) I_4 \qquad .....(5.16)$$

$$V_3 = Z_{31} I_1 + Z_{32}I_2 + Z_{33}I_3 + (Z_{32} - Z_{33}) I_4 \qquad .....(5.17)$$

$$\begin{bmatrix} V_1 \\ V_2 \\ V_3 \end{bmatrix} = \begin{bmatrix} Z_{11} & Z_{12} & Z_{13} \\ Z_{21} & Z_{22} & Z_{23} \\ Z_{31} & Z_{32} & Z_{33} \end{bmatrix} \begin{bmatrix} I_1 \\ I_2 \\ I_3 \end{bmatrix} + \begin{bmatrix} Z_{12} - Z_{13} \\ Z_{22} - Z_{23} \\ Z_{32} - Z_{33} \end{bmatrix} [I_4] \qquad .....(5.18)$$

$$V_3 = Z_b + V_2 I_4 \qquad .....(5.19)$$

Substitute V_2 and V_3 in (5.19)

$\Rightarrow$
$$Z_{31} I_1 + Z_{32}I_2 + Z_{33}I_3 + (Z_{32} - Z_{33})I_4 - Z_{21}I_1 + Z_{22}I_2$$
$$-Z_{23}I_3 - (Z_{22} - Z_{23}) I_4 = Z_b I_4$$

$\Rightarrow$
$$(Z_{31} - Z_{21})I_1 + (Z_{32} - Z_{22})I_2 + (Z_{33} - Z_{23}) I_3$$
$$= (Z_b + Z_{22} - Z_{23} - Z_{32} + Z_{33}) I_4$$

$$I_4 = \frac{(Z_{31} - Z_{21})I_1 + (Z_{32} - Z_{22})I_2 + (Z_{33} - Z_{23}) I_3}{(Z_b + Z_{22} - Z_{23} - Z_{32} + Z_{33})}$$

$\Rightarrow$
$$I_4 = \frac{-1}{(Z_b + Z_{22} - Z_{23} - Z_{32} + Z_{33})} \begin{bmatrix} Z_{21} - Z_{31} \\ Z_{22} - Z_{32} \\ Z_{23} - Z_{33} \end{bmatrix}^T \begin{bmatrix} I_1 \\ I_2 \\ I_3 \end{bmatrix}$$

substitute I_4 in (5.18)

$\Rightarrow$
$$\begin{bmatrix} V_1 \\ V_2 \\ V_3 \end{bmatrix} = \begin{bmatrix} Z_{11} & Z_{12} & Z_{13} \\ Z_{21} & Z_{22} & Z_{23} \\ Z_{31} & Z_{32} & Z_{33} \end{bmatrix} \begin{bmatrix} I_1 \\ I_2 \\ I_3 \end{bmatrix} - \frac{1}{p} \begin{bmatrix} Z_{12} - Z_{13} \\ Z_{22} - Z_{23} \\ Z_{32} - Z_{33} \end{bmatrix} \times \begin{bmatrix} Z_{21} - Z_{31} \\ Z_{22} - Z_{32} \\ Z_{23} - Z_{33} \end{bmatrix}^T \begin{bmatrix} I_1 \\ I_2 \\ I_3 \end{bmatrix}$$

$$Z_{\text{BUS (new)}} = Z_{\text{BUS (old)}} - \frac{1}{p} \begin{bmatrix} Z_{12} - Z_{13} \\ Z_{22} - Z_{23} \\ Z_{32} - Z_{33} \end{bmatrix} \times \begin{bmatrix} Z_{21} - Z_{31} \\ Z_{22} - Z_{32} \\ Z_{23} - Z_{33} \end{bmatrix}^T$$

where
$$P = Z_b + Z_{22} - Z_{23} - Z_{32} + Z_{33}$$

$$Z_{BUS \, (new)} = Z_{BUS \, (old)} - \frac{1}{p} \begin{bmatrix} Z_{1k} - Z_{1l} \\ Z_{2k} - Z_{2l} \\ . \\ . \\ Z_{nk} - Z_{nl} \end{bmatrix} \times \begin{bmatrix} Z_{k1} - Z_{l1} \\ Z_{k2} - Z_{l2} \\ . \\ . \\ Z_{k3} - Z_{l3} \end{bmatrix}^T \qquad .....(5.20)$$

K = old bus (2)

L = old bus (3)

$$P = Z_{kk} + Z_{ll} - Z_{kl} - Z_{lk} + Z_b \qquad .....(5.21)$$

Corresponding to an element

$$Z_{BUS \, (new)} = Z_{BUS \, (old)} - \frac{1}{p} [Z_{ik} - Z_{il}] \times [Z_{kj} - Z_{lj}]^T$$

$i = 1,2,3....n$

$j = 1,2,3....n$

$n \rightarrow$ size of Z_{BUS} matrix

$$Z_{BUS\,(new)} = Z_{BUS\,(old)} - \frac{1}{p}[Z_{ik} - Z_{il}] \times [Z_{jk} - Z_{jl}]^T$$

Q1. Obtain the Z_{Bus} for four bus sample system as given below in figure 5.6. The series impedance of each line is given in the table. Reference bus is denoted by R. No line is mutually coupled.

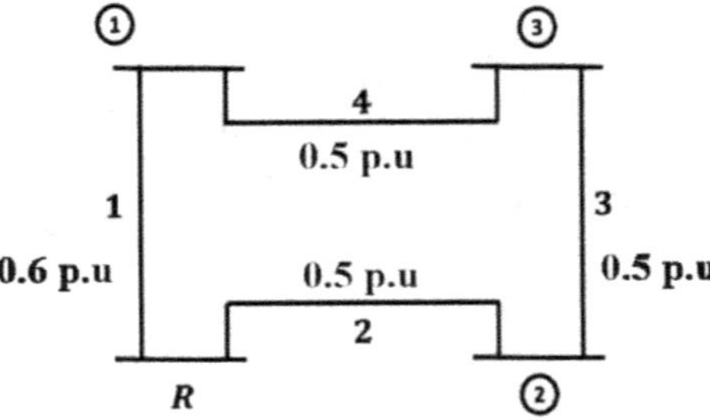

Figure 5.6 Power system network

Solution:

Step 1: An element with impedance 0.6 p.u. is added to reference bus A new bus is created the size of Z_{bus} matrix increases by 1 & the added element is known as branch. i.e Type-1 modification

$$Z_{bus} = \begin{bmatrix} Z_{RR} & Z_{R1} \\ Z_{1R} & Z_{11} \end{bmatrix}$$

The impedance measured with respect to reference bus is zero.

$$Z_{bus} = \begin{bmatrix} 0 & 0 \\ 0 & Z_{11} \end{bmatrix} = [Z_{11}]_{1 \times 1}$$

Calculation of diagonal element

$$Z_{bus} = [Z_{11}]_{1 \times 1}$$

In standard form $Z_{kk} = Z_b$

$$Z_{11} = 0.6 \text{ p.u.}$$

$$[Z_{Bus}] = [0.6]_{1 \times 1} \text{ p.u.}$$

Step 2: Add an element with an impedance of 0.5 p.u. to reference bus. A new bus is created. Size of Z_{bus} increase by one and the added element is known as branch i.e. type 1 modification

$$Z_{bus} = \begin{bmatrix} Z_{11} & Z_{12} \\ Z_{21} & Z_{22} \end{bmatrix}_{2 \times 2}$$

$$= \begin{bmatrix} 0.6 & Z_{12} \\ Z_{21} & Z_{22} \end{bmatrix}_{2 \times 2}$$

Calculation of off diagonal elements:

$K = $ new bus $= 2$

$$Z_{ik} = 0, \quad i = 1,2,3,\ldots n$$

$$Z_{ik} = 0 \Rightarrow Z_{i2} = 0 \quad \text{for} \quad i = 1 \quad \Rightarrow Z_{12} = 0 = Z_{21}$$

$$Z_{bus} = \begin{bmatrix} 0.6 & 0 \\ 0 & Z_{22} \end{bmatrix}_{2 \times 2}$$

Calculation of diagonal elements $Z_{kk} = Z_b$:

$$k = 2 \text{ (new bus)}$$

$$Z_{22} = 0.5 \text{ p.u.} = Z_b$$

$$Z_{bus} = \begin{bmatrix} 0.6 & 0 \\ 0 & 0.5 \end{bmatrix}_{2 \times 2} \text{ p.u.}$$

Step 3: Add an element with impedance 0.5 p.u. is added to bus 2.

A new bus 3 is created the size of matrix increases by one and the added element is known as branch i.e. type-2 modification.

$$Z_{bus} = \begin{bmatrix} Z_{11} & Z_{12} & Z_{13} \\ Z_{21} & Z_{22} & Z_{23} \\ Z_{31} & Z_{32} & Z_{33} \end{bmatrix}$$

$$= \begin{bmatrix} 0.6 & 0 & Z_{13} \\ 0 & 0.5 & Z_{23} \\ Z_{33} & Z_{32} & Z_{33} \end{bmatrix}$$

Calculation of off diagonal elements:

$$k = 3 \text{ (new bus)}$$

$$j = 2 \text{ (old bus)}$$

$$Z_{ik} = Z_{ij} \quad i = 1, 2, \ldots n; \ i \neq k \Rightarrow Z_{i3} = Z_{i2}$$

$$i = 1: Z_{13} = Z_{12} = 0 = Z_{31}$$

$$i = 2: Z_{23} = Z_{22} = 0.5 \text{ p.u.} = Z_{32}$$

$$[Z_{Bus}] = \begin{bmatrix} 0.6 & 0 & 0 \\ 0 & 0.5 & 0.5 \\ 0 & 0.5 & Z_{33} \end{bmatrix}$$

Calculation of diagonal elements:

$$Z_{kk} = Z_{ij} + Z_b$$

$$Z_{33} = Z_{22} + Z_b$$

$$Z_{33} = 0.5 + 0.5 = 1 \text{ p.u.}$$

$$[Z_{Bus}] = \begin{bmatrix} 0.6 & 0 & 0 \\ 0 & 0.5 & 0.5 \\ 0 & 0.5 & 1 \end{bmatrix} \text{ p.u.}$$

Step 4: Add an element between 1 & 3 with impedance 0.25 p.u. The size of Z_{Bus} does not increase by 1 since it is added between existing buses and the added element is known as link. i.e.

type-4 modification.

$$k = \text{old bus} = 1$$

$$l = \text{new bus} = 3$$

$$i = \text{size of } Z_{Bus} \text{ earlier} = 3 \text{ (in previous step)}$$

$$j = \text{size of } Z_{Bus} \text{ earlier} = 3$$

$$\text{Now } P = Z_{kk} + Z_{ll} - Z_{lk} - Z_{kl} + Z_b$$

$$= Z_{11} + Z_{33} - Z_{31} - Z_{13} + Z_b$$

$$= 0.6 + 1 - 0 - 0 + 2.5$$

$$= 1.85 \text{ p.u.}$$

$$Z_{Bus}\text{(new)} = Z_{Bus}\text{(old)} - \frac{1}{P}\begin{bmatrix} Z_{11} - Z_{13} \\ Z_{21} - Z_{23} \\ Z_{31} - Z_{33} \end{bmatrix}\begin{bmatrix} Z_{11} - Z_{13} \\ Z_{21} - Z_{23} \\ Z_{31} - Z_{33} \end{bmatrix}^T$$

$$= \begin{bmatrix} 0.6 & 0 & 0 \\ 0 & 0.5 & 0.5 \\ 0 & 0.5 & 1 \end{bmatrix} - \frac{1}{1.85}\begin{bmatrix} 0.6 \\ -0.5 \\ -1 \end{bmatrix}[0.6 - 0.5]$$

$$= \begin{bmatrix} 0.405 & 0.162 & 0.324 \\ 0.162 & 0.364 & 0.229 \\ 0.324 & 0.229 & 0.459 \end{bmatrix}$$

Q2. Determine the modified Z_{Bus} matrix if the transmission line between 1 & 3 buses is disconnected as in figure 5.6.

Solution:

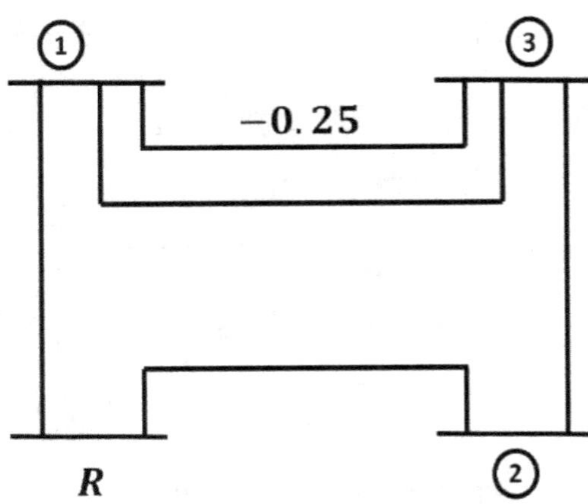

Disconnecting a transmission line between buses 'i' and 'k' is equivalent to adding a transmission line with impedance 'Z_b' between buses i and k

$$p = Z_{kk} + Z_{ll} - Z_{lk} - Z_{kl} + Z_b$$

$$= 0.405 + 0.459 - 0.324 - 0.324 + (-0.25) = -0.033$$

$$Z_{Bus\,(new)} = \begin{bmatrix} 0.405 & 0.162 & 0.324 \\ 0.162 & 0.364 & 0.229 \\ 0.324 & 0.229 & 0.459 \end{bmatrix} + \frac{1}{0.033} \begin{bmatrix} 0.081 \\ -0.06 \\ -0.135 \end{bmatrix} \begin{bmatrix} 0.081 \\ 0.06 \\ -0.135 \end{bmatrix}$$

$$= \begin{bmatrix} 0.6 & 0 & 0 \\ 0 & 0.5 & 0.5 \\ 0 & 0.5 & -1.6 \end{bmatrix}_{3 \times 3}$$

5.3 PERFORMANCE EQUATIONS OF PARTIAL NETWORKS

Consider partial networks as shown in the figures 5.7

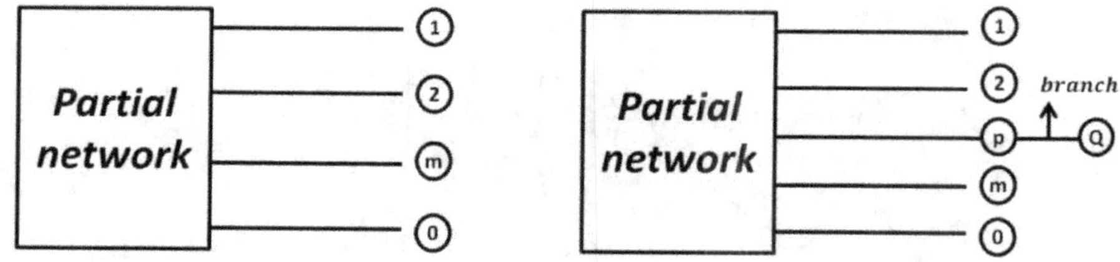

Figure 5.7(a) Partial network   **Figure 5.7(b)** Partial network with branch

Assume that Z_{Bus} is known for partial network consisting of 'm' buses and a reference node. The performance equation of this network is $\overline{E_{Bus}} = \overline{Z_{Bus}}\,\overline{I_{Bus}}$

$\overline{E_{Bus}}$ is m ×1 vector of bus voltages measured with respect to reference bus

$\overline{I_{Bus}}$ is m ×1 vector of impresses bus currents.

If an element p, q is added to the partial network then it may be a branch or a link

If an added element is a branch then a new bus 'q' is added to a partial network then resultant bus impedance matrix is (m+1) × (m+1) & voltage & current vectors are (m + 1) ×1 & (m +1) ×1

Calculate the elements in $(m+1)^{th}$ row & $(m+1)^{th}$ column to determine Z_{Bus} (new)

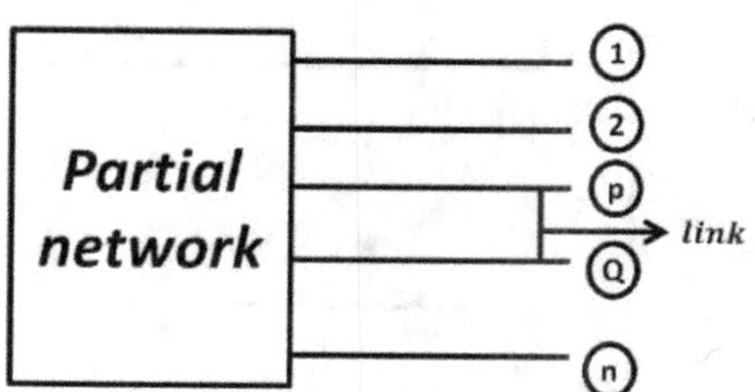

Figure 5.8 Partial network with link

If the added element pq is link then no new bus is not added to the partial network then

(a) Dimension of the matrices do not change

(b) Calculate all the elements of Z_{Bus} to determine Z_{Bus} (new)

5.4   DETERMINATION OF Z_{BUS} BY CONSIDERING MUTUAL COUPLING BETWEEN THE ELEMENTS

5.4.1 Addition of a Branch

The performance equation for the partial network with an added branch pq is shown in figure 5.9.

$$\begin{bmatrix} E_1 \\ E_2 \\ \vdots \\ E_p \\ \vdots \\ E_m \\ E_q \end{bmatrix} = \begin{bmatrix} Z_{11} & Z_{12} & \cdots & Z_{1p} & \cdots & Z_{1m} & Z_{1q} \\ Z_{21} & Z_{22} & \cdots & Z_{2p} & \cdots & Z_{2m} & Z_{2q} \\ \vdots & \vdots & \cdots & \vdots & \cdots & \vdots & \vdots \\ Z_{p1} & Z_{p2} & \cdots & Z_{pp} & \cdots & Z_{pm} & Z_{pq} \\ \vdots & \vdots & \cdots & \vdots & \cdots & \vdots & \vdots \\ Z_{m1} & Z_{m2} & \cdots & Z_{mp} & \cdots & Z_{mm} & Z_{mq} \\ Z_{q1} & Z_{q2} & \cdots & Z_{qp} & \cdots & Z_{qm} & Z_{qq} \end{bmatrix} \begin{bmatrix} I_1 \\ I_2 \\ \vdots \\ I_p \\ \vdots \\ I_m \\ I_q \end{bmatrix} \qquad .....(5.22)$$

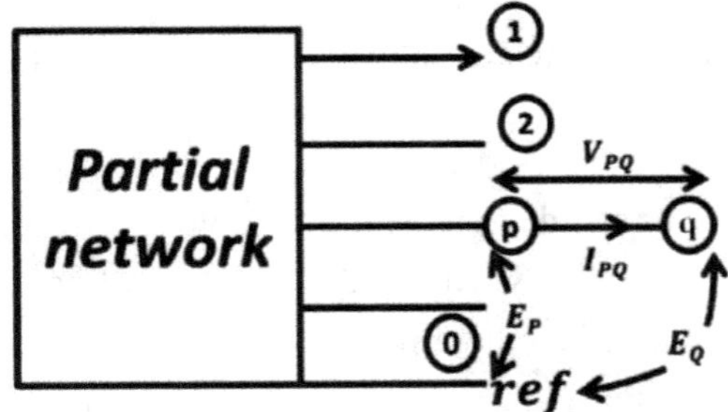

Figure 5.9 Addition of a branch

Assumptions: The network consists of bilateral passive elements

$$Z_{qi} = Z_{iq}, \; i = 1,2,....m$$

And refers to the buses of the partial network & not including the new bus Q.

The added branch p-q is assumed to be mutually coupled with one or more elements of the partial network.

Calculation of off diagonal elements:

The off diagonal element Z_{qi} can be determined by injecting a current at i^{th} bus & calculating voltage at q^{th} bus with reference node

Since all other bus currents are zeros, equation (5.22) is reduced as

$$\left.\begin{matrix} E_1 = Z_{1i}\, I_i \\ E_2 = Z_{2i}\, I_i \\ E_3 = Z_{3i}\, I_i \end{matrix}\right\}$$

$$E_4 = Z_{4i}\, I_i$$

$$\vdots$$

$$E_p = Z_{pi}\, I_i \qquad\qquad .....(5.23)$$

$$\vdots$$

$$E_m = Z_{mi}\, I_i$$
$$E_q = Z_{qi}\, I_i$$
$$I_i = 1 \text{ p.u.}$$
$$Z_{qi} = E_q \qquad \qquad \text{.....(5.24)}$$

The bus voltages associated with added elements & voltage across the added element p-q are related as

$$E_p - E_q = V_{pq}$$
$$E_q = E_p - V_{pq} \qquad \qquad \text{.....(5.25)}$$

The current flowing through elements of the network are expressed in terms of primitive admittances & voltage across the elements

$$\begin{bmatrix} I_{pq} \\ I_{rs} \end{bmatrix} = \begin{bmatrix} y_{pq,pq} & \overline{y_{pq,rs}} \\ y_{rs,pq} & [Y_{rs,pq}] \end{bmatrix} \begin{bmatrix} V_{pq} \\ V_{rs} \end{bmatrix} \qquad \text{.....(5.26)}$$

p-q is a fixed subscript & refers to the added element. It is available subscript & refers to all other elements

Notations

I_{pq} , V_{pq} = current flowing through added element p–q & voltage across added element p–q.

I_{rs} , V_{rs} = current vector & voltage vector of the elements of partial network

$Y_{pq,pq}$ = self admittance of added element p-q.

$\bar{y}_{pq,rs}$ = vector of mutual admittances between added element p–q and elements r-s of the partial network.

$$\bar{y}_{rs,pq} = \text{transpose of } \bar{y}_{pq,rs}$$

$[y_{pq,rs}]$ = primitive admittance matrix of the partial network

Current flowing through the added branch $i_{pq} = 0$ $\qquad \qquad \text{.....(5.27)}$

Voltage across the added branch not equal to zero since the added branch is mutually coupled to one or more elements of partial network.

$$\bar{V}_{rs} = \bar{E}_r - \bar{E}_s \qquad \qquad \text{.....(5.28)}$$

$\bar{E}_r$, $\bar{E}_s$ are voltages at the bus in partial network

From (5.26) and (5.27)

$$0 = y_{pq,pq} * V_{pq} + \bar{y}_{pq,rs}\bar{V}_{rs}$$

$$V_{pq} = \frac{\{\bar{y}_{pq,rs}\bar{V}_{rs}\}}{y_{pq,pq}} \qquad \qquad \text{.....(5.29)}$$

Substitute (5.29) in (5.25)

$$E_p = E_q - V_{pq}$$
$$E_q = E_p + \frac{\bar{y}_{pq,rs}\bar{V}_{rs}}{y_{pq,pq}}$$

$$Z_{qi}I_i = Z_{pi}I_i + \frac{\bar{y}_{pq,rs}[\bar{E}_r - \bar{E}_s]}{y_{pq,pq}}$$

$$Z_{qi}I_i = Z_{pi}I_i + \frac{\bar{y}_{pq,rs}[\bar{Z}_{ri}I_i - \bar{Z}_{si}I_i]}{y_{pq,pq}}$$

$$Z_{qi} = Z_{pi} + \frac{\bar{y}_{pq,rs}[\bar{Z}_{ri} - \bar{Z}_{si}]}{y_{pq,pq}} \qquad \ldots.(5.30)$$

Calculation of diagonal element, Z_{qq}

The element Z_{qq} can be calculated by injecting a current at q^{th} bus and calculating voltage at q^{th} bus as shown in figure 5.10.

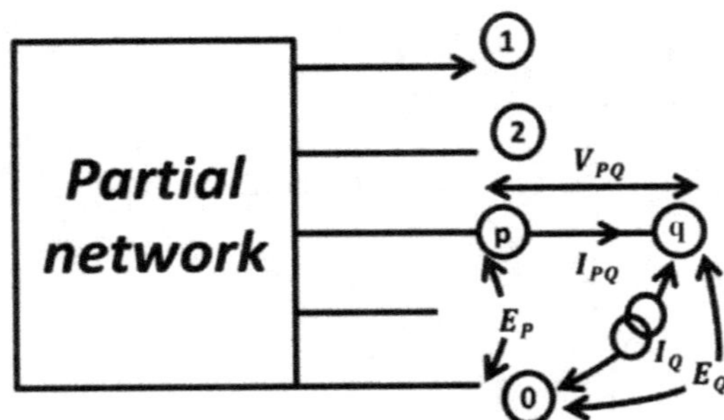

Figure 5.10 To determine diagonal element

Since all other bus currents $= 0$.

$$\left.\begin{array}{l} E_1 = I_q Z_{1q} \\ E_2 = I_q Z_{2q} \\ \vdots \\ E_p = I_q Z_{pq} \\ \vdots \\ E_m = I_q Z_{mq} \\ E_q = I_q Z_{qq} \end{array}\right\} \qquad \ldots.(5.31)$$

Since $I_q = 1\,\text{p.u.}$

$$Z_{qq} = E_q \qquad \ldots.(5.32)$$

Current flowing through the added element, $I_{pq} = -I_q = -1$

From (5.26),

$$-1 = y_{pq,pq} \times V_{pq} + \bar{y}_{pq,rs}\bar{V}_{rs}$$

$$V_{pq} = \frac{-\{1 + \bar{y}_{pq,rs}\bar{V}_{rs}\}}{y_{pq,pq}} \qquad \ldots.(5.33)$$

$$E_q = E_p - V_{pq}$$

$$E_q = E_p + \frac{1 + \bar{y}_{pq,rs}\bar{V}_{rs}}{y_{pq,pq}}$$

Substitute $\quad\quad E_p = Z_{pq}I_q = Z_{pq}$ and $E_q = Z_{qq}$ in above equation

$$Z_{qq} = Z_{pq} + \frac{1+\bar{y}_{pq,rs}\bar{V}_{rs}}{y_{pq,pq}}$$

$$Z_{qq} = Z_{pq} + \frac{1+\bar{y}_{pq,rs}\bar{V}_{rs}}{y_{pq,pq}}$$

$$Z_{qq} = Z_{pq} + \frac{1+\bar{y}_{pq,rs}[\bar{Z}_{rq} - \bar{Z}_{sq}]}{y_{pq,pq}} \quad\quad\quad .....(5.34)$$

Case (i): No mutual coupling between the added branch and other elements of partial network

$$\bar{y}_{pq,rs} = 0$$

$$Z_{qi} = Z_{pi}$$

$$Z_{qq} = Z_{pq} + \frac{1}{y_{pq,pq}} = Z_{pq} + Z_{pq,pq}$$

Case (ii): No mutual coupling between added branch and other elements of the partial network. Bus 'p' is reference bus.

$$\bar{y}_{pq,rs} = 0$$

$$Z_{pi} = 0$$

$$Z_{qi} = 0$$

$$Z_{pq} = 0$$

$$Z_{qq} = Z_{pq,pq}$$

5.4.2 Addition of Link

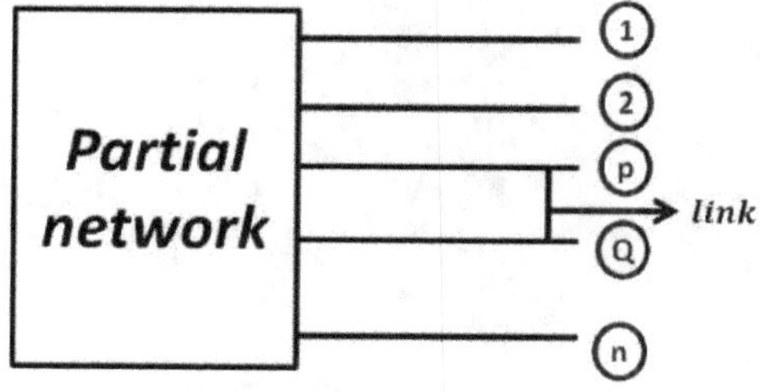

Figure 5.11 Addition of Link between buses p and q.

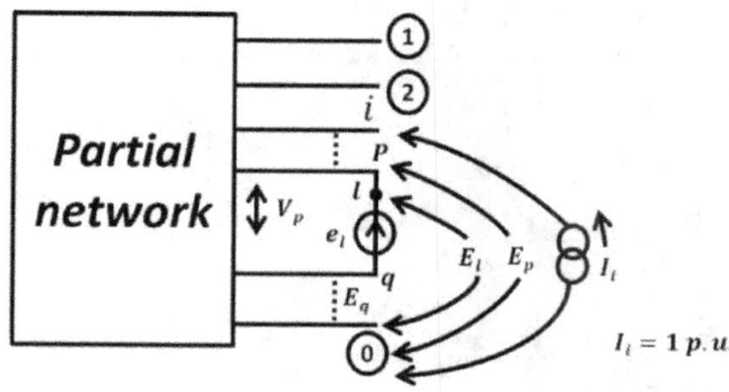

Figure 5.12 To determine off-diagonal and diagonal elements.

Add an element between the existing buses p & q. Therefore, the added element is link.

Connect a voltage source e_l in series with the added element, an imaginary node is created.

The voltage source e_l is such that the current flowing through the added link is zero.

The performance equations for the partial network with added element p and the source voltage e_l is

$$\begin{bmatrix} E_1 \\ E_2 \\ | \\ | \\ E_p \\ | \\ | \\ E_m \\ E_l \end{bmatrix} = \begin{bmatrix} Z_{11} & Z_{12} & - & -Z_{1p} & - & -Z_{1m} & Z_{1l} \\ Z_{21} & Z_{22} & - & -Z_{2p} & - & -Z_{2m} & Z_{2l} \\ | & | & - & - & | & - & - & | & | \\ | & | & - & - & | & - & - & | & | \\ Z_{p1} & Z_{p2} & - & -Z_{pp} & - & -Z_{pm} & Z_{pl} \\ | & | & - & - & | & - & - & | & | \\ | & | & - & - & | & - & - & | & | \\ Z_{m1} & Z_{m2} & - & -Z_{mp} & - & -Z_{mm} & Z_{ml} \\ Z_{l1} & Z_{l2} & - & -Z_{lp} & - & -Z_{lm} & Z_{ll} \end{bmatrix} \begin{bmatrix} I_1 \\ I_2 \\ | \\ | \\ I_p \\ | \\ | \\ I_m \\ I_l \end{bmatrix} \qquad .....(5.35)$$

Calculation of off–diagonal element, Z_{li} .

Z_{li} is calculated by injecting current $I_i = I_p$ at i^{th} bus and calculating voltage at the lth node w.r.t. reference node.

$$e_l = E_l - E_q \qquad .....(5.36)$$

Since all other bus currents are zeros.

$$E_{ki} = Z_{li} I_i \qquad .....(5.37)$$

$$k = 1,2,....p,....m$$

The voltage source $e_l = Z_{li} I_i$ .....(5.38)

As $I_i = 1 p.u.$, $e_l = Z_{li}$ .....(5.39)

Let V_{pl} be the voltage across the element p-l

$$E_p - E_q = V_{pl} + e_l$$

$$E_p - E_q - V_{pl} = e_l \qquad .....(5.40)$$

With imaginary load l size of the $[Z_{bus}]$ increases by one temporarily. Therefore, added element p-l is a branch.

$$\therefore I_{pl} = 0 \rightarrow I_{pq} = 0 \qquad .....(5.41)$$

The current flowing through the element p-l in terms of primitive admittances and voltage across the element is

$$i_{pl} = y_{pl,pl} V_{pl} + \bar{y}_{pl,rs} \overline{V}_{rs} = 0$$

$$i_{pl} = \frac{-\{\bar{y}_{pl,rs} \overline{V}_{rs}\}}{y_{pl,pl}}$$

$$V_{pl} = \frac{-\{\bar{y}_{pq,rs} \overline{V}_{rs}\}}{y_{pq,pq}} \qquad .....(5.42)$$

As admittance offered by ideal voltage source is zero.

Substitute (5.42) in (5.40)

$$e_l = E_p - E_q + \frac{\{\bar{y}_{pq,rs}\bar{V}_{rs}\}}{y_{pq,pq}}$$

$$e_l = Z_{pi} I_i - Z_{qi} I_i + \frac{\{\bar{y}_{pq,rs}[\bar{E}_r - \bar{E}_s]\}}{y_{pq,pq}}$$

$$Z_{li} = Z_{pi} - Z_{qi} + \frac{\{\bar{y}_{pq,rs}[\bar{Z}_{ri} - \bar{Z}_{si}]\}}{y_{pq,pq}} \qquad .....(5.43)$$

$$i = 1,2,.....m \quad i \neq l$$

Calculation of diagonal element: Z_{ll} can be calculated by injecting current I_l at i^{th} bus and calculating voltage at i^{th} bus w.r.t. bus Q.

Since all other bus currents are zero

$$E_{kl} = Z_{kl} I_l \qquad .....(5.44)$$

$$k = 1,2,....m$$

$$E_l = z_{ll} I_l \qquad .....(5.45)$$

$$E_l = z_{ll} \qquad .....(5.46)$$

The current flowing through the element p-l, $I_{pl} = -I_l = -1$ p.u.

This current interms of primitive admittances and voltage across the elements

$$-1 = y_{pl,pl}V_{pl} + \bar{y}_{pl,rs}\bar{V}_{rs}$$

$$V_{pl} = \frac{-\{1 + \bar{y}_{pl,rs}\bar{V}_{rs}\}}{y_{pl,pl}}$$

As admittance of ideal voltage source = 0, p-l can be replaced with p-q

$$V_{pl} = \frac{-\{1 + \bar{y}_{pq,rs}\bar{V}_{rs}\}}{y_{pq,pq}} \qquad .....(5.47)$$

Substitute (5.47) in (5.40)

$$E_p - E_q - V_{pl} = e_l$$

$$Z_{pl} I_l - Z_{ql} I_l \frac{\{1 + \bar{y}_{pq,rs}\bar{V}_{rs}\}}{y_{pq,pq}} = Z_{ll} I_l$$

$$Z_{ll} = Z_{pl} - Z_{ql} + \frac{\{1 + \bar{y}_{pq,rs}[\bar{Z}_{ri} - \bar{Z}_{si}]\}}{y_{pq,pq}} \qquad .....(5.48)$$

Case (i): No mutual coupling between added element and other elements od partial network, $\bar{y}_{pq,rs} = 0$

$$(5.43) \rightarrow Z_{li} = Z_{pi} - Z_{qi} \ , i = 1,2..........p,..........m \quad i \neq l$$

$$(5.48) \rightarrow Z_{ll} = Z_{pl} - Z_{ql} + \frac{1}{y_{pq,pq}}$$

$$= Z_{pl} - Z_{ql} + Z_{pq,pq}$$

Case (ii): There is no mutual coupling between the added and other elements of network and bus 'p' is reference bus.

$$Z_{pi} = 0 \ , Z_{pl} = 0$$

$$Z_{li} = -Z_{qi}$$

$$Z_{ll} = -Z_{ql} + \frac{1}{y_{pq,pq}}$$

Elimination of node l

$$\overline{E_{BUS}} = \overline{Z_{BUS}}\,\overline{I_{BUS}} + Z_{ll}I_l \qquad\qquad .....(5.49)$$

The series voltage source e_l

$$e_l = Z_{lj}\overline{I_{BUS}} + Z_{ll}I_l$$

$$0 = Z_{lj}\overline{I_{BUS}} + Z_{ll}I_l$$

$$I_l = -\frac{Z_{lj}\overline{I_{BUS}}}{Z_{ll}} \qquad\qquad .....(5.50)$$

Substituting (5.48) in (5.49)

$$\overline{E_{BUS}} = Z_{BUS}\overline{I_{BUS}} - Z_{li}\frac{Z_{lj}\overline{I_{BUS}}}{Z_{ll}}$$

$$\overline{E_{BUS}} = \left[Z_{BUS} - \frac{Z_{li}Z_{lj}}{Z_{ll}}\right]\overline{I_{BUS}} \qquad\qquad .....(5.51)$$

$$\overline{E_{BUS}} = Z_{BUS}(before\ elimination) - \frac{Z_{li}Z_{lj}}{Z_{ll}} \qquad\qquad .....(5.52)$$

$$Z_{ij(mod)} = Z_{ij}(before\ elimination) - \frac{Z_{il}Z_{lj}}{Z_{ll}} \qquad\qquad .....(5.53)$$

Modification of $[Z_{BUS}]$ to make changes in network:

The performance equation for the partial network is

$$[\overline{Z_{BUS}}][\overline{I_{BUS}}] = [\overline{E_{BUS}}] \qquad\qquad .....(5.54)$$

The ith bus voltage from the above equation

$$E_i = \sum_{k=i}^{n} Z_{ik}I_k \qquad\qquad .....(5.55)$$

Upon removal of one element or due to change in impedance of a coupled element

$$E_{BUS\,(N)} = Z_{BUS(N)}I_{BUS(N)} \qquad\qquad .....(5.56)$$

Now the ith bus voltage now changes to

$$E_i = \sum_{k=i}^{n} Z_{ik(n)}I_{k(n)} \qquad\qquad .....(5.57)$$

When a mutually coupled element is removed from the network, only those $[Z_{BUS}]$ elements corresponding to elements removed gets effected while rest of the matrix elements remains same. However, all bus voltages and currents change in the network and in order to get the desired bus voltages according to the changes in the network, suitable increments to bus currents are provided by retaining the same elements.

$$\overline{E_{BUS\,(N)}} = \overline{E_{BUS\,(0)}} + \Delta\overline{E_{BUS}} \qquad\qquad .....(5.58)$$

$$= Z_{BUS}I_{BUS} + Z_{BUS}\,\Delta I_{BUS}$$

$$E_{i\,(n)} = \sum_{k=i}^{n} Z_{ik} \left(I_k + \Delta I_k\right) \qquad \qquad .....(5.59)$$

Where ΔI_k is the change in the kth bus current.

Suppose an element p-q is to be removed or its impedance value is to be changed and is mutually coupled to an element r-s, changes in element currents are

$$k = p, \Delta I_k = \Delta I_{pq}$$
$$k = q, \Delta I_k = -\Delta I_{pq} \qquad \qquad .....(5.60)$$
$$l = r, \Delta I_k = \Delta I_{rs}$$
$$k = s, \Delta I_k = -\Delta I_{rs}$$

Let a current source, $I_j = 1$p.u. is connected to j^{th} bus with all other bus currents $= 0$.

$$I_k = 0\;, I_j = 1\text{p.u.} \qquad \qquad .....(5.61)$$

$$k = 1,2............n \; k \neq j$$

$$\therefore E_{i\,(N)} = Z_{ij}I_j + Z_{i\,p}\Delta I_{pq} - Z_{i\,q}\Delta I_{pq} + Z_{i\,r}\Delta I_{rs} - Z_{i\,s}\Delta I_{rs}$$

$$E_{i\,(N)} = Z_{ij} + \left(Z_{i\,p} - Z_{i\,q}\right)\Delta I_{pq} + \left(Z_{i\,r} - Z_{i\,s}\right)\Delta I_{rs} \qquad .....(5.62)$$

Let group of subscripts pq, rs be denoted as a &b

$$a = [p \; r]\;;\; b = [q\; s]$$
$$E_{i\,(N)} = Z_{ij} + \left(\overline{Z_{ia}} - \overline{Z_{ib}}\right)\Delta I_{ab} \qquad \qquad .....(5.63)$$

Let Y' (before) be the primitive admittance matrix before removing the element.

Y' (after) be the primitive admittance matrix after removing the element.

$$\Delta I_{ab} = (Y'(before) - Y'(after))\,\Delta V_{cd} \qquad \qquad .....(5.64)$$

Let 'cd' be the subscripts of [Y' (before) $-$ Y' (after)]

As $\Delta V_{cd} = E_c - E_d$,

$$\overline{E_c} = \overline{Z_{cj}} + ([\overline{Z_{ca}}] - [\overline{Z_{cb}}])\overline{\Delta I_{ab}} \qquad \qquad .....(5.65)$$
$$\overline{E_d} = \overline{Z_{dj}} + ([\overline{Z_{da}}] - [\overline{Z_{db}}])\overline{\Delta I_{ab}}$$

Substituting (5.65) in (5.64)

$$\Delta I_{ab} = \left(Y'(before) - Y'(after)\right)[E_c - E_d]$$
$$= [Y'_{(B)} - Y'_{(A)}]\,[\overline{Z_{cj}} + ([\overline{Z_{ca}}] - [\overline{Z_{cb}}])\overline{\Delta I_{ab}} - \overline{Z_{dj}} + ([\overline{Z_{da}}] - [\overline{Z_{db}}])\overline{\Delta I_{ab}}]$$
$$\Delta I_{ab} = (Y'_{(B)} - Y'_{(A)})(\overline{Z_{cj}} - \overline{Z_{dj}} + \overline{Z_{ca}} - \overline{Z_{cb}} - \overline{Z_{da}} + \overline{Z_{db}})\,\overline{\Delta I_{ab}}$$
$$\Delta I_{ab}\{ U - (Y'_{(B)} - Y'_{(A)}) + \overline{Z_{ca}} - \overline{Z_{cb}} - \overline{Z_{da}} + \overline{Z_{db}} \} = (Y'_{(B)} - Y'_{(A)})\,\overline{Z_{cj}} - \overline{Z_{dj}}$$
$$\Delta I_{ab} = m^{-1}(Y'_{(B)} - Y'_{(A)})\,\overline{Z_{cj}} - \overline{Z_{dj}} \qquad (13)$$
$$E_{i\,(new)} = Z_{ij} + (\overline{Z_{ia}} - \overline{Z_{ib}})m^{-1}(Y'_{(B)} - Y'_{(A)})\,\overline{Z_{cj}} - \overline{Z_{dj}} \qquad .....(5.66)$$

Equation (5.66) gives the voltage at bus 'i' by injecting a current at bus 'j' such that $I_j = 1$ p.u. and appropriate changes in current at buses p,q,r,s to determine the effect of changes in the elements p,q.

From the definition of the bus impedance matrix, the ith and j^{th} elements of the modified bus impedance matrix are

$$Z_{ij} + (\overline{Z_{ia}} - \overline{Z_{ib}})m^{-1}\,(Y'_{(B)} - Y'_{(A)})\,\overline{Z_{cj}} - \overline{Z_{dj}} \qquad\qquad .....(5.67)$$

Where $i = 1, 2,.....n$

The process is repeated for each $j = 1,2,....n$ to obtain all the elements of Z_{BUS} or $Z_{BUS\,(N)}$.

Q1. Obtain Z_{bus} for the bus sample system given in the figure below in 5.13. The self and mutual impedance of each line are given and reference is R.

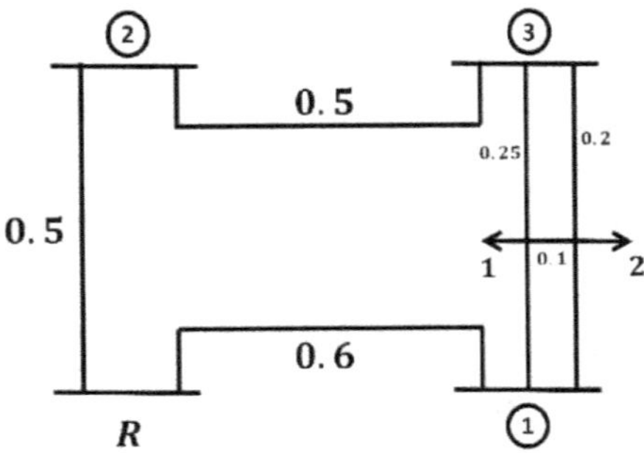

Figure 5.13 Power system network

Line no	1	2	3	4	5
Bus code p–q	R–1	R–2	2–3	1–3(1)	1–3(2)
Self Impedance, Zpq	0.6	0.5	0.5	0.25	0.2
Mutual Impedance, Zpq					0.1

Solution:

Step1: Add an element with impedance 0.6p.u. to reference bus

A new bus1 is created and size of $[Z_{BUS}]$ increases by one and the added element is known as branch

i.e. type -1 modification and $Z_{kk} = Z_b = 0.6$

$$[Z_{BUS}] = \begin{array}{c} R \\ 1 \end{array}\begin{bmatrix} Z_{RR} & Z_{Rq} \\ Z_{1R} & Z_{11} \end{bmatrix} = \begin{array}{c} R \\ 1 \end{array}\begin{bmatrix} 0 & 0 \\ 0 & 0.6 \end{bmatrix} p.u. = [0.6]$$

As impedance measured w.r.t. reference bus is 0, $Z_{00} = Z_{01} = Z_{10} = 0$

Step 2: An element is added to reference bus. A new bus 2 is created and size of $[Z_{BUS}]$ increases by one and the added element is known as branch

Type – 2 Modification

$$[Z_{BUS}] = \begin{array}{c} 1 \\ 2 \end{array}\begin{bmatrix} Z_{11} & Z_{12} \\ Z_{21} & Z_{22} \end{bmatrix} = \begin{bmatrix} 0.6 & Z_{12} \\ Z_{21} & Z_{22} \end{bmatrix}$$

Where $Z_{kk} = Z_b$ and $Z_{ki} = Z_{ji}$ where $k = 2, j = 1$

$Z_{12} = Z_{21} = 0$

$$Z_{22} = 0.5$$

$$[Z_{BUS}] = \begin{bmatrix} 0.6 & 0 \\ 0 & 0.5 \end{bmatrix}$$

Step 3: Add an element called branch to bus 2. A new bus '3' is created and size of $[Z_{BUS}]$ increases by 1 and the added element is known as branch.

i.e. Type – 2 modification

$$[Z_{BUS}] = \begin{bmatrix} Z_{11} & Z_{12} & Z_{13} \\ Z_{21} & Z_{22} & Z_{23} \\ Z_{31} & Z_{32} & Z_{33} \end{bmatrix} = \begin{bmatrix} 0.6 & 0 & Z_{13} \\ 0 & 0.5 & Z_{23} \\ Z_{31} & Z_{32} & Z_{33} \end{bmatrix}$$

Calculation of off–diagonal elements:

$$k = 3, i = 1, 2 \quad i \neq k \text{ and } j = 2$$

$$Z_{ki} = Z_{ji}$$

$$i = 1; \therefore \quad Z_{3i} = Z_{2i}$$

$$\Rightarrow Z_{31} = Z_{21} = 0$$

$$i = 2: Z_{32} = Z_{22} = 0.3$$

$$= \begin{bmatrix} 0.6 & 0 & 0 \\ 0 & 0.5 & 0.5 \\ 0 & 0.5 & Z_{22} \end{bmatrix}$$

Calculation of diagonal elements:

$$Z_{kk} = Z_b + Z_{jj}$$

$$Z_{33} = Z_b + Z_{22} = 0.5 + 0.5 = 1\,p.u.$$

$$[Z_{BUS}] = \begin{bmatrix} 0.6 & 0 & 0 \\ 0 & 0.5 & 0.5 \\ 0 & 0.5 & 1 \end{bmatrix}$$

Step 4: An element called link, $Z_{1-3(1)}$ is added between existing buses 1 and 3 $Z_{1-3(1)}$ i.e. Type–4 modification. The size of Z_{BUS} matrix do not increase by 'i' and the added element is known as 'link'.

$$Z_{BUS\,(N)} = Z_{BUS\,(old)} - \frac{1}{P}[Z_{ik} - Z_{il}][Z_{jk} - Z_{jl}]^T$$

$$P = Z_{kk} + Z_{ll} - Z_{kl} - Z_{lk} + Z_b$$

$$K = 3, l = 1$$

$$= Z_{33} + Z_{11} - Z_{13} - Z_{31} + Z_b$$

$$P = 0.1 + 0.6 - 0 - 0 + 0.25$$

$$P = 1.6 + 0.25$$

$$P = 1.85\ \text{p.u.}$$

$$[Z_{BUS\,(N)}] = \begin{bmatrix} 0.6 & 0 & 0 \\ 0 & 0.5 & 0.5 \\ 0 & 0.5 & 1 \end{bmatrix} - \frac{1}{1.85}\begin{bmatrix} Z_{13} - Z_{11} \\ Z_{23} - Z_{21} \\ Z_{33} - Z_{31} \end{bmatrix}\begin{bmatrix} Z_{13} - Z_{11} \\ Z_{23} - Z_{21} \\ Z_{33} - Z_{31} \end{bmatrix}^T$$

$$= \begin{bmatrix} 0.6 & 0 & 0 \\ 0 & 0.5 & 0.5 \\ 0 & 0.5 & 1 \end{bmatrix} - \frac{1}{1.85} \begin{bmatrix} -0.6 \\ -0.5 \\ -1 \end{bmatrix} \begin{bmatrix} -0.6 & -0.7 \end{bmatrix}$$

$$= \begin{bmatrix} 0.6 & 0 & 0 \\ 0 & 0.5 & 0.5 \\ 0 & 0.5 & 1 \end{bmatrix} - \begin{bmatrix} 0.19 & 0 & 0.32 \\ 0 & 0 & 0 \\ -0.32 & 0 & 0.54 \end{bmatrix}$$

$$= \begin{bmatrix} 0.41 & 0.16 & 0.32 \\ 0.16 & 0.36 & 0.229 \\ 0.32 & 0.22 & 0.46 \end{bmatrix}$$

Step 5: An element is added between existing buses 1 and 3, $Z_{1-3}(2)$

$Z_{1-3(2)} = 0.2$ p.u. $Z_{1-3(1)}$ is added of link

The matrix corresponding to pq and rs

$$Z_{pq,rs} = \begin{array}{cc} & \begin{array}{cc} 1-3(1) & 1-3(2) \end{array} \\ \begin{array}{c} 1-3(1) \\ 1-3(2) \end{array} & \begin{bmatrix} 0.25 & 0.1 \\ 0.1 & 0.2 \end{bmatrix} \end{array}$$

$$\bar{Y}_{pq,rs} = \left[\overline{Z_{pq,rs}} \right]^{-1}$$

$$= \begin{bmatrix} 0.25 & 0.1 \\ 0.1 & 0.2 \end{bmatrix}^{-1}$$

$$= \begin{bmatrix} 5 & -2.5 \\ -2.5 & 6.25 \end{bmatrix}$$

Calculation of off diagonal elements

$$Z_{li} = Z_{pi} - Z_{qi} + \frac{\{\bar{y}_{pq,rs}[\bar{Z}_{ri} - \bar{Z}_{si}\}}{y_{pq,pq}}$$

$i = 1,2,\ldots.n \qquad i \neq l$

$n = 3$

<u>$i = 1$</u>

$$Z_{l1} = Z_{p1} - Z_{q1} + \frac{\{\bar{y}_{pq,rs}[\bar{Z}_{r1} - \bar{Z}_{s1}\}}{y_{pq,pq}}$$

$r = 1, s = 3$

$$Z_{l1} = Z_{p1} - Z_{q1} + \frac{\{\bar{y}_{1-3(2),1-3(1)}[\bar{Z}_{1,1} - \bar{Z}_{3,1}\}}{y_{1-3(2),1-3(2)}}$$

$p = 1, q = 3$

$$Z_{l1} = Z_{11} - Z_{31} + \frac{\{\bar{y}_{1-3(2),1-3(1)}[\bar{Z}_{1,1} - \bar{Z}_{3,1}\}}{y_{1-3(2),1-3(2)}}$$

$$Z_{l1} = 0.45 - 0.324 + \frac{(-2.5)[\bar{Z}_{1,1} - \bar{Z}_{3,1}]}{6.25}$$

$$Z_{l1} = 0.45 - 0.324 + \frac{(-2.5)[0.405 - 0.324]}{6.25}$$

$$Z_{l1} = 0.048 = Z_{l1}$$

<u>$i = 2$</u>

$$Z_{21} = Z_{p2} - Z_{q2} + \frac{\{\bar{y}_{pq,rs}[\bar{Z}_{r2} - \bar{Z}_{s2}\}}{y_{pq,pq}}$$

$$= Z_{11} - Z_{32} + \frac{-2.5\,[\bar{Z}_{1,2} - \bar{Z}_{3,2}]}{63.25}$$

$$= 0.16 - 0.229 + \frac{(-2.5)[0.16 - 0.229]}{6.25}$$

$$Z_{21} = -0.0405 = Z_{12}$$

$$\underline{i = 3}$$

$$Z_{31} = Z_{p3} - Z_{q3} + \frac{\{\bar{y}_{pq,rs}[\bar{Z}_{r3} - \bar{Z}_{s3}\}}{y_{pq,pq}}$$

$$= Z_{13} - Z_{33} + \frac{-2.5\,[\bar{Z}_{1,3} - \bar{Z}_{3,3}]}{63.25}$$

$$= 0.32 - 0.46 + \frac{(-2.5)[0.32 - 0.46]}{6.25}$$

$$Z_{31} = -0.084 = Z_{13}$$

Calculation of diagonal element

$$Z_{ll} = Z_{11} - Z_{31} + \frac{\{1 + \bar{y}_{(1-3)2,1-3(1)}[\bar{Z}_{il} - \bar{Z}_{3l}\}}{y_{(1-3)2,1-3(1)}}$$

$$= 0.048 - (-0.0081) + \frac{1 + (-0.25)[0.048 - 0.08]}{6.25}$$

$$Z_{ll} = 0.237$$

Considering the imaginary node of Z_{BUS} matrix by 'l'

$$[Z_{BUS}] = \begin{array}{c} 1 \\ 2 \\ 3 \\ l \end{array} \begin{bmatrix} 0.405 & 0.162 & 0.32 & 0.048 \\ 0.162 & 0.36 & 0.229 & -0.0405 \\ 0.324 & 0.229 & 0.459 & -0.081 \\ 0.048 & -0.0405 & -0.081 & 0.237 \end{bmatrix}$$

Eliminating the node l and the modified $[Z_{\text{BUS}}]$ is determined as follows

$$Z_{ij(N)} = Z_{ij(0)} - \frac{\overline{Z_{il(0)}}\,\overline{Z_{jl(0)}}}{Z_{ll}}$$

$$j = i = 1,2,\ldots\ldots.n \quad n = 3$$

$i = 1, j = 1:$

$$Z_{11(N)} = Z_{11(0)} - \frac{Z_{1l(0)}\overline{Z_{1l(0)}}}{Z_{ll}}$$

$$= 0.0405 - \frac{0.048^2}{0.237}$$

$$= 0.395$$

$i = 1, j = 2:$

$$Z_{12(N)} = Z_{12(0)} - \frac{Z_{1l(0)}\overline{Z_{2l(0)}}}{Z_{ll}}$$

$$= 0.171$$

$i = 1, j = 3$:

$$Z_{13(N)} = Z_{13(0)} - \frac{Z_{1l(0)}\overline{Z_{3l(0)}}}{Z_{ll}}$$

$$= 0.324 - \frac{(0.048)(-0.081)}{0.237}$$

$$= 0.341$$

$i = 2, j = 2$:

$$Z_{22(N)} = Z_{22(0)} - \frac{Z_{2l(0)}\overline{Z_{2l(0)}}}{Z_{ll}}$$

$$= 0.36 - \frac{(-0.0405)^2}{0.237}$$

$$= 0.35$$

$i = 2, j = 3$:

$$Z_{23(N)} = Z_{23(0)} - \frac{Z_{2l(0)}\overline{Z_{3l(0)}}}{Z_{ll}}$$

$$= 0.229 - \frac{(-0.0405)(-0.081)}{0.237}$$

$$= 0.215$$

$i = 3, j = 3$:

$$Z_{33(N)} = Z_{33(0)} - \frac{Z_{3l(0)}\overline{Z_{3l(0)}}}{Z_{ll}}$$

$$= 0.459 - \frac{(-0.081)^2}{0.237}$$

$$= 0.431$$

$$Z_{BUS} = \begin{bmatrix} 0.395 & 0.171 & 0.341 \\ 0.171 & 0.358 & 0.215 \\ 0.341 & 0.215 & 0.431 \end{bmatrix} \text{p.u.}$$

OBJECTIVE QUESTIONS

1. Define primitive network?

    Ans: A set of unconnected elements is defined as a primitive network.

2. What is the performance equation of primitive element in impedance form?

    Ans: The performance equation of a primitive element in impedance form is

$$v_{pq} + e_{pq} = z_{pq}i_{pq}$$

3. What is the performance equation of primitive element in admittance form?

    Ans: The performance equation of a primitive element in admittance form is

    $$i_{pq} + j_{pq} = y_{pq}v_{pq}$$

4. Define partial network?

    Ans: A subset of the existing power system network is known as partial network.

5. Define branch and it's significance?

    Ans: If an element is added to a bus and results in the formation of a new bus then the added element is known as branch and the size of Z_{bus} matrix increases by 1.

6. Define link and it's significance?

    Ans: If an element is added to a bus and do not result in the formation of a new bus then the added element is known as link and the size of Z_{bus} matrix do not increase by 1.

7. What are the four types of modifications in Z_{bus} formulation?

    Ans:

    Type–1 modification refers to addition of an element between a reference bus and a new bus.

    Type–2 modification refers to addition of an element between an old bus and a new bus.

    Type–3 modification refers to addition of an element between a reference bus and an old bus.

    Type–4 modification refers to addition of an element between two old buses.

8. What is the impedance of a reference bus or measured with respect to reference bus?

    Ans: The impedance of a reference bus or measured with respect to reference bus is zero.

9. What is bus impedance matrix?

    Ans: The matrix consisting of driving point impedances and transfer impedances of a power system network is called bus impedance matrix.

10. What is primitive impedance matrix?

    Ans: The primitive impedance matrix consists of self and mutual impedance of individual network elements neglecting their interconnection.

11. How Z_{loop} is determined with all the network elements uncoupled?

    Ans: All the diagonal elements are determined as the sum of the impedances of elements in the basic loop corresponding to diagonal element. An off–diagonal element is determined as the negative of the sum of the impedances of the network elements common to basic loops pertaining to this off–diagonal element.

12. What are the advantages of Z_{bus} building algorithm?

    Ans: The calculation of Z_{bus} by singular transformation involves inversions and multiplications. This is a time–consuming process for a large system consisting of hundreds of buses. But the Z_{bus} algorithm is systematically formulated by adding the elements in sequence.

13. What is the modification of Z_{bus} if one link is removed from original network and the link is having no mutual coupling with the other element?

   Ans: The modified impedance matrix can be obtained by adding a link equal in magnitude and opposite in sign to the link to be disconnected.

6 Load Flow Analysis

6.1 INTRODUCTION

Load flow solution is the solution of the network under steady state condition subjected to certain inequality constraints under which system operates.

Load flow solution gives the nodal voltages and phase angles and hence the power injected at all the buses and power flows through line.

The load flow study is essential for designing a new power system and for planning extension of the existing power system network for the increased load demands.

A load flow study also gives the initial conditions of the system to study the transient behavior of the power system network

A load flow solution of the power system network requires

1. Formulating network equation

2. Suitable mathematical technique for solution of the equations

3. Operating constraints

All equipment in power system network are designed to operate at fixed voltages

(a) Consumers require operating voltage to lie within certain range.

$$|V_{i,min}| \leq |V_i| \leq |V_{i,\,max}|$$

(b) The mechanical power input and rating of alternators impose a limit on the active power generation

$$P_{Gi,min} \leq P_{Gi} \leq P_{Gi,max}$$

(c) Limitations of the generator field excitation constraints the reactive power generation

$$Q_{Gi,min} \leq Q_{Gi} \leq Q_{Gi,max}$$

(d) Consideration of stability impose constraints on the maximum permissible power angle between buses i and j

$$|\delta i - \delta j|_{min} \leq |\delta i - \delta j| \leq |\,\delta i - \delta j|_{\,max}$$

(e) The active power and reactive power should satisfy the condition

$$\sum_{i=1}^{n} P_{Gi} = \sum_{i=1}^{n} P_{Di} + P_L$$

$$\sum_{i=1}^{n} Q_{Gi} = \sum_{i=1}^{n} Q_{Di} + Q_L$$

6.2   CLASSIFICATION OF BUSES

At each bus, generated active power P_{Gi}, active load demand. P_{Di}, generated reactive power Q_{Gi} and reactive power demand Q_{Di} is specified.

Knowing generation and load demand the net power injected at the ith bus are

$$P_i = P_{Gi} - P_{Di}$$
$$P_i = Q_{Gi} - Q_{Di}$$

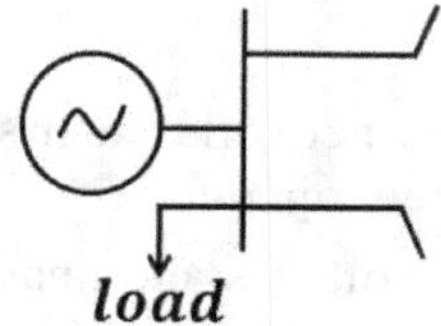

PV bus: The below figure represents an exporting bus

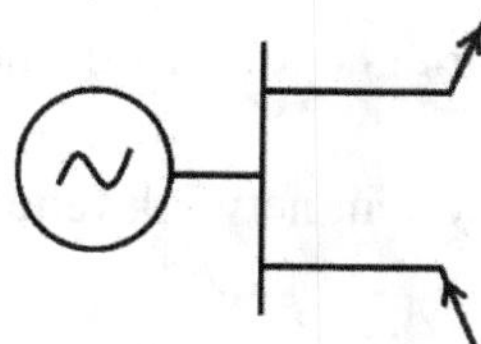

$$P_i = P_{Gi}$$
$$Q_i = Q_{Gi}$$

PQ bus: The below figure represents an importing bus

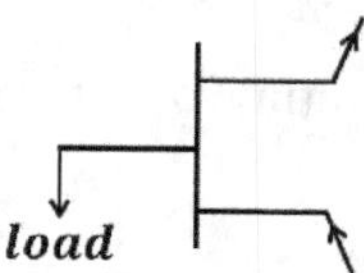

$$P_i = P_{Di}$$
$$Q_i = Q_{Di}$$

Transmission line:

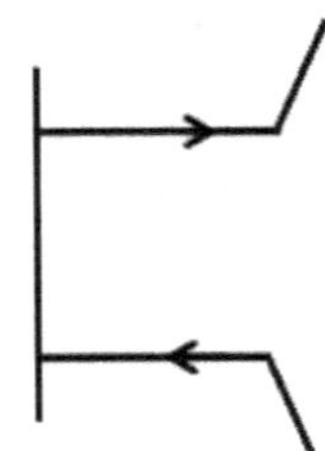

$$P_i = 0$$
$$Q_i = 0$$

1. **PQ or load Bus**

   Specified are P_i and Q_i

   Unspecified are $|V_i|$ and δ

2. **PV bus or Generator bus:** A bus can be designated as a PV Bus only when a generator is connected to it.

   Specified injected active power P_i, $|V_i|$

   P_{Gi} can be said to the desired value based on the requirements of active power at importing bus with permissible control action of generator

   PV bus is an exporting bus

   Within the permissible control of exciter, voltage magnitude can be controlled at this bus.

   As excitation has limits, generator at this bus can generate reactive power within the limits

   $$Q_{Gi(min)} < Q_{Gi} < Q_{Gi(max)}$$

   (a) Information about reactive power limits is required

   (b) Within the limits of reactive power, PV bus or generator bus is referred as voltage control bus

3. **Voltage controlled bus:**

   (a) Because of the capabilities in maintaining the bus voltage to the specified value, a PV bus or generator bus is designated as voltage–controlled bus.

   (b) However, a pure voltage controlled bus is connected to the voltage–controlled equipment but not to generators $\rightarrow P_{Gi} = 0$ $Q_{Gi} = 0$. Active and reactive powers injected into i^{th} bus

   $$P_i = P_{Gi} - P_{Di} = - P_{Di}$$
   $$Q_i = Q_{Gi} - Q_{Di} = - Q_{Di}$$

   (c) The phase angle δ_i is an unknown parameter.

4. **Reference Bus/Swing bus/Slack bus**

   (i) In a power system network, if one bus is connected to an alternator with high generation capacities P and Q, it is designated a ref bus

   (ii) Specified $|V_i|$, δ_i

   Unspecified P_i, Q_i.

(iii) Justification for specifying V_i and δ_i. As one of the bus voltage vectors is selected as a ref vector to find phase angle δ of the remaining n bus voltage network (a) Assign $|V_i|=1$p.u. and $\delta = 0$ representing flat start

(iv) Justification for not specifying P_i and Q_i

(a) The power balance equations are

$$\Sigma P_{Gi} - \Sigma P_{Di} - \Sigma P_{losses} = 0$$

$$\sum_{i=1}^{n} Q_{Gi} - Q_{Di} - P_{losses} = 0$$

(b) In a power flow study, the generation of active and reactive power cannot be set to correct values since loss in lines are unknown. Therefore, it is necessary to have one bus as slack bus at which complex power generation is not initially set.

Therefore, generation at slack bus is such that it supplies the difference in total system load and losses.

Sum of complex power specified at the remaining buses.

Bus Type	Specified	Unspecified	No. of Bus
Slack	$|V_i|$, δ	P_i, Q_i	1
PV	P_i, $|V_i|$	$|Q_i|$, δ_i	15%
PQ bus	P_i, Q_i	$|V_i|$, δ_i	85%

6.3 DETERMINATION OF STATIC LOAD FLOW EQUATIONS

1. Complex power injected into ith bus

$$S_i = S_{Gi} - S_{Di} = (P_{Gi} + jQ_{Gi}) - (P_{Di} + jQ_{Di})$$
$$= (P_{Gi} - P_{Di}) + j(Q_{Gi} - Q_{Di})$$
$$= P_i + j Q_i = V_i I_i \qquad\qquad .....(6.1)$$

Current injecting into the ith bus

$$I_i = \sum_{k=1}^{n} Y_{ik} V_k \qquad\qquad .....(6.2)$$

Substitute (6.2) in (6.1)

$$S_i = V_i \left\{ \sum_{k=1}^{n} Y_{ik} V_k \right\}^*$$

$$S_i^* = P_i - Q_i = V_i^* \left\{ \sum_{k=1}^{n} Y_{ik} V_k \right\} \qquad\qquad .....(6.3)$$

Expressing the power system network quantities in polar form

$$V_i = |V_i| \angle \delta_i$$

$$V_k = |V_k| \angle \delta_k$$

Series admittance Y_{ik} between transmission lines between 'I' and 'k' $= G_{ik} - jB_{ik}$

$$= \sqrt{G^2_{ik} + B^2_{ik}} \angle -\tau_{ik}$$

$$= |Y_{ik}| \angle -\tan^{-1}\left(\frac{-B_{ik}}{G_{ik}}\right)$$

$$= |Y_{ik}| \angle -\gamma_{ik}$$

$$S^*_i = P_i - jQ_i = |V_i| \angle -\delta_i \sum_{k=1}^{n} |Y_{ik}| \angle -\gamma_{ik} |V_k| \angle \delta_k$$

$$P_i - jQ_i = \sum_{k=1}^{n} |V_i||Y_{ik}||V_k| \left\{ \angle -\delta_i - \gamma_{ik} + \delta_k \right\}$$

$$P_i - jQ_i = \sum_{k=1}^{n} |V_i||Y_{ik}||V_k| \left\{ -\angle \delta_i + \gamma_{ik} - \delta_k \right\}$$

$$P_i = \sum_{k=1}^{n} |V_i||Y_{ik}||V_k| \cos\left(\delta_i + \gamma_{ik} - \delta_K \right) \qquad \qquad .....(6.4)$$

$$Q_i = \sum_{k=1}^{n} |V_i||Y_{ik}||V_k| \sin\left(\delta_i - \delta_K + \gamma_{ik} \right) \qquad \qquad .....(6.5)$$

Equations 6.4 and 6.5 are known as static load flow equations

6.4   VOLTAGE AT i^{th} BUS

The complex power injected into i^{th} bus

$$S_i = P_i + jQ_i = V^*_i I_i$$

$$S^*_i \quad P_i - jQ_i = V^*_i I_i$$

$$V^*_i \left\{ \sum_{K=1}^{n} Y_{ik} V_k \right\} = P_i - jQ^*_j$$

$$\sum_{K=1}^{n} Y_{ik} V_k = \frac{P_i - jQ_i}{V^*_i}$$

$$Y_{ii}V_i + \sum_{\substack{K=1 \\ K \neq i}}^{n} Y_{ik} V_k = \frac{P_i - jQ_i}{V_i^*}$$

$$Y_{ii}V_i = \left\{ \frac{P_i jQ_i}{V_i^*} - \sum_{\substack{K=1 \\ K \neq i}}^{n} Y_{ik} V_k \right\}$$

$$V_i = \frac{1}{Y_{ii}} \left\{ \frac{P_i - jQ_i}{V_i^*} - \sum_{\substack{K=1 \\ K \neq i}}^{n} Y_{ik} V_k \right\} \qquad \qquad .....(6.6)$$

6.5 DETERMINATION OF POWER TRANSMISSION IN THE LOAD FLOW ANALYSIS

Consider a transmission line between buses i and k as shown below in figure 6.1.

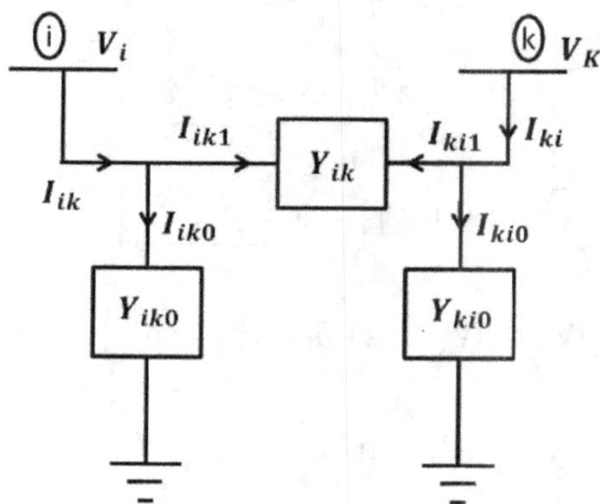

Figure 6.1 Transmission line between buses 'i' and 'k'

Y_{ik} = series admittance of transmission line between 'i' and 'k'

Y_{iko}, Y_{kio} are shunt admittances

I_{ik}, I_{ki} are currents flowing through transmission line from buses i to k and k to i

V_i and V_k are voltages at bus i and k current fed by bus i in to the transmission line

$$I_{ik} = I_{ik1} + I_{iK0} = Y_{ik}(V_i - V_k) + Y_{ik0} V_i$$

Power injected into the transmission line from bus

Bus i to bus k

$$S_{ik} = P_{ik} + jQ_{ik} = V_i I_{ik}^* = V_i\{y_{ik}^*(V_i^* - V_k^*) + y_{iko}^* V_i^*\}$$

power injected into line from bus k to bus i

$$S_{ki} = P_{ki} + jQ_{ki} = V_k I_{ki}^*$$

Current fed by $I_{ki}^* = I_{ki1} + I_{kio}$

$$I_{ki} = Y_{ki}(V_k - V_i) + Y_{ki0}\, V_k$$

$$S_{ki} = V_k(y_{ki}{}^*(V_k - V_i) + y_{ki0}\, V_k)$$

Total power transmitted through the line $= S_{ik} = S_{ki}$

Power losses between the buses (i) and (k) in line,

$$P_{loss,ik} = S_{ik} + S_{ki}$$

$$\text{Total losses} = \sum_{l=1}^{n} P_{loss}$$

6.6 NUMERICAL METHODS

The unknown electrical quantities can be determined by numerical methods

1. Gauss seidel
2. Newton Raphson
3. Decoupled load flow
4. Fast decoupled load flow

6.6.1 Gauss Seidal Method

Consider three algebraic equations

$$2x_1 + x_2 + x_3 = 5$$

$$3x_1 + 5x_2 + 2x_3 = 15$$

$$2x_1 + x_2 + 4x_3 = 8$$

Assume $x_1^{(0)} = 0$, $x_2^{(0)} = 0$, $x_3^{(0)} = 0$

Iteration 1:

$$x_1^{(1)} = \frac{5 - x_2^{(0)} - x_3^{(0)}}{2} = \frac{5 - 0 - 0}{2} = 2.5$$

$$x_2^{(1)} = \frac{15 - 3x_1^{(1)} - 5\,x_2^{(0)}}{3} = \frac{15 - 3(2.5) - 0}{3} = 1.5$$

$$x_3^{(1)} = \frac{8 - 2x_1^{(1)} - x_2^{(1)}}{4} = \frac{8 - 2 \times 2.5 - 1.5}{4} = 0.37$$

Iteration 2:

$$x_1^{(2)} = \frac{5 - x_2^{(1)} - x_3^{(1)}}{2} = \frac{15 - 1.5 - 0.4}{2} = 1.6$$

$$x_2^{(2)} = \frac{15 - 3x_1^{(2)} - x_3^{(1)}}{5} = 1.9$$

$$x_3^{(2)} = \frac{8 - 2x_1^{(2)} - x_3^{(1)}}{4} = 0.7$$

Iteration 3:

$$x_1^{(3)} = \frac{5 - x_2^{(2)} - x_3^{(2)}}{2} = \frac{5 - 1.9 - 0.7}{2} = 1.2$$

$$x_2^{(3)} = \frac{15 - 3x_1^{(3)} - 2x_3^{(2)}}{5} = \frac{15 - 3(1.2) - 2(0.7)}{5} = 2$$

$$x_3^{(3)} = \frac{8 - 2x_1^{(3)} - x_2^{(3)}}{4} = \frac{8 - 2(1.2) - 2}{4} = 0.9$$

Iteration 4:

$$x_1^{(4)} = \frac{5 - x_2^{(3)} - x_3^{(3)}}{2} = \frac{5 - 2 - 0.9}{2} = 1.05$$

$$x_2^{(4)} = \frac{15 - 3x_1^{(4)} - x_3^{(3)}}{5} = \frac{15 - 3(1.05) - 2(0.9)}{5} = 2.01$$

$$x_3^{(4)} = \frac{8 - 2x_1^{(4)} - x_2^{(4)}}{4} = \frac{8 - 2(1.05) - 2.01}{4} = 0.972$$

Iteration 5:

$$x_1^{(5)} = \frac{5 - x_2^{(4)} - x_3^{(4)}}{2} = \frac{5 - 1.05 - 2.01}{2} = 1.008$$

$$x_2^{(5)} = \frac{15 - 3x_1^{(5)} - 2x_3^{(4)}}{5} = 2.0062$$

$$x_3^{(5)} = \frac{8 - 2x_1^{(5)} - x_2^{(5)}}{4} 0.994$$

The iteration process continues till convergence is obtained.

Q1. Obtain the voltage at bus 2 for the simple power system network shown in the figure using Gauss seidel method if $V_1 = 1\lfloor 0^\circ$ $p.u.$

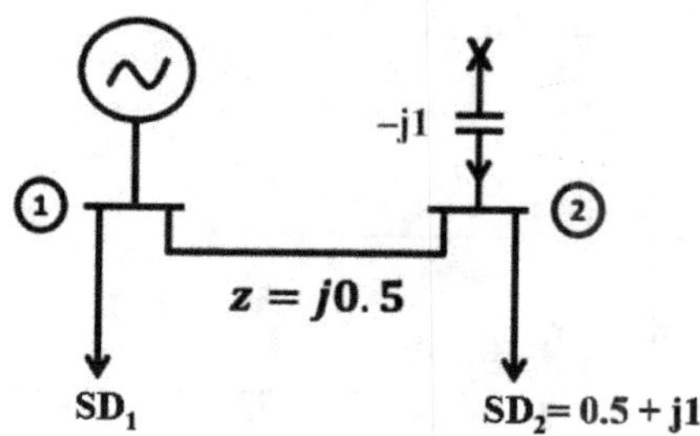

Solution:

Capacitor at bus 2 injects reactive power of 1p.u.

Complex power injected into bus 2.

$$S_2 = j1 - (j0.5 + j1) = -j0.5 \text{p.u.}$$

Bus admittance matrix for the two–bus network

$$Y_{Bus} = \begin{matrix} 1 \\ 2 \end{matrix} \begin{bmatrix} \dfrac{1}{j0.5} & \dfrac{-1}{j0.5} \\ \dfrac{-1}{j0.5} & \dfrac{1}{j0.5} \end{bmatrix} = \begin{bmatrix} -j0.2 & j0.2 \\ j0.2 & -j0.2 \end{bmatrix}$$

Assume voltage at the bus 2 for zeroth iteration $V_2^0 = 1\angle 0^0 \text{p.u.}$

Voltage at bus 2 for 1st iteration.

$$V_2^1 = \frac{1}{Y_{22}} \left\{ \frac{(P_2 - jQ_2)}{V_2^{0*}} - (Y_{21}V_1) \right\}$$

$$= \frac{1}{-j0.2} \left\{ \frac{-0.5}{1\underline{|0^0}} - 2j1\underline{|0^0} \right\}$$

$$= j0.5\{-0.5 - 0.2j\}$$

$$= j0.5\{0.5 + j0.2\} = -j0.25 + 1$$

$$= 1.03\underline{|-14.03^0} \text{ p.u.}$$

Q2.   Calculate the total losses in transmission line between buses 1 and 2 for Q1.

Solution:

Current flowing from bus 1 to bus 2

$$I_{12} = \frac{V_1 - V_2}{Z} = \frac{1\underline{|0^0} - 1.03\underline{|-14.0^0}}{j0.5}$$

$$= \frac{1 - (0.999 - 0.24j)}{j0.5} = +\frac{0.001 + j0.24}{j0.5} = 0.5\underline{|-0.16^0}$$

Complex power injected from bus 1 to line

$$S_{12} = V_1 \, I_{12}^* = 1\underline{|0^0} \left(0.5\underline{|0.16^0} \right)$$

$$= 0.5\underline{|0.16^0} \ p.u.$$

Current injected from bus 2 to bus 1

$$I_{21} = \frac{(V_2 - V_1)}{Z} = \frac{(1.261 - j0.3) - 1}{j0.5}$$

$$= 0.5\underline{|179.8^0} \text{ p.u.}$$

Complex power injected from bus 2 to line

$$S_{21} = V_2^1 I_{21}^* = 0.5\lfloor 179.8^0 \times 1.03 \lfloor -14.03^0$$

$$= 0.515 \lfloor -143.83^0$$

Total losses = $S_{12} + S_{21} = (-7.17 \times 10^{-5} + j0.1245)$p.u.

Acceleration factor

The no. of iterations corresponding method corresponding to Gauss–seidel method can be minimized by Acceleration factor

Voltage at i^{th} bus for (r+1) iteration

$$V_i^{r+1} = V_i^r + \alpha\left\{V_i^{r+1} - V_i^r\right\}$$

By default, $\alpha = 1.6$

Q3. The following is the system data for a load flow solution

Bus code	Admittance	Schedule of active and reaction			
1–2	2–j8	Bus code	P	Q	V	Remark
1–3	1–j4	1	–	–	1.06	Slack
2–3	0.66–j2.66	2	0.5	0.5	1+j0	PQ
2–4	1–j4	3	0.4	0.4	1	PQ
3–4	2–j8	4	0.3	0.3	1	PQ

Determine the voltages at the end of 1st iteration using Gauss Seidel method

Solution:

$$Y_{Bus} = \begin{bmatrix} 3-j12 & -2+j8 & -1+j4 & 0 \\ -2+j8 & 366-j14.66 & -0.66+j26 & -1+j4 \\ -1+j4 & -0.66+j2.6 & 3.66-j14.6 & -2+j8 \\ 0 & -1+j4 & -2+j8 & 3-j12 \end{bmatrix}$$

The powers for load buses are to be taken as negative and that for generator bus are to be taken as positive.

Voltage at the ith bus

$$V_i = \frac{1}{Y_{ii}}\left\{\frac{(P_1-jQ_i)}{V_i^*} - \sum_{\substack{K=1 \\ K\neq i}}^{n} Y_{ik}V_k\right\}$$

Assume the voltages for 0th iteration

$$V_2^0 = 1+j0 = 1\angle 0^0 \text{ p.u.}$$

$$V_3^0 = 1 = 1\angle 0^0 \text{ p.u.}$$

$$V_4^0 = 1 = 1\angle 0 \, \text{p.u.}$$

Voltage at bus 2 at the end of first iteration

$$V_2^1 = \frac{1}{Y_{22}}\left\{\frac{(P_2 - jQ_2)}{V_2^{*0}} - \left(Y_{21}V_1 + Y_{23}V_3^0 + Y_{24}V_4^0\right)\right\}.$$

$$V_2^1 = \frac{1}{15.1\angle -759}\left\{\frac{(-0.5)-(-0.2)j}{1\angle 0^0} - \begin{bmatrix}(-2+j8)(1.06)\\+(-0.66+2.0)1+(-1+i4)(1)\end{bmatrix}\right\}$$

$$= 0.9938 - j0.0295 \, \text{p.u}$$

By using acceleration factor

$$V_i^{r+1} = V_i^r + \alpha\left\{V_i^{r+1} - V_i^r\right\}$$

$$V_3^1 = V_3^0 + 1.6\left\{V_3^1 - V_3^0\right\}$$

$$r = 0, i = 3$$

$$V_3^1 = 0.9900 - j0.0472$$

$$r = 0, i = 4$$

$$V_4^1 = \frac{1}{Y_{44}}\left\{\frac{P_4 - Q_4 j}{V_4^{0*}} - \sum_{\substack{K=1\\K\,'4}}^{4} Y_{4k}V_k\right\}$$

$$= \frac{1}{3-j12}\left\{\frac{-0.3+j0.1}{1} - \left(Y_{41}V_1 + Y_{42}V_2 + Y_{43}V_3\right)\right\}$$

$$= \frac{1}{3-j12}\left\{(-0.3+j0.1) - \begin{pmatrix}0+(-1+j4)(1.0189-j0.046)\\+(-2+j8)(0.9900-j0.0472)\end{pmatrix}\right\}$$

$$= \frac{2.1372 - 12.036j}{3 - j12}$$

$$= (0.9859 - j0.0683)\,\text{p.u.}$$

By using acceleration factor,

$$V_4^1 = V_4^0 + 1.6\left\{V_4^1 - V_4^0\right\}$$

$$= 1 + 1.6\{0.9859 - j0.0683 - 1\}$$

$$= (0.9774 - j0.10928)\,\text{p.u.}$$

Q4.   If in the last question bus 2 is taken as generator bus with $|V_2| = 1.04$ and reactive power constraint $0.1 \leq Q_2 \leq 2$ determine the voltages starting with flat voltage profile is assuming $\alpha = 1$

Solution:

The reactive power at bus 2 is written as follows.

$$S_i^* = P_i - jQ_i = V_i^* \sum_{\substack{K=1 \\ K \neq i}}^{n} Y_{ik} V_k$$

$i = 2$

$$P_2 - jQ_2 = V_2^* \sum_{k=1}^{n} Y_{2k} V_k$$

$$= V_2^* \left\{ Y_{21} V_1 + Y_{23} V_3 + Y_{24} V_4 + Y_{22} V_2 \right\}$$

$$= 1.04 \left\{ \begin{array}{l} (-2+j8)(1.06)+(-0.666+j2.6)(1+j0)(-1+j4)(1+j0) \\ +(3.666-1Q66)\times(1.04+j0) \end{array} \right\}$$

$$= (0.0212 - j0.17)$$

$$Q_2 = IP(0.0212 - j0.17)$$

$$Q_2 = 0.173 \text{ p.u.}$$

Since Q_2 lies within the limits $(0.1 \leq Q_2 \leq 1)$ bus 2 is considered as PV bus for 1st iteration

$i = 2$

$$V_2^1 = \frac{1}{Y_{22}} \left[\frac{P_2 - jQ_2}{V_2^0} - \left(Y_{21} V_1 + Y_{23} V_3^0 + Y_{24} V_4^0 \right) \right]$$

$$= \frac{1}{3.66 - j14.66} \left[\frac{0.5 - j0.17}{(1+4)} - \left[(-2+j8)(1.06)+(-0.66+j2.6)(1)+(-1+j4)(1) \right] \right]$$

$$= \frac{4.26 - j15.24}{3.66 - j14.66}$$

$$= (1.04 + j0.029) \text{ p.u.}$$

$$= 1.04 \lfloor 1.59^0 \text{ p.u.}$$

Phase angle for bus 2 is 1.59^0

$i = 3$

$$V_3^1 = \frac{1}{Y_{33}} \left[\frac{P_3 - jQ_3}{V_3^{0*}} - \left(Y_{31} V_1 + Y_{23} V_2 + Y_{34} V_4 \right) \right]$$

$$V_3^1 = \frac{1}{3.66 - j14.66}\left[\frac{-0.4 + j0.3}{1} - \left[(-1 + j4)(1.06) + (-0.66 + j2.6)(1.04 + 0.02) + (-2 + j8)(1)\right]\right]$$

$$= \frac{3.4218 - j14.62}{3.66 - j14.62} = 0.9977 - j0.01575 \text{ p.u}$$

$i = 4$

$$V_4^1 = \frac{1}{Y_{44}}\left[\frac{P_4 - jQ_4}{V_4^{*0}} - \left(Y_{41}V_1 + Y_{42}V_2^1 + Y_{43}V_3^1\right)\right]$$

$$= \frac{1}{3 - j12}\left[\frac{-0.3 + j0.1}{1} - \left(0 + (-1 + j4)(1.047 + j0.029) + (-2 + j8)(0.9977 - 0.05)\right)\right]$$

$$= \frac{2.7324 - j12.072}{2} = 1.0004 - j0.0224 \text{ p.u.}$$

Q5. If the reactive power constrains on generator 2 is $0.2 \leq Q_2 \leq 2$. Solve the previous question at the end of 1st iteration.

$$Q_2 = 0.17$$

As the reactive power constraints are not satisfied, the generator / PV is considered as PQ bus with reactive power Q_2 = reactive power limit that is limited $Q_2 = 0.2$ p.u.

Solution:

As generator bus is connected as load bus, P and Q are specified and $|V|$; δ are specified. But P and Q are considered as positive as in case of generator bus is contrast to load bus where P and Q are considered as negative.

$i = 2$:

$$V_2' = \frac{1}{Y_{22}}\left[\frac{(P_2 - jQ_2)}{V_2^{*0}} - (Y_{21}V_1 + Y_{23}V_3^0 + Y_{24}V_4^0)\right]$$

$$= \frac{1}{3.66 - j14.66}\left[\frac{0.5 - j0.2}{1 + j0} - ((-2 + j8)(1.06) + (-0.66 + j2.664)1 + (-1 + j4)(1 + j0))\right]$$

$$= \frac{4.286 - j15.34}{3.66 - j14.664} = 1.0535 + j0.0288$$

Acceleration value = $1.0857 + j0.0462$

$i = 3$

$$V_3^1 = \frac{1}{3.66 - j14.6}\left[\frac{(-0.4 + j0.3)}{1} - \left[\begin{array}{l}(-1 + j4)(1.06) + (-0.66 + j2.6)\\ \times(1.05659 + j0.0462) + (-2 + j8)(1)\end{array}\right]\right]$$

$$= \frac{3.496 - j14.734}{3.66 - j14.6} = 1.005 - j0.0126$$

Acceleration value,

$$V_3^1 = V_3^0 + 1.6\left(V_3^1 - V_3^0\right)$$

$$V_3^1 = 1.009 - 0.0202j$$

$$i = 4$$

$$V_4^1 = \frac{1}{Y_{44}}\left[\frac{P_4 - jQ_4}{V_4^{0*}} - \left(Y_{41}V_1 + Y_{42}V_2^1 + Y_{43}V_3^1\right)\right]$$

$$= \frac{1}{3-12j}\left[\frac{(-0.3+j0.1)}{1} - \left(0 + (-1+j4)(1.005-0.0126j) + (-2+8j)(1009-0.0202)\right)\right]$$

$$= \frac{2.511 - j12.045}{-3 - j12} = 0.999 - j0.0392$$

Acceleration value,

$$V_4^1 = 1 + 1.6\left(0.999 - j0.0392 - 1\right)$$

$$= \left(0.99030 - j0.0627\right)\text{P.u.}$$

6.6.2 Newton Raphson Method

Newton Raphson for single valued function

1. Consider a single valued function f(x)= 0
2. The solution is the value of x which satisfies f(x) = 0
3. The duration process begins with initial guess $x^0 = 0$ & assume $f(x^0) = 0$
4. Assume 1st iteration value of x i.e

   x^1 be the solution of f(x) = 0

   where $x^1 = x^0 + \Delta x^0$

   & $f(x^1) = 0 \Rightarrow + f(x^0 + \Delta x^0) = 0$
5. The increment Δx^0 is unknown and can be estimated form taylor series of expansion,

$$f(x) = f(x^0 + \Delta x^0) = f(x^0) + \left(\frac{\partial f}{\partial x}\right)^0 \Delta x^0 + \frac{1}{21}\left(\frac{\partial^2 f}{\partial x^2}\right)(\Delta x^0)^2 + = 0 \qquad \text{.....(6.7)}$$

Assumptions:

As Δx^0 is negligible (small) Δx^{02}, Δx^{03} are neglected

The initial guess Δx^0 though is not the exact solution but is nearer to the exact solution

$$f(x^1) = 0 \Rightarrow f\left(x^0\right) + \left(\frac{\partial f}{\partial x}\right)^0 \Delta x^0 = 0$$

6. $\Delta x^0 = -\dfrac{f\left(x^0\right)}{\left(\dfrac{\partial f}{\partial x}\right)^0}$ .....(6.8)

7. $x^1 = x^0 + \Delta x^r = x^0 - \dfrac{f\left(x^0\right)}{\left(\dfrac{\partial f}{\partial x}\right)^0}$ .....(6.9)

For $(r+1)^{th}$ iteration

$$x^{r+1} = x^r + \Delta x^r = x^r - \dfrac{f(x^r)}{\left(\dfrac{\partial f}{\partial x}\right)^r}$$ .....(6.10)

8. The iteration procedure continue till the convergence is obtained
   i.e $|x^{r+1} - x^r| \leq \epsilon$ .....(6.11)

Example 1: Find the root the equation $f(x) = x^2 - 3x + 2$ using new to Raphson method
$$f(x) = x^2 - 3x + 2$$

Iteration 0:
$$x^0 = 0 \Rightarrow + (x^0) = 0 - 0 + 2 = 2$$

Iteration 1:

$$x^1 = x^0 - \dfrac{f(x^0)}{\left(\dfrac{\partial f}{\partial x}\right)^0}$$

$$= 0 - \dfrac{2}{\dfrac{\partial(2x-3)}{\partial x^0}} = \dfrac{+2}{+3} = 0.667$$

Check $x^2 - 3x + 2$
$$= (0.667)^2 - 3(0.667) + 2$$
$$= 0.443889$$

Iteration 2:
$$x^1 = 0.667$$

$$\left(\dfrac{\partial f}{\partial x}\right)^1 = 2x - 3 = 2(0.667) - 3 = -1.666$$

$$x^2 = x^1 - \frac{f(x^1)}{f^1(x^1}$$

$$= 0.667 - \frac{0.443889}{-1.666}$$

$$= 0.9334$$

$$\text{Check} \Rightarrow (0.9334)^2 - 3(0.9334) + 2$$

$$= 0.07099$$

Iteration 3:

$$x_2 = 0.9334$$

$$f(x_2) = 2(0.9334) - 3 = -1.1332$$

$$x^{(3)} = x^2 = \frac{f(x^2)}{f(x^2)}$$

$$= 0.9334 - \frac{0.07099}{-1.1332}$$

$$x^{(3)} = 0.996$$

$$\text{check: } f\left(x^{(3)}\right) = 4.016 \times 10^{-3}$$

Iteration 4:

$$x_2 = 0.996$$

$$f'(x^{(3)}) = 2(0.996) - 3 = -1.008$$

$$x^4 = 0.996 - \frac{3.9700 \times 10^{-3}}{-1.008}$$

$$x^4 = 0.1$$

$$\text{check: } 1.587 \times 10^{-5}$$

Newton Raphson method for multivalued functions

1. Consider two algebraic equations with 2 unknown functions $f_1(x_1,x_2) = 0$, $f_2(x_1,x_2) = 0$
2. The solution for $f_1(x_1,x_2) = 0$ & $f_2(x_1, x_2) = 0$ are the values of x_1 & x_2 which satisfies the equitation $f_1(x_1, x_2) = 0$ & $f_2(x_1,x_2) = 0$
3. The iteration process begins with initial guess $x_1^0 = 0$ & $x_2^0 = 0$ & assumed that
$$f_1(x_1^0, x_2^0) \neq 0$$
4. Assume the 1st iteration values of x_1 and x_2
Let $x_1^1 = x_1^0 + \Delta x_1^0$ and $x_2^1 = x_2^0 + \Delta x_2^0$ be the solution for $f_1(x_1, x_2) = 0$, $f_2(x_1 x_2) = 0$ then

$$f_1\left(x_1^1, x_2^1\right) = 0 \Rightarrow f_1\left(x_1^0 + \Delta x_1^0, x_2^0 + \Delta x_2^0\right)$$

$$f_1\left(x_1^1, x_2^1\right) = 0 \Rightarrow f_1\left(x_1^0 + \Delta x_1^0, x_2^0 + \Delta x_2^0\right)$$

5. Δx_1^0 & Δx_2^0 are unknowns and can be obtained from Taylor series of expansion neglecting higher order terms.

6. $f_1\left(x_1^1, x_2^1\right) = 0 \Rightarrow f_1\left(x_1^0 + \Delta x_1^0, x_2^0 + \Delta x_2^0\right)$

$$\left.\begin{array}{l} = f_1\left(x_1^0, x_2^0\right) + \left(\dfrac{\partial f_1}{\partial x_1}\right)^0 \Delta x_1^0 \\[2em] \left(\dfrac{\partial f_1}{\partial x_2}\right)^0 \Delta x_2^0 = 0 \end{array}\right\} \to 1$$

$$f_2\left(x_1^1, x_2^1\right) = f_2\left(x_1^0, x_2^0\right) + \left(\dfrac{\partial f_1}{\partial x_1}\right)^0 \Delta x_1^0 + \left(\dfrac{\partial f_1}{\partial x_2}\right)^0 \Delta x_2^0 = 0$$

7. Representing equation in the matrix form

$$\underbrace{\begin{bmatrix} \dfrac{\delta f_1}{\delta x_1} & \dfrac{\delta f_1}{\delta x_2} \\[1em] \dfrac{\delta f_2}{\delta x_1} & \dfrac{\delta f_1}{\delta x_2} \end{bmatrix}^0}_{\text{Jacobian matrix}} \quad \underbrace{\begin{bmatrix} \Delta x_1 \\ \Delta x_2 \end{bmatrix}^0}_{\substack{\text{Increment}\\ \text{matrix}}} = \underbrace{\begin{bmatrix} f_1(x_1, x_2) \\ f_2(x_1, x_2) \end{bmatrix}^0}_{\substack{\text{Residual \ column}\\ \text{vector}}}$$

$$[J]^0[\Delta x]^0 = [f]^0$$
$$[\Delta x]^0 = -[f]^0[J]^0 \to (2)$$

Therefore, x_1 and x_2 for first iterations are

$$x_1^1 = x_1^0 + \Delta x_1^0; x_2^1 = x_2^0 + \Delta x_2^0$$
$$x_1^{r+1} = x_1^r + \Delta x_1^r; x_2^{r+1} = x_2^r + \Delta x_2^r$$
$$[\Delta x^r] = -[f]^r[J]^r$$
$$[x_i^{r+1} - x_i^r] \leq \epsilon$$

The iteration procedure continuous till the convergence is obtained.

Example 2: Use newton Raphson method to solve algebraic equations

$$f(x_1, x_2) = x_1^2 - x_2^2 - 4$$
$$g(x_1, x_2) = x_1^2 - x_2^2 - 1$$
$$x_1^0 = 2, \quad x_2^0 = -1$$

Update the values of x_1 and x_2 from iteration.

Solution:

The Jacobian matrix elements are

$$\left(\frac{\delta f}{\delta x_1}\right)^0 = 2x_1^0 = 2 \times 2 = 4$$

$$\left(\frac{\delta f}{\delta x_2}\right)^0 = -2 \times (-1) = 2$$

$$\left(\frac{\partial g}{\partial x_1}\right)^0 = 2x_1 = 4$$

$$\left(\frac{\partial g}{\partial x_2}\right)^0 = 2x_2 = 2(-1) = -2$$

$$\begin{bmatrix} 4 & 2 \\ 4 & -2 \end{bmatrix}\begin{bmatrix} \Delta x_1 \\ \Delta x_2 \end{bmatrix}^0 - \begin{bmatrix} f(x_1,x_2) \\ g(x_1,x_2) \end{bmatrix}^0$$

$$f_1(x_1,x_2)^0 = x_1^{2^0} - x_2^{2^0} + 4 = 4 - 1 - 4 = -1$$

$$g_1(x_1,x_2)^0 = x_1^{2^0} + x_2^{2^0} - 1 = 4 + 1 - 1 = 4$$

$$\Rightarrow \begin{bmatrix} 4 & 2 \\ 4 & -2 \end{bmatrix}\begin{bmatrix} \Delta x_1 \\ \Delta x_2 \end{bmatrix}^0 = \begin{bmatrix} -1 \\ 4 \end{bmatrix}$$

$$[\Delta x_1]^0 = -[+]^0 [J]^{-1^0}$$

$$\begin{bmatrix} \Delta x_1 \\ \Delta x_2 \end{bmatrix}^0 = \begin{bmatrix} -1 \\ 4 \end{bmatrix}\left[\frac{1}{-8-8}\begin{bmatrix} +2 & 2 \\ 4 & -4 \end{bmatrix}\right]$$

$$= \begin{bmatrix} -1 \\ 4 \end{bmatrix}\left[\frac{1}{-16}\begin{bmatrix} 2 & 2 \\ 4 & -4 \end{bmatrix}\right]$$

$$= \begin{bmatrix} \dfrac{-1}{8} & \dfrac{-1}{8} \\ \dfrac{-1}{4} & \dfrac{1}{4} \end{bmatrix}\begin{bmatrix} -1 \\ 4 \end{bmatrix}$$

$$= \begin{bmatrix} \dfrac{+1}{8} - \dfrac{4}{8} \\ \dfrac{1}{4} + \dfrac{4}{4} \end{bmatrix} = \begin{bmatrix} \dfrac{-3}{8} \\ \dfrac{5}{4} \end{bmatrix} = \begin{bmatrix} -0.375 \\ 1.25 \end{bmatrix}$$

$$\Delta x_1^0 = -0.375$$

$$\Delta x_2^0 = 1.25$$

Representing equation in the matrix form

$$\begin{bmatrix} \dfrac{\partial f_1}{\partial x_1} & \dfrac{\partial f_1}{\partial x_2} \\[2mm] \dfrac{\partial f_2}{\partial x_1} & \dfrac{\partial f_2}{\partial x_2} \end{bmatrix}^0 \begin{bmatrix} \Delta x_1 \\ \Delta x_2 \end{bmatrix}^0 = \begin{bmatrix} f_1(x_1,x_2) \\ f_2(x_1,x_2) \end{bmatrix}^0$$

$$[J][\Delta x]^0 = [f]^0$$

$$[\Delta x]^0 = -[f]^0 [J]^{0^{-1}} \qquad\qquad .....(6.12)$$

Therefore, x_1 and x_2 for first iteration are

$$x_1^1 = x_1^0 + \Delta x_1^0 \,; x_2^1 = x_2^0 + \Delta x_2^0$$

$$x_1^{r+1} = x_1^r + \Delta x_1^r \,; x_2^{r+1} = x_2^r + \Delta x_2^r$$

$$\left[\Delta x^r\right] = -[f]^r \left[J^r\right]^{-1}$$

$$\left| x_i^{r+1} - x_r^r \right| \le \epsilon$$

$$i = 1, 2 ....n$$

The iteration procedure continuous till the convergence is obtained.

Newton Raphson method is an iterative method which approximates the set of nonlinear simultaneous equations to a set of simultaneous equations using Taylor of expansion and the terms are limited to first approximation.

Newton Raphson method (rectangular form):

At any bus 'p', complex power injected is

$$S_p = P_p + jQ_p = V_p \left\{ \sum_{Q=1}^{n} Y_{pQ} V_Q \right\}^*$$

$$S_p^* = P_p - jQ_p = V_p^* \sum_{Q=1}^{n} Y_{pQ} V_Q$$

$$\text{let } V_p = e_p + jf_p$$

$$V_q = e_q + jf_q$$

$$V_{pq} = e_{pq} + jB_{pq}$$

$$S_p^* = (e_p - jf_p) \sum_{Q=1}^{n} (e_{pq} - jB_{pq})(e_p + jfq)$$

$$= (e_p - jf_p) \sum_{Q=1}^{n} (G_{pq}e_p + B_{pq}f_q) + j(G_{pq}f_q + B_{pq}e_p)$$

$$P_p = \sum_{Q=1}^{n} e_p\left(G_{pq}e_P + B_{pq}f_q\right) + fP\left(G_{pq}f_P - B_{pq}e_p\right)$$

$$Q_p = \sum_{Q=1}^{n} f_p\left(G_{pq}e_P + B_{pq}f_q\right) + e_p\left(G_{pq}f_p - B_{pq}e_p\right) \qquad \ldots\ldots(6.13)$$

$$\left|V_p\right|^2 = \sqrt{e_p^{\,2} + f_p^{\,2}} \qquad \ldots\ldots(6.14)$$

Equations 12 and 13, 14 are load flow equations and are nonlinear equations consisting of real and imaginary components of nodal voltages. The LHS terms P_p and Q_p are for load bus.

P_p and V_p for generator bus are known then e_p and f_p are unknowns.

For n bus system no. of unknowns $= 2(n-1)$ because voltage at the slack bus is known and fixed in both magnitude and phase considering bus as reference unknown variables are $e_2, e_3, \ldots e_{n-1}, e_n, f_2, f_3 \ldots\ldots f_{n-1}, f_n$.

To solve $2(n-1)$ variables $2(n-1)$ equations are required.

Assuming bus as the slack bus, and remaining $(n-1)$ buses as load buses

$$\begin{bmatrix} \Delta P_2 \\ \Delta P_3 \\ \Delta P_4 \\ \vdots \\ \Delta P_n \\ \Delta Q_2 \\ \Delta Q_3 \\ \vdots \\ \Delta Q_n \end{bmatrix}_{2(n-1)\times 1} = \begin{bmatrix} \dfrac{\delta P_2}{\delta e_2} & \dfrac{\delta P_2}{\delta e_3} & \cdots & \dfrac{\delta P_2}{\delta e_n} & \dfrac{\delta P_2}{\delta f_2} & \dfrac{\delta P_2}{\delta f_3} & \cdots & \dfrac{\delta P_3}{\delta f_n} \\[2ex] \dfrac{\delta P_3}{\delta e_2} & \dfrac{\delta P_3}{\delta e_3} & \cdots & \dfrac{\delta P_3}{\delta e_n} & \dfrac{\delta P_3}{\delta f_2} & \dfrac{\delta P_3}{\delta f_3} & \cdots & \dfrac{\delta P_3}{\delta f_n} \\[1ex] \vdots & \vdots & \vdots & \vdots & \vdots & \vdots & \vdots & \vdots \\[1ex] \dfrac{\delta P_n}{\delta e_2} & \dfrac{\delta P_n}{\delta e_3} & \cdots & \dfrac{\delta P_n}{\delta e_n} & \dfrac{\delta P_n}{\delta f_2} & \dfrac{\delta P_n}{\delta f_3} & \cdots & \dfrac{\delta P_n}{\delta f_n} \\[2ex] \dfrac{\delta Q_2}{\delta e_2} & \dfrac{\delta Q_2}{\delta e_3} & \cdots & \dfrac{\delta Q_2}{\delta e_n} & \dfrac{\delta Q_2}{\delta f_2} & \dfrac{\delta Q_2}{\delta f_3} & \cdots & \dfrac{\delta Q_n}{\delta f_n} \\[1ex] \vdots & \vdots & \vdots & \vdots & \vdots & \vdots & \vdots & \vdots \\[1ex] \dfrac{\delta Q_n}{\delta e_2} & \dfrac{\delta Q_n}{\delta e_3} & \cdots & \dfrac{\delta Q_n}{\delta e_n} & \dfrac{\delta Q_n}{\delta f_2} & \dfrac{\delta Q_n}{\delta f_3} & \cdots & \dfrac{\delta Q_n}{\delta f_n} \end{bmatrix}_{2(n-1)\times 2(n-1)} \begin{bmatrix} \Delta e_2 \\ \Delta e_3 \\ \Delta e_4 \\ \vdots \\ \Delta e_n \\ \Delta f_2 \\ \Delta f_3 \\ \vdots \\ \Delta f_n \end{bmatrix}_{2(n-1)\times 1}$$

Off diagonal elements of J_1

$$\frac{\partial P_P}{\partial e_Q} = e_P G_{PQ} - f_P B_{PQ}; Q \neq P \rightarrow 3$$

Diagonal elements of J_1

$$P_p = \left[\left(e_p^2 G_{pp} + e_p f_p B_{pp}\right) + \left(f_p^2 G_{pp} + f_p e_p B_{pp}\right)\right]$$

$$+ \sum_{\substack{Q=1 \\ Q \neq P}}^{n} e_p\left(e_Q G_{pQ} + f_Q B_{pQ}\right) + f_p\left(f_q G_{pQ} + e_q B_{pQ}\right)$$

$$\frac{\partial P_p}{\partial e_p} = \left(2e_p G_{pp} + f_p B_{pp} - f_p B_{pp}\right) + \sum_{\substack{Q=1 \\ Q \neq P}}^{n} \left(e_Q G_{pQ} + f_Q B_{pq}\right)$$

$$\frac{\partial P_p}{\partial e_p} = 2e_p G_{pp} + \sum_{\substack{Q=1 \\ Q \neq P}}^{n} \left(e_Q G_{pQ} + f_Q B_{pq}\right) \rightarrow 4$$

Off diagonal elements of J_2

$$\frac{\partial P_p}{\partial f_Q} = e_p B_{pQ} + f_p G_{pQ} \quad Q \neq P \rightarrow 3$$

Diagonal elements of J_2

$$\frac{\partial P_p}{\partial f_p} = \left(f_p B_{pp} + 2f_p G_{pp} - e_p B_{pp}\right) + \sum \left(f_Q G_{pQ} - e_Q B_{pQ}\right) \rightarrow 4$$

Off diagonal element of J_3

$$\frac{\partial Q_p}{\partial e_Q} = f_p G_{pQ} + e_p B_{pQ} \quad Q \neq P \rightarrow 5$$

Diagonal element of J_3

$$Q_p = f_p e_p G_{pp} + f_p f_p B_{pp} - e_p f_p G_{pp} + e_p^{\,2} B_{pp} + \sum_{\substack{Q=1 \\ Q \neq P}}^{n} f_p \left(e_Q G_{pQ} + f_Q B_{pQ}\right) - e_p \left(f_p G_{pQ} - e_Q B_{pQ}\right)$$

$$\frac{\partial Q_p}{\partial e_p} = f_p G_{pp} - f_p G_{pp} + 2e_p B_{pp} + \sum_{\substack{Q=1 \\ Q \neq P}}^{n} -\left(f_p G_{pQ} + e_Q B_{pQ}\right) \rightarrow 6$$

Off diagonal element of J_4

$$\frac{\partial Q_p}{\partial f_Q} = f_p B_{pq} - e_p G_{pq}; Q \neq P \rightarrow 7$$

Diagonal element of J_4

$$\frac{\partial Q_p}{\partial f_p} = 2f_{pp} B_{pp} + \sum_{\substack{Q=1 \\ Q \neq 1}}^{n} \left(e_Q G_{pQ} + f_Q B_{pQ}\right) \rightarrow 8$$

Off diagonal elements of J_5

$$|V_p|^2 = e_p^{\,2} + f_p^{\,2}$$

$$\frac{\partial |V_p|^2}{\partial e_Q} = 0 \rightarrow 9$$

Diagonal elements of J_5

$$\frac{\partial |V_p|^2}{\partial e_p} = 2e_p \to 10$$

off diagonal elements of J_6

$$\frac{\partial |V_p|^2}{\partial f_Q} = 0 \to 11$$

Diagonal elements of J_6

$$\frac{\partial |V_p|^2}{\partial f_p} = 2f_p \to 12$$

Calculation of residual column vector ΔP, ΔQ, $\Delta |V|^2$

Let P_{sp}, Q_{sp}, $|V|_{2sp}$ be the specified quantities at bus(p).

Assuming that flat voltage profile, P, Q and $|V|^2$ at various bus are calculated

$$\Delta P_p = P_{sp} - P_p^0$$

$$\Delta Q_p = Q_{sp} - Q_p^0$$

$$\begin{bmatrix} \Delta P \\ \Delta Q \end{bmatrix} = \begin{bmatrix} J_1 & J_2 \\ J_3 & J_4 \end{bmatrix} \begin{bmatrix} \Delta e \\ \Delta f \end{bmatrix} \to A$$

$$J_1 = \frac{\partial P}{\partial e} \quad J_2 = \frac{\partial P}{\partial f}$$

$$J_3 = \frac{\partial Q}{\partial e} \quad J_2 = \left.\frac{\partial Q}{\partial e}\right|_{(n-1)\times(n-1)}$$

Each including generator buses in (n–1) bus

$$\begin{bmatrix} \Delta P \\ \Delta Q \\ \Delta |V_p|^2 \end{bmatrix} = \begin{bmatrix} J_1 & J_2 \\ J_3 & J_4 \\ J_5 & J_6 \end{bmatrix} \begin{bmatrix} \Delta e \\ \Delta f \end{bmatrix} \to B$$

$$\Delta |V|_p^2 = |V|_{sp}^2 - |V|^2{}_p^0$$

Where superscript indicates the value obtained corresponding to 0th iteration. Calculation of increment voltage vector Δe and Δf

$$e_p{}^1 = e_p{}^0 + \Delta e_p{}^0$$

$$f_p{}^1 = f_p{}^0 + \Delta f_p{}^0$$

These values are used for next iteration

Assume that new better estimates obtained for $(k+1)^{th}$ iteration

$$e_p^{k+1} = e_p^k + \Delta e_p^k$$

$$f_p^{k+1} = f_p^k + \Delta f_p^k$$

The process is repeated until largest column in the residual column vector is less than specified value.

6.6.3 Decoupled Load Flow Method

It is the simplified version of Newton Raphson method.

The active power changes ΔP are less sensitive to the changes in voltage magnitude and more sensitive to δ.

The reactive power changes ΔQ are less sensitive to δ and more sensitive to voltage magnitude

$$\begin{bmatrix} \Delta P \\ \Delta Q \end{bmatrix} = \begin{bmatrix} H & O \\ O & L \end{bmatrix} \begin{bmatrix} \Delta\delta \\ \dfrac{\Delta|V|}{|V|} \end{bmatrix} \rightarrow (14)$$

Equation (14) represents decoupled equations. From (14)

$$[\Delta P] = [H][\Delta\delta] \rightarrow (15)$$

$$[\Delta Q] = [L]\left[\Delta\frac{|V|}{|V|}\right] \rightarrow (16)$$

The diagonal and off diagonal elements of 'H' and 'L' can be calculated from equations (6, 9).

$\Delta\delta$ can be calculated from equation (15). The updated values of δ are used in equations 16 to find $\Delta|V|$

6.6.4 Fast Decoupled Load Flow Method

Assumptions:

1. $\delta_p - \delta_q = 0 \Rightarrow \cos(\delta_p - \delta_q) = 1 \Rightarrow \sin(\delta_p - \delta_q) = 0$

2. As the power system network has more reactance with respect to resistor, resistance of transmission line is neglected.

   i.e, $G_{pq} <<<< B_{pq}$

     $G_{pq}\sin(\delta_p - \delta_q) << B_{pq}\cos(\delta_p - \delta_q) << B_P$

3. The reactive power $Q_p << B_{pp}\left|V_p\right|^2$

$$\text{From } 10 \Rightarrow H_{pp} = \frac{\partial P_p}{\partial\delta_p} = -B_{pp}\left|V_p\right|^2$$

$$6 \Rightarrow H_{pq} = \frac{\partial P_p}{\partial \delta_q} = -\left|V_p\right|\left|V_q\right| B_{pq}$$

$$13 \Rightarrow L_{pp} = \frac{\dfrac{\partial Q_p}{\partial \left|V_p\right|}}{\left|V_p\right|} = -B_{pp}\left|V_p\right|^2$$

$$9 \rightarrow L_{pq} = \frac{\dfrac{\partial Q_p}{\partial \left|V_q\right|}}{\left|V_q\right|} = -\left|V_p\right|\left|V_q\right| B_{pq}$$

From above 17 $H_{pp} = L_{pp}$ and $H_{pq} = L_{pq}$

The unknown elements $\Delta\delta$ and $\Delta|V|$ are calculated as follows.

$$\left.\begin{array}{l}\left[B'\right]\left[\Delta\delta\right] = \left[\dfrac{\Delta P}{|V|}\right] \\[2em] \left[B''\right]\left[\Delta|V|\right] = \left[\dfrac{\Delta Q}{|V|}\right]\end{array}\right\} \rightarrow (18)$$

$[B']$ is the susceptance matrix having elements $B_{pq}(p=2,3\ldots n\,;\,q=2,3,\ldots.n)$.

$[B'']$ is a part of susceptance matrix, having the elements B_{pq} corresponding to the elements pq

Fast decoupled method load flow method is popular method and Newton Raphson method is accurate method.

Q8. Consider a three–bus system as shown. The line data and bus data are given the reactive power limit for $Q_{2,min} = 0$, $Q_{2,max} = 0.6$ update the voltages and phase angle using Newton Raphson method for 1 iteration neglect line charging admittance.

Line	L_1	L_2	L_3
Series impedance	0.025+j0.1	0.025+j0.1	0.025+j0.1

Bus No	Generation		Load		V_i	δ_i	Type
	P_{Gi}	Q_{Gi}	P_{Di}	Q_{Di}			
1	–	–	0.9	0.4	1.02	0	Slack
2	1.4	–	–	–	1.03	–	PV
3	–	–	1.1	0.4	–	–	PQ

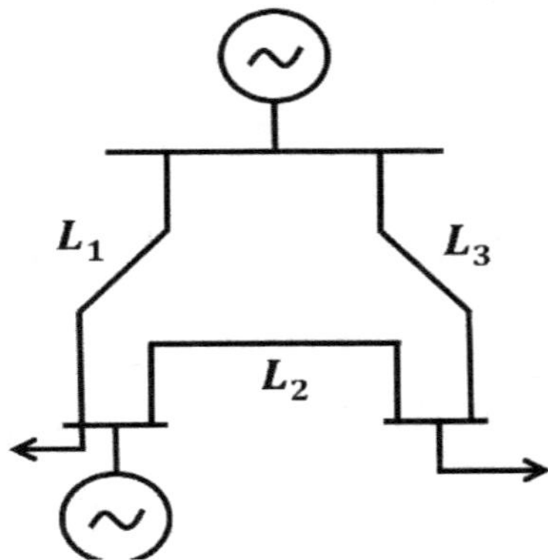

Solution:

Step 1:

Series Impedance of all lines = 0.025+j0.1p.u.

Series Admittance, $y = \dfrac{1}{z} - \dfrac{1}{0.025 + j0.1}$

Diagonal elements of Y_{Bus} = 2.35–j9.41p.u.

$$Y_{11} = Y_{22} = Y_{33} = 2(2.35 - j9.41)$$

$$= 4.7 - j18.82 \, p.u.$$

Off diagonal element

$$Y_{12} = Y_{21} = Y_{23} = Y_{32} = Y_{13} = Y_{31}$$

$$= -2.35 + j9.41 \ p.u.$$

$$[Y_{Bus}] = \begin{bmatrix} 4.7 - j18.8 & -2.3 + j9.4 & -2.3 + j9.4 \\ -2.3 + j9.4 & 4.7 - j18.8 & -2.3 + j9.4 \\ -2.3 + j9.4 & -2.3 + j9.4 & 4.7 + j - 8.8 \end{bmatrix} = G - Bj$$

$$G_{11} = G_{22} = G_{33} = 4.7$$

$$G_{22} = G_{21} = G_{23} = G_{32} = G_{31} = G_{13} = -23$$

$$B_{11} = B_{22} = B_{33} = 18.8$$

$$B_{12} = B_{21} = B_{23} = B_{32} = B_{31} = B_{13} = -9.4$$

Step 2: Computation of powers

Summaries of the specified bus quantity are:

Bus 1: $V_1 = 1.02$p.u., $\delta_1 = 0$ →slack bus

Bus 2: $P_2 = 1.4 \ V_2 = 1.03$ →PV bus

Bus 3: $P_3 = 1.1 \ Q_3 = -0.4$ →PQ bus

Assuming flat start for bus voltages

$$V_3^0 = 1 \ p.u. \quad \text{and} \quad \delta_2^0 = \delta_3^0 = 0 \ rad$$

Active power injected at bus 2

$$P_p = \left|V_p\right|^2 \left|Y_{pp}\right| \text{Cos}\,\theta_{pp} + \sum_{\substack{Q=1 \\ Q \neq P}}^{n} \left|V_p\right|\left|Y_{pQ}\right|\left|V_Q\right| \text{Cos}\left(\theta_{pq} + \delta_p - \delta_q\right)$$

$$= \left|V_p\right|^2 G_{pp} + \sum_{\substack{Q=1 \\ Q \neq P}}^{n} \left|V_p\right|\left|Y_{pQ}\right|\left|V_Q\right| \text{Cos}\left(\theta_{pq} + \delta_p - \delta_q\right)$$

$$P = 2$$

$$P_2 = \left|V_2\right|^2 G_{22} + \sum_{\substack{Q=1 \\ Q \neq 2}}^{n} \left|V_2\right|\left|Y_{2Q}\right|\left|V_Q\right| \text{Cos}\left(Q_{2q} + \delta_2 - \delta_2\right)$$

$$P_2 = |V_2|^2 G_{22} + \{|V_2||Y_{21}||V_1|\text{Cos}(\theta_{21} + \delta_2 - \delta_1) + |V_2||Y_{23}||V_3|\text{Cos}(\theta_{23} + \delta_2 - \delta_3)\}$$

$$= 1.03^2 \times 4.7 + 1.03(1.02)[|Y_{21}|\text{Cos}(\delta_2 - \delta_1)\text{Cos}\,\theta_{21} - |Y_{21}|\text{Sin}\,\theta_{21}\text{Sin}(\delta_2 - \delta_1)]$$
$$+ |V_2||V_3|[|Y_{23}|\text{Cos}\,\theta_{23}\text{Cos}(\delta_2 - \delta_3) - |Y_{23}|\text{Sin}\,\theta_{23}\text{Sin}(\delta_2 - \delta_3)]$$

$$= 1.03^2 \times 4.7 + 1.03 \times 1.02\{G_{21}\text{Cos}(\delta_2 - \delta_1) + B_{21}\text{Sin}(\delta_2 - \delta_1)\}$$
$$+ 1.03 \times 1\{G_{23}\text{Cos}(\delta_2 - \delta_3) + B_{23}\text{Sin}(\delta_2 - \delta_3)\}$$

$$= 1.03^2 \times 4.7 + 1.03 \times 1.02\{(-2.3)\text{Cos}(0) + (-2.3)\text{Sin}(0)\}$$
$$+ 1.03\{(-2.3)\text{Cos}0 + (-9.4)\text{Sin}0\}$$

$$P_2 = 0.103 \ p.u.$$
$$P = 3$$

$$P_3 = \left|V_3\right|^2 G_{33} + \sum_{\substack{Q=1 \\ Q \neq 3}}^{3} \left|V_3\right|\left|Y_{3Q}\right|\left|V_Q\right| \text{Cos}\left(\theta_{3q} + \delta_3 - \delta_q\right)$$

$$= \left|V_3\right|^2 G_{33} + \left|V_3\right|\left\{\left|Y_{31}\right|\left|V_1\right|\text{Cos}\left(\theta_{31} + \delta_3 - \delta_1\right) + \left|Y_{32}\right|\left|V_2\right|\text{Cos}\left(\theta_{32} + \delta_3 - \delta_2\right)\right\}$$

$$= \left|V_3\right|^2 G_{33} + \left|V_3\right|\left\{\left|V_1\right|\left|Y_{31}\right|\text{Cos}\,\theta_{31}\,\text{Cos}\left(\delta_3 - \delta_1\right) - \left|V_1\right|\left|Y_{31}\right|\text{Sin}\,\theta_{32}\,\text{Sin}\left(\delta_3 - \delta_2\right)\right\}$$

$$\left|V_2\right|\left|Y_{32}\right|\text{Cos}\,\theta_{32}\,\text{Cos}\left(\delta_3 - \delta_2\right) - \left|V_2\right|\left|Y_{32}\right|\text{Sin}\,\theta_{32}\,\text{Sin}\left(\delta_3 - \delta_2\right)$$

$$= |V_3|^2 G_{33} + |V_3|\{|V_1|(G_{31}\text{Cos}(\delta_3 - \delta_1) + B_{31}\text{Sin}(\delta_3 - \delta_1) + |V_2|G_{32}\text{Cos}(\delta_3 - \delta_2)$$
$$+ B_{32}\text{Sin}(\delta_3 - \delta_2))\}$$

$$= 1 \times (4.7) + 1\{1.02(2.3) + 1.03(-2.3)\} = -0.0212 \ p.u.$$

The reactive power injected at bus P

$$Q_P = |V_P|^2 |Y_{pp}| \sin\theta_{pp} + \sum_{\substack{Q=1 \\ Q \neq P}}^{n} |V_P||Y_{PQ}||V_Q| \times \sin\left(\theta_{pq} + \delta_p - \delta_Q\right)$$

$$= |V_P|^2 \left(-B_{pp}\right) + \sum_{\substack{Q=1 \\ Q \neq P}}^{n} |V_P||V_Q|\left\{ |Y_{PQ}| \sin\theta_{pq} \cos\left(\delta_p - \delta_Q\right) + |Y_{PQ}| \cos\theta_{pq} \sin\left(\delta_p - \delta_Q\right) \right\}$$

$$= |V_P|^2 \left(-B_{pp}\right) + \sum_{\substack{Q=1 \\ Q \neq P}}^{n} |V_P||V_Q|\left\{ \left(-B_{pq}\right) \cos\left(\delta_p - \delta_Q\right) + G_{pq} \sin\left(\delta_p - \delta_q\right) \right\}$$

$P = 2$

$$Q_2^0 = -|V_2|^2 \left(B_{22}\right) + \left\{ \begin{array}{l} |V_2||V_1|\left\{ \left(-B_{21} \cos\left(\delta_2 - \delta_1\right) + G_{21} \sin\left(\delta_2 - \delta_1\right)\right)\right\} + |V_2||V_3|\left\{ \left(-B_{23}\right)\cos\left(\delta_2 - \delta_3\right)\right\} \\ + G_{23} \sin\left(\delta_2 - \delta_3\right) \end{array} \right\}$$

$$= -(1.03)^2(1.88) + (1.03)\{(1.02)(+9.4) + (1)(9.4)\}$$

$$= -0.38728 \ p.u.$$

$P = 3$

$$Q_3^0 = -|V_B|^2 B_{33} + \left\{ \begin{array}{l} |V_3||V_1|(-B_{31}\cos(\delta_3 - \delta_1) + G_{31}\sin(\delta_3 - \delta_1)) + \\ |V_3||V_2|(-B_{32}\cos(\delta_3 - \delta_2) + G_{32}\sin(\delta_3 - \delta_2)) \end{array} \right\}$$

$$= -(1)^2(18.8) + 1\left\{1.02\left(+9.4(1) + (-2.3)(0)1.03(+2.4 + (-2.3)(0))\right)\right\} = 0.47 \ p.u.$$

As Q_p is negative of right–hand side terms

$$Q_2^0 = +0.3872 \ p.u.$$

$$Q_3^0 = -0.47 \ p.u.$$

Step 3: Verification of reactive power limits for PV bus

Therefore, PV bus is treated as PV bus for the next iteration

Step 4: Computation of power mismatches

$$\Delta P_2 = P_{2,s} - P_{2,cal}, \Delta P_3 = P_{3s} - P_{3,cal}$$

$$\Delta Q_3 = Q_{3,s} - Q_{3,cal}$$

$$\Delta P_2 = +1.4 - 0.1948 = 1.2052$$

$$\Delta P_3 = -1.1 - (-0.0212) = -1.0788$$

$$\Delta Q_3 = -0.4 - (-0.47) = 0.07$$

5 Calculation of Jacobian elements

$$\begin{bmatrix} \Delta P \\ \Delta Q \end{bmatrix} = \begin{bmatrix} H & M \\ N & L \end{bmatrix} \begin{bmatrix} \Delta\delta \\ \dfrac{|\Delta V|}{|V|} \end{bmatrix}$$

$$\begin{bmatrix} \Delta P_2 \\ \Delta P_3 \\ \Delta q_3 \end{bmatrix} = \begin{bmatrix} H_{22} & H_{23} & N_{23} \\ H_{32} & H_{33} & N_{33} \\ M_{32} & M_{33} & L_{33} \end{bmatrix} \begin{bmatrix} S_2 \\ S_3 \\ \dfrac{|\Delta V_1|}{|V_3|} \end{bmatrix}$$

Calculation of matrix 'H' elements:

Diagonal elements:

$$H_{pp} = -Q_p - |V_p|^2 B_{pp}$$

$$P = 2$$

$$H_{pp} = -Q_2 - |V_2|^2 B_{22}$$

$$= -0.38 - (1.03)^2(-18.8)$$

$$19.55 \, p.u.$$

$$P = 3$$

$$H_{33} = -Q_3 - |V_3|^2 B_{33}$$

$$= 0.47 - (1)^2 (-18.8)$$

$$= 19.27$$

Off diagonal elements

$$H_{pq} = |V_p||V_q|\{-B_{pq}\text{Cos}(\delta_p - \delta_Q) + G_{pQ}\text{Sin}(\delta_p - \delta_Q)\}$$

$$P = 2$$

$$P = 3$$

$$\Rightarrow H_{23} = |V_2||V_3|\{-B_{23}\text{Cos}(\delta_2 - \delta_3) + G_{23}\text{Sin}(\delta_2 - \delta_3)\}$$

$$= 1.03 \times \{-9.4(1) + G_{23} \times 0\}$$

$$= -9.682 \, p.u.$$

$$p = 3, q = 2$$

$$H_{32} = |V_3||V_2|\{-B_{32}\text{Cos}(\delta_3 - \delta_2) + G_{32}\text{Sin}(\delta_3 - \delta_2)\}$$

$$= 1(1.03)\{-9.4\}$$

$$= -9.682 \, p.u.$$

Calculation of matrix 'M' elements

Diagonal elements

$$M_{pp} = P_p - |V_p|^2 G_{pp}$$

$$M_{22} = P_2 - |V_2|^2 G_{22} = 0.199 - (1.03)^2 (4.7) = -9.787$$

$$M_{33} = P_3 - \left|V_3\right|^2 G_{33} = 0 - 18.33 - (1)^2 (4.7) = -23.03$$

Off diagonal elements

$$M_{pq} = \left|V_p\right|\left|V_q\right|\{-G_{pq}\text{Cos}(\delta_p - \delta_q) - B_{pq}\text{Sin}(\delta_p - \delta_q)\}$$

p = 3, q = 2

$$M_{32} = |V_3||V_2|\{-G_{32}\text{Cos}(\delta_3 - \delta_2) - B_{32}\text{Sin}(\delta_3 - \delta_2)\}$$
$$= 1(1.03)\ \{-(-2.3)\ (1)\}$$
$$= 2.369\ p.u.$$

Calculation of matrix 'N' elements

Diagonal elements

$$N_{pp} = P_p - \left|V_p\right|^2 G_{pp}$$

$$P = 3 \rightarrow N_{33} = P_3 - \left|V_3\right|^2 G_{33} = -18.33 + (1)(4.7) = -13.63\ p.u.$$

Off diagonal elements

$$N_{pq} = \left|V_p\right|\left|V_q\right|\{G_{pq}\text{Cos}(\delta_p - \delta_q) + B_{pq}\text{Sin}(\delta_p - \delta_q)\}$$
$$p = 2,\ q = 3,\ N_{23} = |V_2||V_3|\{G_{23}\text{Cos}(\delta_2 - \delta_3) + B_{23}\text{Sin}(\delta_2 - \delta_3)\}$$
$$= 1.03\ (1)\ (-2.3)$$
$$= -2.369$$

Calculation of matrix 'L' elements

Diagonal elements

$$L_{pp} = -\left|V_p\right|^2 B_{pp} + Q_p$$

$$P = 3 \rightarrow L_{33} = -\left|V_3\right|^2 B_{33} + Q_3 = -(1)(-18.8) + 0.47 = 19.26\ p.u.$$

$$\begin{bmatrix} 0.199 \\ -0.0212 \\ -0.47 \end{bmatrix} = \begin{bmatrix} 19.55 & -9.682 & -2.39 \\ -9.682 & 19.27 & -13.63 \\ 2.39 & -23.03 & 19.27 \end{bmatrix} \begin{bmatrix} \Delta\delta_2 \\ \Delta\delta_3 \\ \Delta|V_3| \end{bmatrix}$$

$$\Delta\delta_2^0 = 0.0978 = 5.6^0$$

$$\Delta\delta_3^0 = 0.1437 = 8.23^0$$

$$\frac{\Delta|V_3|^0}{|V_3|} = 0.1350 \rightarrow \Delta|V_3|^0 = 0.1350$$

$$\delta_2^1 = \delta_2^0 + \Delta\delta_2^0 = 5.6^0 + 0 = 5.6^0$$

$$\delta_3^1 = \delta_3^0 + \Delta\delta_3^0 = 0 + 8.23^0 = 8.23^0$$

$$|V_3|^1 = |V_3|^0 + \Delta V_3^0 = 1 + 0.1350 = 1.350\ p.u.$$

Q9. In a 15–bus power system network with 3 voltage–controlled bus the size of Jacobian matrix power system.

Total no. of buses given = 15 Bus

$$\begin{bmatrix} \Delta P \\ \Delta Q \\ \Delta |V| \end{bmatrix} = \begin{bmatrix} J_1 & J_2 \\ J_3 & J_4 \end{bmatrix} \begin{bmatrix} \Delta \delta \\ \Delta |V| \end{bmatrix}$$

There are three voltage control busses

$$\Rightarrow 15 - 3 = 12$$

Among 12 buses one is the Slack bus

Therefore Total no.of PQ buses 12–1 = 11

No. of variables contributed by PQ bus

$$\Rightarrow 11 \times 2 = 22$$

Plus voltage buses = 22 + 3 = 25

$$Size = 25 \times 25$$

Comparison of load flow methods:

Description	Gauss seidel method	Newton Raphson method	Decoupled load flow
Arithmetic operations	Least in number to complete one iteration	Elements of Jacobian matrix are to be calculated	Less than newton Raphson
Time	Requires less time per iteration bus increases with no. of buses	Time per iteration is 7 times of Gauss seidal and increases with no. of buses	Less time when compared to Gauss seidal and newton Raphson
Convergence	Linear convergence	Quadratic	Geometeric
No. of iterations	Large in no. and increases with buses	Very less for large systems (3 to 5)	Only 2 to 3 iterations
Acceleration	External acceleration due to α	Internal acceleration	Internal acceleration
Slack bus selection	Effects convergence	Effect is minimal	Effect is moderate
Accuracy	Less accurate	More	Moderate
Memory	Least due to sparsity of [Ybus]	More memory due to Jacobian matrix	Less (around 60% of memory as compare to NR
Application	Small size system	For all large size systems	Large system
Programming logic	Easiest	Most difficult	Moderan difficult
Reliability	For small systems	For both small and large	More reliable than NR

DC load flow method:

DC load flow method is formulated by Stott and Alsac in 1974

Real power flow $R_{DC} = \theta_{DC}B_{DC}$

DC includes phase angles of voltages and BDC includes susceptance of transmission system.

Consider the power flow equations

$$P_K = \sum_{j=1}^{N} |V_K||V_j|\left\{G_{kj}\cos\left(\theta_k - \theta_j\right) + B_{kj}\sin\left(\theta_k - \theta_j\right)\right\}$$

$$Q_K = \sum_{j=1}^{N} |V_K||V_j|\left\{G_{kj}\sin\left(\theta_k - \theta_j\right) - B_{kj}\cos\left(\theta_k - \theta_j\right)\right\}$$

Approximation 1: Series admittance of transmission line,

$$y_{ik} = -\frac{1}{Z_{ik}} = \frac{1}{r_{ik} + Z_{ikj}} \times \frac{R_{ik} - X_{ikj}}{R_{ik} = X_{ikj}}$$

$$Y_{ik} = \frac{R_{ik}}{R_{ik^2} + X_{ik^2}} - \frac{jX_{ik}}{R_{ik^2} + X_{ik^2}}$$

tr line $x \gg R$ $Y_{ik} = G_{ik} - B_{ikj}$

$d/s\, R \gg x$

where

$$\frac{R_{ik}}{R_{ik^2} + X_{ik^2}} = G$$

$$B = \frac{X_{ik}}{R_{ik} + X_{ik^2}}$$

For a transmission line, $X = R$ or $R = 0$

$$\therefore G = 0, B = \frac{1}{X_{ik}}$$

$$\therefore P_k = \sum_{j=1}^{N} |V_k||V_j|\left\{B_k j\sin\left(\theta_k - \theta_j\right)\right\}$$

$$Q_k = \sum_{j=1}^{N} |V_k||V_j|\left\{B_k j\cos\left(\theta_k - \theta_j\right)\right\}$$

Approximation 2: For long transmission line

$$\sin\left(\theta_k - \theta_j\right) \cong \left(\theta_k - \theta_j\right)$$

$$\cos\left(\theta_k - \theta_j\right) \cong 1$$

$$P_k = \sum_{j=1}^{N} |V_k||V_j|B_{kj}\left(\theta_k - \theta_j\right)$$

$$Q_k = \sum_{j=1}^{N} |V_k||V_j|B_{kj}$$

OBJECTIVE QUESTIONS

1. State load flow analysis?

   Ans: Load flow solution is a solution of the power system network under steady state condition subject to certain inequality constraints under which the system operates.

2. What is the information obtained from a load flow study?

   Ans: The information obtained from a load flow study are magnitude and phase of bus voltages, real and reactive power flowing in each line and the line losses.The load flow solution also gives the information regarding initial conditions of the system when transient behavior of the system has to be studied.

3. What is the need for load flow study?

   Ans: The load flow study of a power system is essential to decide the best operation of existing system and for planning the future expansion of the system.It is also essential for designing a new power system.

4. How a load flow study is performed?

   Ans: (a)Representation of the system by single line diagram (b)Determining the impedance diagram using the information of single line diagram. (c) Formulation of network equations (d) Solution of network equations

5. What are the static load flow equations?

   Ans:
   $$P = \sum_{k=1}^{n} |V_i|\,|Y_{ik}||V_k|\cos(\theta_{ik} + \delta_i - \delta_k)$$
   $$Q = \sum_{k=1}^{n} |V_i|\,|Y_{ik}||V_k|\sin(\theta_{ik} + \delta_i - \delta_k)$$

6. Define PV or Generator bus?

   Ans: A bus is said to be voltage–controlled bus if the magnitude of voltage and real power are specified. The magnitude of voltage is not allowed to change. Any bus to which a generator is connected is known as a Generator bus or PV bus.

7. What is a PQ bus?

   Ans: A bus is said to be PQ bus or load bus when real and reactive components of power are specified for the bus. Voltage is allowed to change within the permissible limits.

8. What is a swing bus or slack bus?

   Ans: A bus is said to be a swing bus or slack bus when the magnitude and phase of bus voltage are specified.

9. What is the need for Slack bus?

   Ans: In a power system network, the total power generation must be equal to load demand. In a power system, only the generated power and load power are specified for buses. A generator bus considering the additional real and reactive power to supply transmission losses is known as slack or swing or reference bus.

10. What are the operating constraints imposed in the load flow studies?

    Ans: (a) Reactive power limits for generator buses (b) Allowable change in voltage magnitude for load buses.

11. Define acceleration factor in power flow studies?

    Ans: In power flow solution by iterative methods, the number of iterations can be reduced if the correction voltage at each bus is multiplied by some constant. The multiplication of the constant increases the correction to bring the voltage closer to the value it is approaching. The multipliers that accomplish this improved convergence are called acceleration factors.

12. Why iterative methods are required for power flow studies?

    Ans: The power flow equations are non–linear algebraic equations and explicit solution is not possible. The solution of non–linear equations can be obtained only by numerical techniques.

13. What is Flat voltage start in power flow studies?

    Ans: In iterative methods of power flow studies, the initial voltages of all buses except slack bus are assumed as $1 + j0$ p.u. and is referred as flat start.

14. What are the advantages of Gauss–Seidel Method?

    Ans:
    (a) Calculations are simple and programming task is less.
    (b) Memory requirement is less.
    (c) Useful for small systems.

15. What are the disadvantages of Gauss–Seidel Method?

    Ans:
    (a) Requires more number of iterations to reach convergence.
    (b) Not suitable for large systems.
    (c) Convergence time increases with time.

16. How approximation is performed in Newton Raphson Method?

    Ans: The set of non–linear power flow equations are approximated to a set of linear simultaneous equations using Taylor series of expansion and the terms are limited to only first approximation.

17. What is Jacobian Matrix? How the elements of Jacobian matrix are calculated?

    Ans: The matrix formed from the first derivatives of power flow equations is called Jacobian matrix,J.Jacobian matrix is a first order partial derivative matrix relating the elements of the power mismatch matrix and elements of the unknown vector.

    The elements of Jacobian matrix changes in every iteration. In each iteration, the elements of Jacobian matrix are obtained by partially differentiating the power flow equations with respect to an unknown variable and then evaluating the first derivatives using the solution of the previous iteration.

18. What are the advantages of Newton–Raphson Method?

    Ans:

    (a) Faster, reliable and accurate.

    (b) Requires less number of iterations for convergence.

    (c) Number of iterations are independent of size of the system.

    (d) Suitable for large size system

19. What are the disadvantages of Newton–Raphson Method?

    Ans:

    (a) Programming is complex.

    (b) Memory requirement is more.

    (c) Computational time per iteration is higher due to more calculations.

20. Mention the advantages of Newton Raphson method over Gauss Seidel Method?

    Ans:

    (a) Newton Raphson method has quadratic convergence and hence converges faster than Gauss Seidel method.

    (b) The number of iterations for convergence is independent of system size.

    (c) The convergence is not affected by the choice of the slack bus.

21. How the convergence of Newton–Raphson method is speeded up?

    Ans: The convergence can be speeded up in NR method by using fast decoupled method algorithm. The weak coupling between active power, P–Voltage, V and reactive power, Q– load angle, δ are decoupled and then the equations are further simplified using the knowledge of practical operating conditions of a power system.

22. How the disadvantages of Newton–Raphson method are overcome?

    Ans: The disadvantage of large memory requirement can be overcome by decoupling the weak coupling between active power, P–Voltage, V and reactive power, Q– load angle, δ. The disadvantage of large computational time per iteration can be reduced by simplifying the decoupled load flow equations. The simplifications are made based on the practical operating conditions of a power system.

23. When the iterative process in the Gauss–Seidel method terminates?

    Ans: The iterative process terminates when the difference between the magnitudes of the voltage during the latest two successive iterations at every bus is less than prescribed tolerance.

24. Why the time required per iteration is more in Newton–Raphson method?

    Ans: As the Jacobian matrix elements are to be evaluated in each iteration, Newton–Raphson method takes more time for each iteration.

25. Why Ybus is used in load flow solution instead of Zbus?

    Ans: As Ybus matrix consists of more zero elements than non–zero elements, the memory required is less and is preferred over Zbus matrix.

26. How many terms can be stored in Ybus matrix for load flow solution?
    Ans: Number of terms=n(n+1)/2

27. What is the sensitivity of Jacobian matrices?

    Ans: The sensitivity refers to decoupling the weak coupling between active power, P – Voltage, V and reactive power, Q – load angle, δ.

28. Why Bbus matrix is used in Fast decoupled load flow method than Ybus matrix?

    Ans: As susceptance of the transmission line is greater than it's conductance, Bbus matrix is used in Fast decoupled load flow method than Ybus matrix

29. What are the constraints to be satisfied for PV or generator bus?

    Ans: The reactive power constraints are to be considered for PV bus or generator bus.

30. What is the sign assigned for P and Q for PV bus?

    Ans: P and Q values are considered as positive for PV bus.

31. Is PV bus an exporting or importing bus?

    Ans: PV bus is also known as exporting bus.

32. Is PQ bus an exporting or importing bus?

    Ans: PQ bus is also known as importing bus.

33. What is the sign assigned for P and Q for PQ bus?

    Ans: P and Q values are considered as negative for PQ bus.

34. What happens if the reactive power limits are satisfied by PV bus?

    Ans: If the reactive power limits are satisfied by PV bus then PV bus is considered as PV bus for that iteration.

35. What happens if the reactive power limits are not satisfied by PV bus?

    Ans: If the reactive power limits are not satisfied by PV bus then PV bus is considered as PQ bus for that iteration with Q being the reactive power limit violated but P and Q are still considered positive.

36. What is the significance of Slack bus?

    Ans: A generator bus considering the additional real and reactive power to supply transmission losses is known as slack or swing or reference bus.

37. What are the dimensions of Jacobian matrix?

    Ans: The dimensions of the Jacobian matrix is $(2n–2) \times (2n–2)$

38. What are the dimensions of Jacobian matrix?

    Ans: The dimensions of the Jacobian sub matrix is $(n–1) \times (n–1)$

39. How many Jacobian sub–matrices exist in a Jacobian matrix?

    Ans: The number of Jacobian sub–matrices with only PQ buses=4

    The number of Jacobian sub–matrices with PV and PQ buses=6

40. What are the assumptions of Decoupled Load flow method?

    Ans: The assumptions of Decoupled Load flow method are

    The real power changes are less sensitive to changes in voltage magnitude and are mainly sensitive to changes in load angle.

    The reactive power changes are less sensitive to changes in load angle and are mainly sensitive to changes in voltage magnitude.

41. What are the assumptions of Fast Decoupled Load flow method?

    Ans: The assumptions of Fast Decoupled Load flow method are

    $$\text{Cos}(\delta_p – \delta_q) = 1 \text{ and } \sin(\delta_p – \delta_q) = 1$$
    $$\text{Gpq } \sin(\delta_p – \delta_q) < B_{pq}$$
    $$Q_p < B_{pp} V_p^2$$

    The phase angle effects of phase shifters from B" are neglected

    The representation of the network elements that affect MVAR flows from B" are neglected.

    The series resistance of the line is neglected from B'.

42. What is the accurate power flow method?

    Ans: The most accurate load flow method is Newton Raphson method as it has internal acceleration and no assumptions.

43. What is the most popular power flow method?

    Ans: The most popular load flow method is Fast Decoupled Load Flow Method.

44. What are the assumptions of DC load flow method?

    Ans:

    (a)   Resistance of the transmission line is negligible.

    (b)   Change in the phase angles of voltages is neglected.

    $$\delta p – \delta q = 0, \cos(\delta p – \delta q) = 1 \text{ and } \sin(\delta p – \delta q) = 0$$

    PV bus is treated as <u>PQ</u> bus if the reactive power limits are violated.

    In Newton Raphson method, non–linear equations are approximated using _______________

    The elements of _______________ are first order derivatives of load flow equations.

In _______________, convergence is not affected by the choice of Slack bus.

The Gauss Seidel method has _______________ convergence and Newton Raphson method has _______________ convergence.

The voltage variations in a bus are directly related to _______________ power.

_______________ method is used for small system and _______________ method is used for large systems.

7   Fault Analysis

7.1   INTRODUCTION

Fault is defined as an abnormal condition resulting in change in electrical parameters like voltage, current, power factor and frequency.

Whenever fault occurs power system network becomes unbalanced. Before analyzing an unbalanced power system network, a three phase power system network can be represented as follows in figure 7.1.

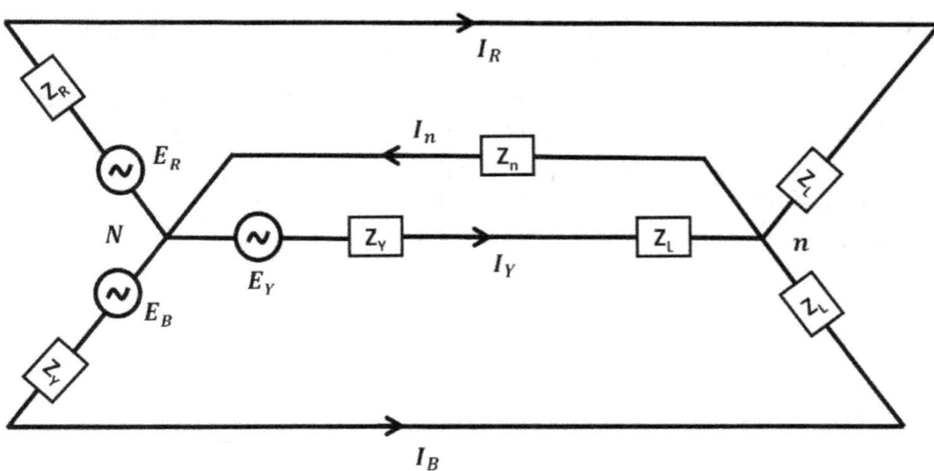

Figure 7.1 Representation of a Balanced three phase Network

Features:

1. Source and load are balanced for a balanced three phase network. Source is balanced & load is unbalanced for three phase unbalanced network.
2. 'N' is fixed neutral & n is floating neutral.

3. The unbalanced load voltages are obtained from Milliman's theorem.
4. For a balanced three phase power system network single phase equivalent can be represented as follows:

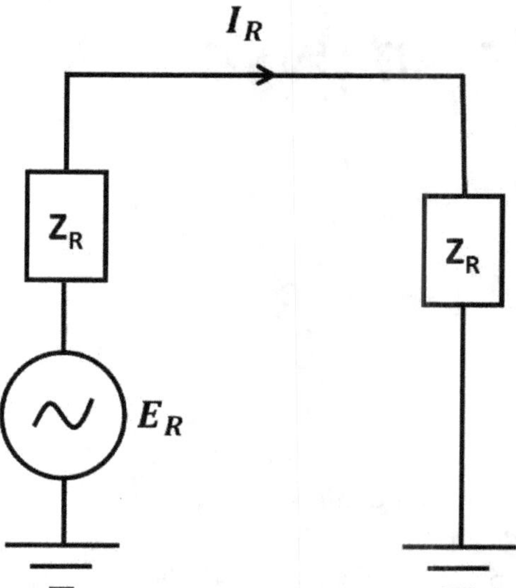

Figure 7.2 Per unit impedance and per unit reactance diagram

Assumptions

1. A generator can be represented by a voltage source in series with an inductive reactance & the internal resistance of the generator is negligible when compared to that of reactance.
2. The loads are inductive in nature. Transformer is ideal and represented by reactance.
3. Transmission line is medium and can be represented by T or π circuit.
4. The $\Delta - Y$ connecting transformer can be represented by an equivalent Y–Y connected transformer. Such impedance diagram may be drawn on per phase bases.

Per unit impedance diagram

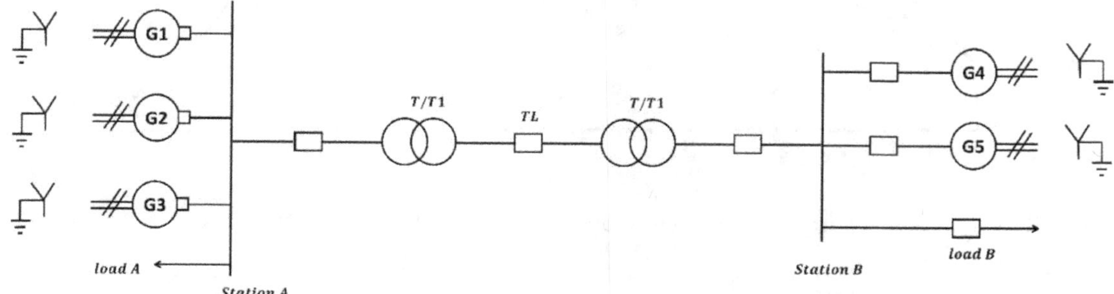

Figure 7.3 Power system network

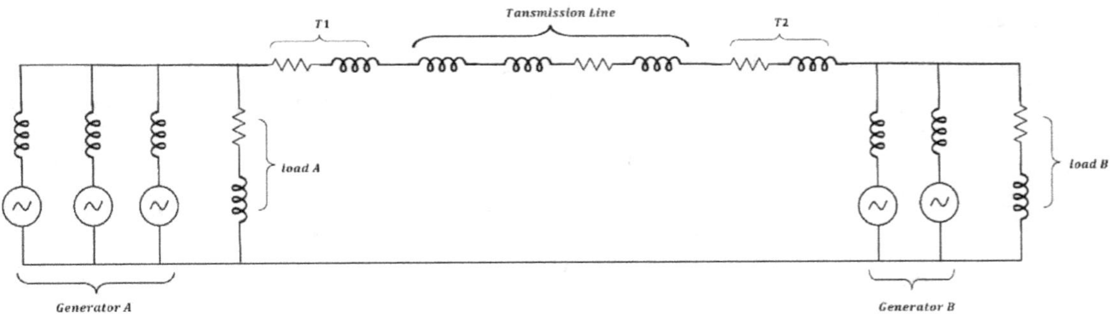

Figure 7.4 Per unit reactance diagram

The per unit reactance diagram of power system network in figure 7.3 is shown in figure 7.4.

7.2   REACTANCE CONTROL OF SHORT CIRCUIT CURRENTS

1. As fault current, $I_f \propto \dfrac{1}{\text{Sub transient reactance, } X_d''}$, the magnitude of fault current can be decreased by connecting reactance at suitable locations in power system network.
2. A reactor is coil with high inductive reactance and low resistance.
3. As resistance is low, I^2R losses are low and efficiency of power system is not affected by location of reactance.

7.2.1 Generator Reactors

1. Reactors are connected in series with each generator as shown in figure 7.5.
2. This reactance is considered as part of leakage reactance of generator.
3. The reactance of an alternator is 2 p.u.
4. Even though a dead short circuit at alternator terminals $I_{sc} \leq I_{FL}$ i.e. no external reactor is required to limit short circuit current.

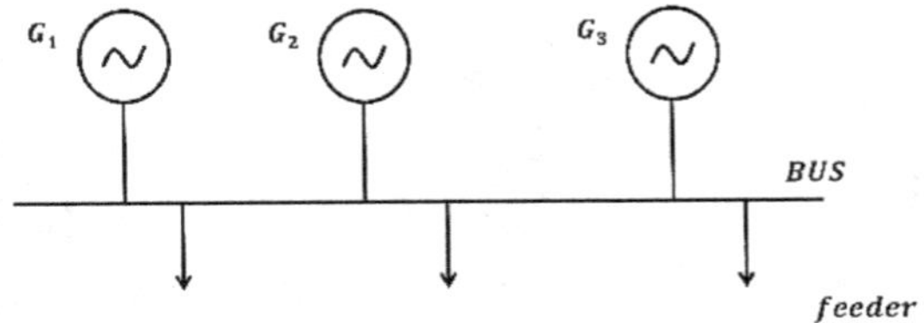

Figure 7.5 Generator reactors

Demerits

1. Reactive voltage drop & reactive power losses occurs
2. If fault occurs on a feeder near bus bar or on the bus bar then voltage at the bus bar is zero and alternator falls outs of synchronism.

7.2.2 Feeder Reactors

1. Reactors are connected in series with each feeder as shown in figure 5.6.
2. As 99% of power system network is transmission line or feeders, faults occur commonly on transmission lines.
3. The perunit reactance of feeder is high and hence external reactors are not required to limit short circuit.

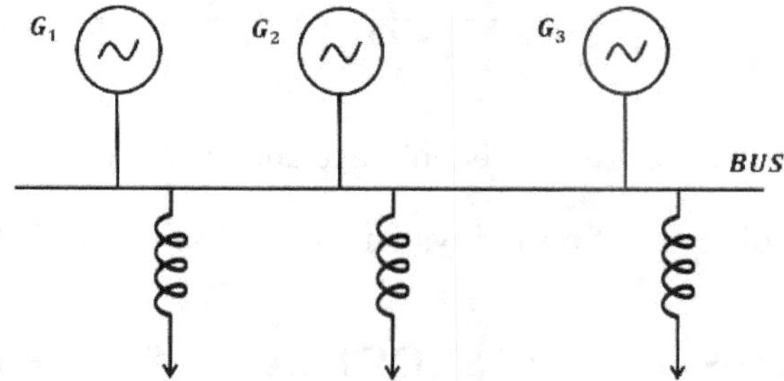

Figure 7.6 Feeder reactors

Merit: Fault on a feeder does not affect other feeders

Demerit: If fault occurs on a feeder near bus bar, voltage decreases & generators fall out of step.

7.2.3 Bus Bar Reactors

(a) **Ring main system:** Ring main system is divided into several sections & each provided with reactance. One feeder is fed from only one generator.

Under normal conditions, each generator supplies its own section of load thereby decreasing reactive voltage drops & reactive power loss.

Merit: If fault occurs on any feeder, only one generator feeds the fault current and only that section of bus bar is affected.

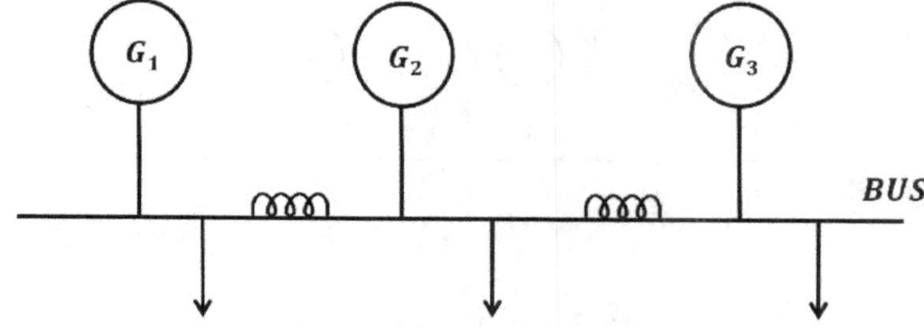

Figure 7.7 Ring main reactors

(b) Tie bar system

1. There are two reactors in series between sections so that each has X/2 reactance.
2. Additional generators may be connected to the system for further expansion.

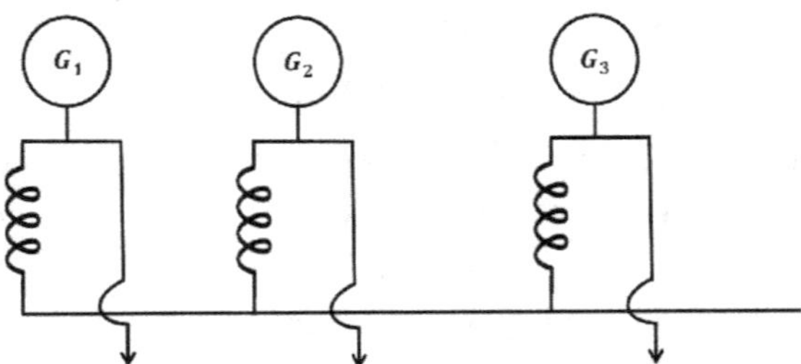

Figure 7.8 Tie bar reactors

7.3 SHORT CIRCUIT CAPACITY OR FAULT LEVEL OF A BUS

1. Consider a part of power system network.
2. Assume that fault occurs on Bus 3 as shown in figure 7.8.

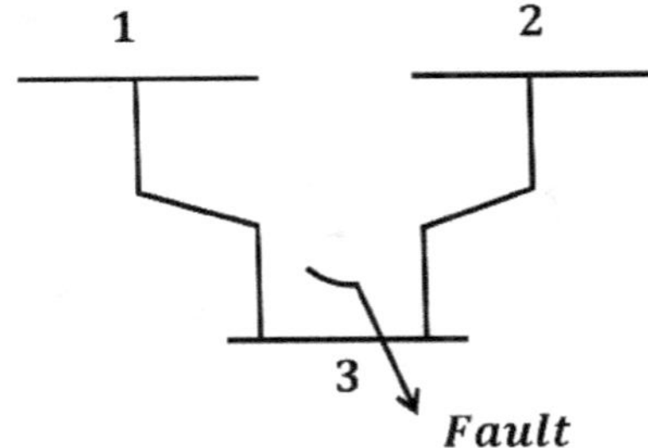

Figure 7.9 Short Circuit capacity of a bus

3. Voltage at bus-3 during fault, $V_3 = 0$.
4. The voltage at the other bus bars decreases and deduction in voltage at other bus is indication of short circuit capacity of buses '1' & '2'.
5. Higher the short circuit capacity, better is the capacity to maintain constant voltage.

Infinite bus: A bus for which pre fault voltage = post fault voltage is infinite bus and infinite bus has zero impedance.

Although the potential at the fault bus is zero, short circuit current is expressed in terms of short circuit KVA based on normal system. Voltage at the point of fault short circuit capacity equal to pre–fault voltage multiplied with post fault current.

$$I_{sc(Amp)} = \frac{V_b}{X_\Omega} = \frac{V_b}{\frac{\%X V_b}{100\, I_b}} = I_{b(Amp)} \left(\frac{100}{X\%}\right) \qquad (1)$$

Short circuit volt ampere for a single phase system is

$$V_b I_{sc(A)} = V_b I_b \left(\frac{100}{X\%}\right) = (VA)_b \left(\frac{100}{X\%}\right)$$

Short circuit volt ampere for a three phase system is

$$3 V_b I_{sc(A)} = 3 V_b I_b \left(\frac{100}{X\%}\right) = (VA)_{b(3-\Phi)} \left(\frac{100}{X\%}\right)$$

The zero sequence diagram can be drawn as follows:

1. An alternator is represented by $E_{R0} = 0$ and X_{R0} connected in series.
2. If the transformer winding is connected in star then series switch must be operated.
   2a. Close series switch with $X_n = 0$ for soild neutral
   2b. Close series switch with $3X_n$ for reactance neutral.
   2c. Open series switch for isolated neutral.
3. If the transformer winding is connected in delta then shunt switch must be closed.
4. Remaining power system components are shown by their zero sequence reactances.

Q1. Determine the equivalent zero sequence per unit reactance for the power system network in figure.

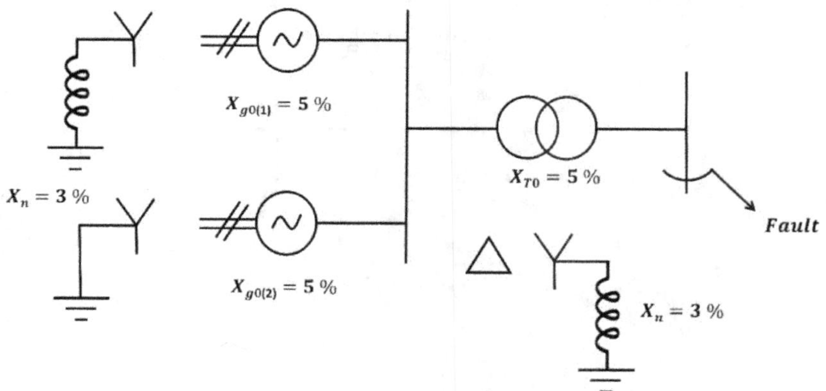

Solution:

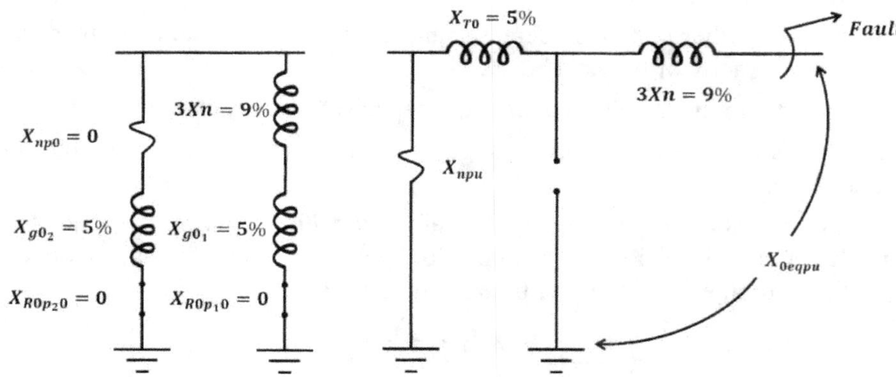

$$X_{0eqp.u} = j(3 \times n + X_{T0} + X_{npu})$$
$$= j(0.09 + 0.05 + 0)$$
$$= j0.14 \ p.u.$$

Q2.   Determine $X_{0eqp.u.}$ with respect to point P or 3 for the given network.

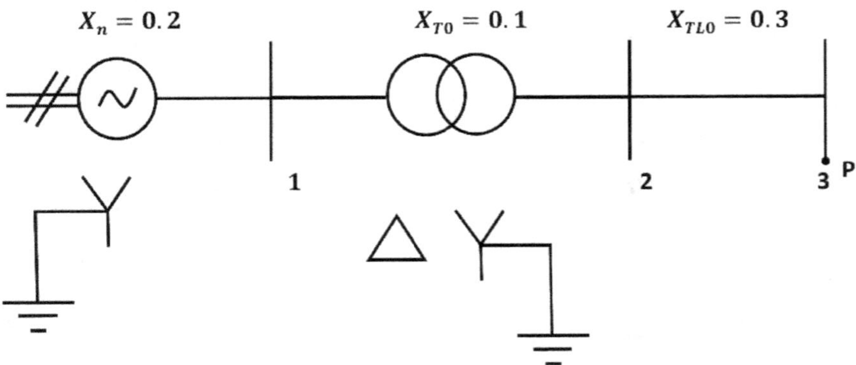

Solution:

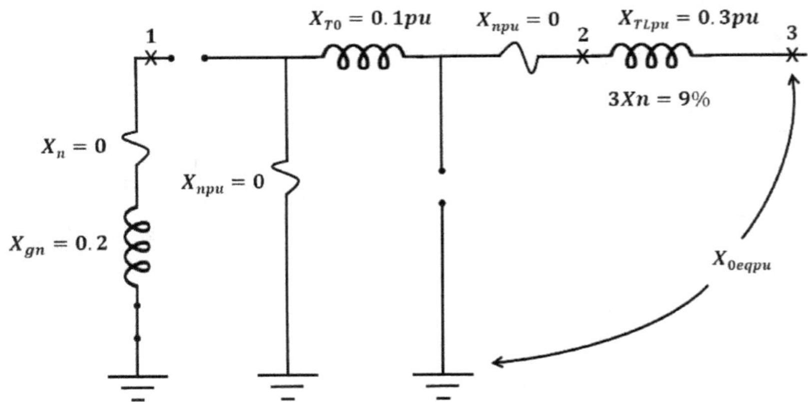

$$X_{0eq \ pu} = j(0.3 + 0 + 0.1 + 0) \ p.u.$$
$$= j0.4 \ p.u.$$

Q3.   Determine $X_{0eqp.u}$ with respect to fault point at the middle of the transmission line.

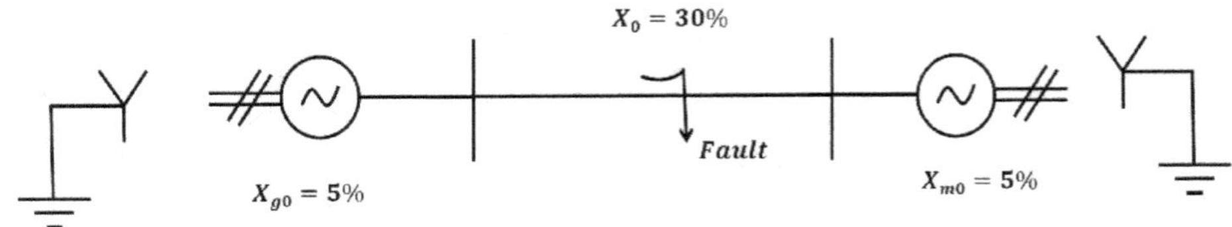

Solution:

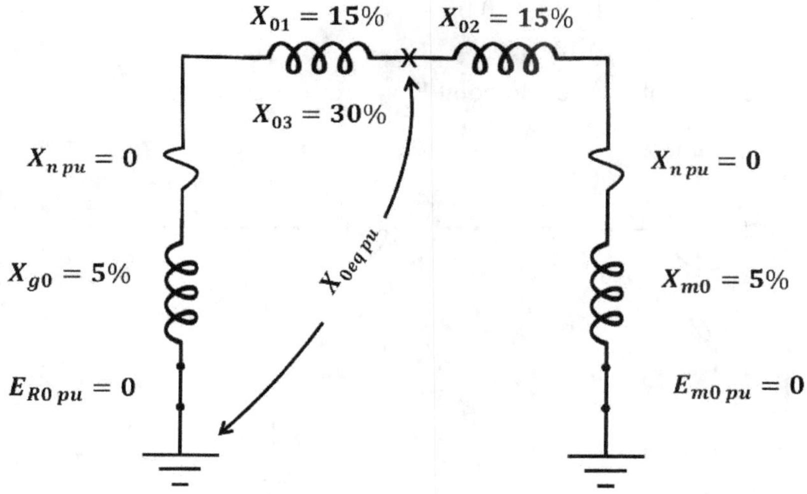

$$X_{0eq\ p.u.} = (0.15 + 0.05 + 0 + 0.2)j\ pu\ //\ (0.15 + 0.05 + 0)j\ p.u.$$
$$= 0.2\ j\ p.u.\ //\ 0.2\ j\ p.u.$$
$$= 0.1\ j\ p.u.$$

Q4. Determine X_{0eq} at the bus 3 for the given network.

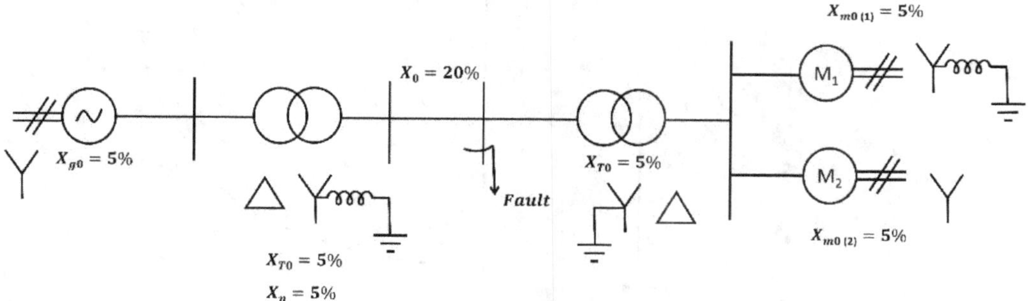

Solution:

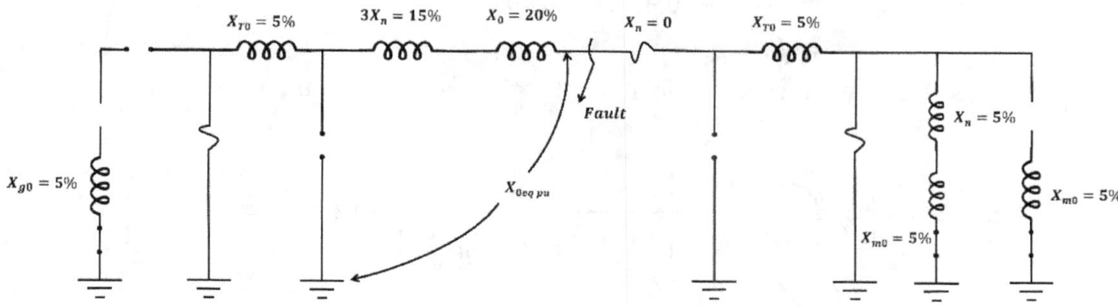

$$X_{ep\ p.u.} = j(0.2 + 0.15 + 0.05)\ //j\ (0.15 + 0.05)\ p.u.$$
$$= j0.4\ p.u.//j0.2\ p.u.$$
$$= \frac{(0.4 \times 0.2)(-1)}{j(0.4 + 0.2)}$$
$$= \frac{\frac{8}{100}(-1)}{j0.6}$$
$$= \frac{j\frac{8}{100}}{\frac{6}{10}}$$
$$= j0.133$$

Q5. Determine the equivalent per unit reactance for the power system network given below.

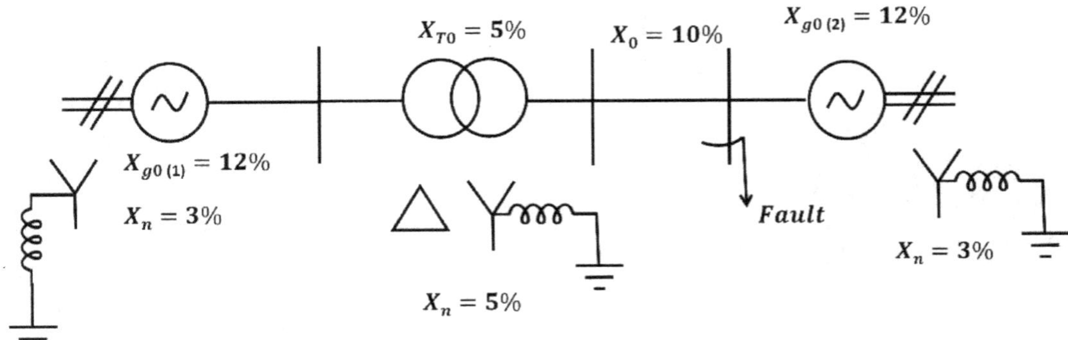

Solution:

$$X_{0eqp.u.} = j(0.05 + 0.15 + 0.10)\ p.u.\ //\ (0.12 + 0.09)$$
$$= j\,0.12\ p.u.$$

Q6. The transmission line impedance on its own base 110 KV, 100 MVA is j 0.5 p.u. What is the line impedance on p.u. new line values 116 KV, 20 MVA?

Solution:

$$X_{p.u.\ (new)} = X_{p.u.\ (old)}\left[\frac{MVA\ (New)}{MVA\ (old)} * \left(\frac{KV\ (old)}{KV\ (New)}\right)^2\right]$$

$$= j0.5\left[\frac{200}{100} * \left(\frac{110}{116}\right)^2\right]$$

$$= j0.89$$

Q7. The single line diagram of a three phase system is shown below. The percentage reactance of each alternator is based on its own capacity. Find the short circuit current that flows into complete three phase at fault point?

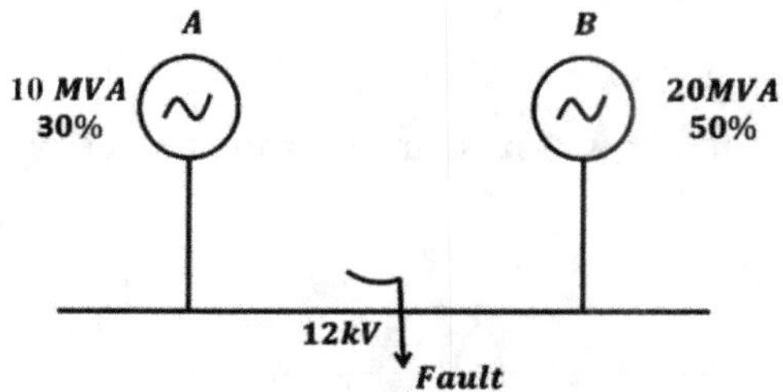

Solution:

Let base MVA = 20 MVA

    p.u. reactance diagram is as follows.

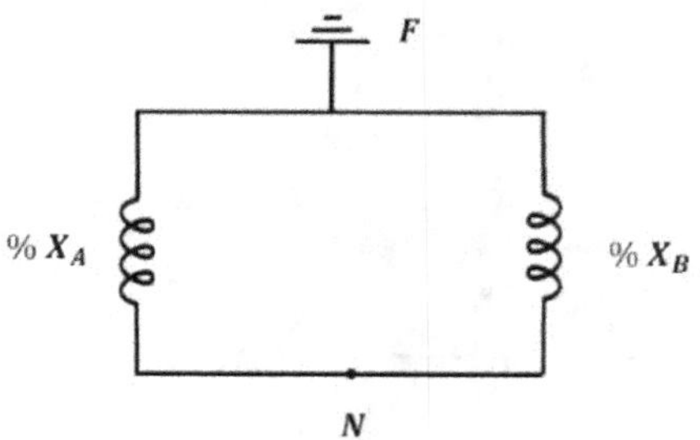

Percentage reactance of alternator A,

$$\%X = 100\ Xp.u.$$

$$= 100 * \frac{MVA_b}{KV_b^{\ 2}}$$

$$\%\ X_A \propto MVA_{base}$$

$$30\% \propto 15\ MVA_b$$

$$\%\ X_A \propto 20\ MVA_b$$

$$\%\ X_A = \frac{20*30}{15} = 60\%$$

Percentage reactance of alternator B,

As the base MVA and rated MVA of alternator B are same,

$$\%X_B = 50\%$$

Equivalent percentage reactance from the generator neutral to the fault point

$$\% \ X = 60 \ \% \parallel 50\%$$

$$= \frac{60 \times 50}{60 + 50} = 27.27\%$$

The base current corresponding to base MVA and 12 KV

$$VA_b = \sqrt{3} \ V_b I_b$$

$$20 \ MVA = \sqrt{3} * 12 \ K * I_b$$

Base current, $I_b = 962.25$ A

Short Circuit current, $I_{sc} = \frac{100*I_b}{\%X} = \frac{100*962.25}{22.2} = 3528.6$ A

Q8. A three phase, 20 MVA, 10KV alternator has internal reactance of 5 % and negligible resistance find the external reactance per phase to be connected in series with the on SC does not exceed 10 times the full load current.

Solution:

$$I_{sc} = 10 \ I_{FL}$$

Full load current, $I_{FL} = \frac{20 \ MVA}{\sqrt{3}*10} = 1154.7$ A

To satisfy the $I_{sc} = 10 \ I_{FL}$

The total % reactance is

$$I_{sc} = \frac{100*I_{FL}}{\%X} = 10 \ I_{FL}$$

$$\% \ X = 10 \ \%$$

External reactance required $= 10 - 5 = 5\%$

And $\%X = 100 \ X_{p.u.} = X_{(\Omega)} * 10 * \frac{MVA_b}{(KV_b)2} = 5\%$

$$100 \ X_{(\Omega)} * \frac{20}{10*10} = 5$$

$$X_{(\Omega)} = \frac{5}{20} = 0.25\Omega$$

Q9. A three phase power system operating at 10 KV having resistance of 1 Ω and reactance of 4 Ω is connected to the generating station bus bars through 5 MVA step up transformer having a reactance of 5 %. The bus bars are supplied by 10 MVA alternators having 20% reactance. Calculate short circuit KVA fed to the symmetrical fault between phases if it occurs.

1. at load end of the transmission line
2. at high voltage terminals of the transformer

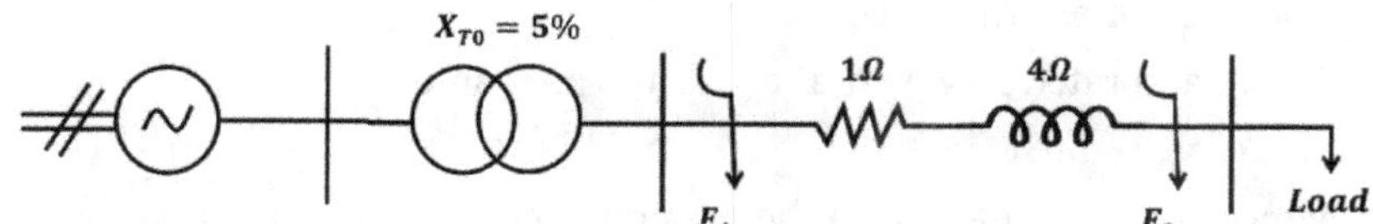

Solution:

Let base MVA = 10 MVA

Percentage reactance of alternator on base MVA,

$$\% X \propto MVA$$

$$10 \% \propto 10$$

$$\% X_{alt.} \propto 10 \Rightarrow \% X_{alt} = 20\%$$

Percentage reactance of transformer on base MVA,

$$5 \% \propto 5$$

$$\% X_{TF} \propto 10$$

$$\% X_{TF} = 10 \%$$

Percentage reactance of transmission line on base MVA,

$$\% X_{TL} = 100 X_{p.u.}$$

$$= 100 X_{(\Omega)} * \frac{MVA_b}{(KV_b)^2}$$

$$= 100*4* \frac{10}{10^2} = 40\%$$

Percentage resistance of transmission line based on MVA_b and KV_b

$$\% R_{TL} = 100 R_{TLp.u.}$$

$$= 100 R_{TL(\Omega)} * \frac{MVA_b}{(KV_b)^2}$$

$$= 100 * \frac{10}{10^2} = 10\%$$

Percentage impedance diagram

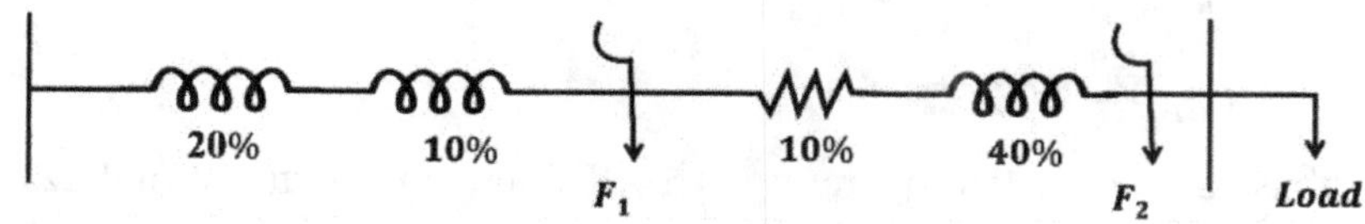

Fault at high voltage terminals of transformer, F_1

$$\% X_{F1} = \% X_{G1} + \% X_{TF} = 20 \% + 10 \% = 30\%$$

$$\text{Short circuit MVA} = \frac{100}{\% X_{F1}} * MVA_b$$

$$= \frac{100}{30} * 10 = 33.33 \text{ MVA}$$

$$= 33333.3 \text{ KVA}$$

Fault at load end of transmission line, F_2

$$\% \ X_{F2} = j \ (\% \ X_{GI} + \% \ X_{TF} + \% \ X_{TL}) + (\% \ R_{T1})$$
$$= j \ (20\% + 10\% + 40\%) + 10\%$$
$$= (10 + j \ 70) \ \% = 70.71\%$$

Short circuit MVA $_{F2} = \dfrac{100}{\% \ Z_{F2}} \ MVA_b$

$$= \dfrac{100}{70.71} * 10 = 14.14 \ MVA$$

Note: As the fault point from the alternator increases, short circuit MVA decreases

Q10. The estimating short circuit MVA at bus bars of a generating station A is 1500 MVA and generating station B is 1200 MVA. The generated voltage at each station is 33 KV. If these stations are interconnecting through a transmission line having 1 Ω reactance and negligible resistance, calculate the possible short circuit MVA at both stations with base MVA = 100 MVA.

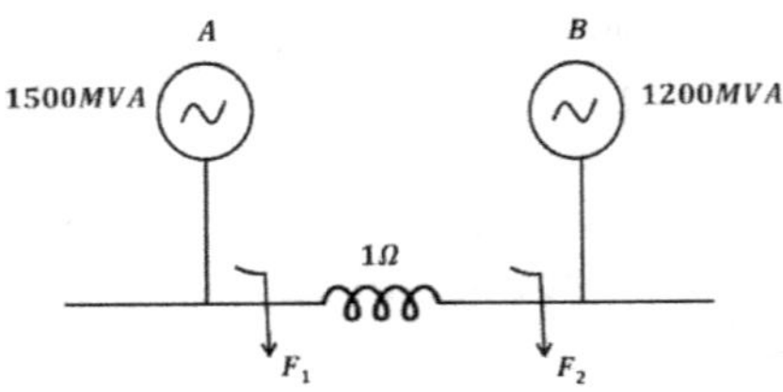

Solution:

Base MVA = 100 MVA

% reactance of generating station A on base MVA, $\% \ X_A = \dfrac{MVA_b}{MVA_{SC}} * 100$

$$= \dfrac{200}{1500} * 100 = 13.34\%$$

Percentage reactance of station B on base MVA, $\% \ X_B = \dfrac{200}{1200} * 100 = 16.67\%$

Percentage reactance of transmission line having reactance of 1Ω is,

$$\% \ X = 100 \ X_{p.u.} = 100 * X\Omega \ \dfrac{MVA_b}{(KV_b)2} = 100 * X\Omega \ \dfrac{100}{33^2} = 18.362\%$$

Percentage reactance diagram for the given network is shown below.

Fault on generating station A

$$\% \ X_{F1} = (\% \ X_A) \ \| (\% \ X_{TL} + \% \ X_B) = 13.34 \ // \ (16.67 + 18.362)$$
$$= \dfrac{6.67 \ (9.18+8.33)}{6.67+9.18+8.33} = 9.66\%$$

Short circuit MVA, $MVA_{sc} = \dfrac{100}{\% \ X_{F1}} * MVA_b$

$$= \dfrac{100}{9.66} * 100 = 1035 \ MVA$$

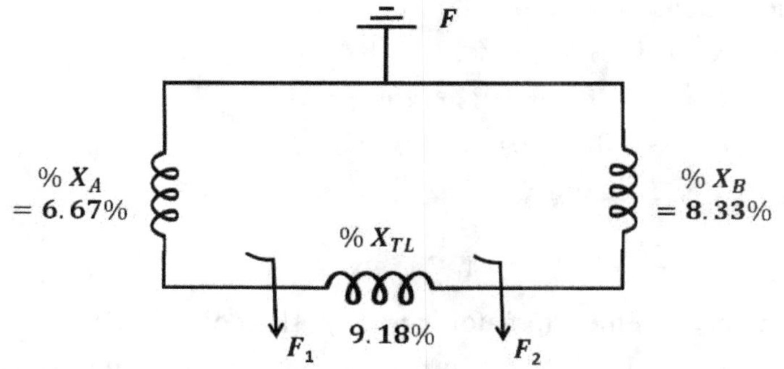

Fault on generating station B, % $X_{F2} = \dfrac{8.33(6.67+9.18)}{8.33+6.67+9.18} = 12\%$

Short circuit MVA, $\text{MVA}_{sc} = \dfrac{100}{\% X_{F2}} * \text{MVA}_b$

$$= \dfrac{100^2}{10.92} = 3662.78 \text{ MVA}$$

Q11. The figure shows a single diagram of a power system. The percentage resistance of generator are calculated by taking ratings as base value. Calculate the short circuit KVA, if three phase fault occurs on a feeder at the beginning if the SC KVA has to be reduced to 60% of that obtained earlier. Find% reactance of feeder reactor?

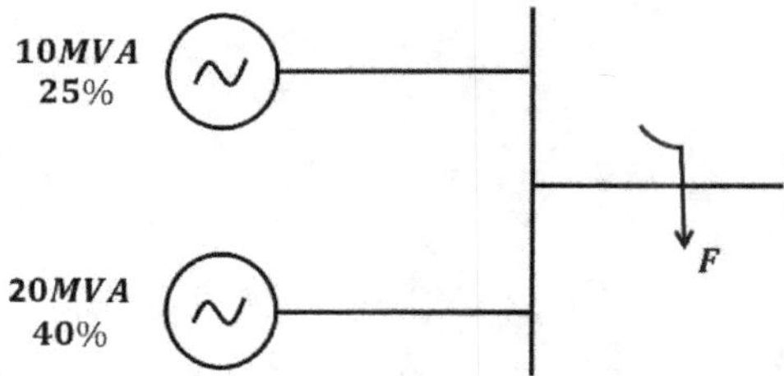

Solution:

Percentage reactance of station G_1,

$\% X_{G1} \propto \text{MVA}$

$25 \% \propto 10$

$\% X_{G1} \propto \text{MVA}_b$

$\% X_{G1} \propto 20 \longrightarrow \dfrac{20*25}{10}$

$\% X_{G1} = 50\%$

Percentage reactance of station G_2,

$\% X_{G2} \propto \text{MVA}$

$40 \% \propto 20$

$\% X_{G2} \propto \text{MVA}_b$

$$\% \, X_{G2} \propto 20$$

$$\% \, X_{G2} = \frac{20*40}{20} = 40\%$$

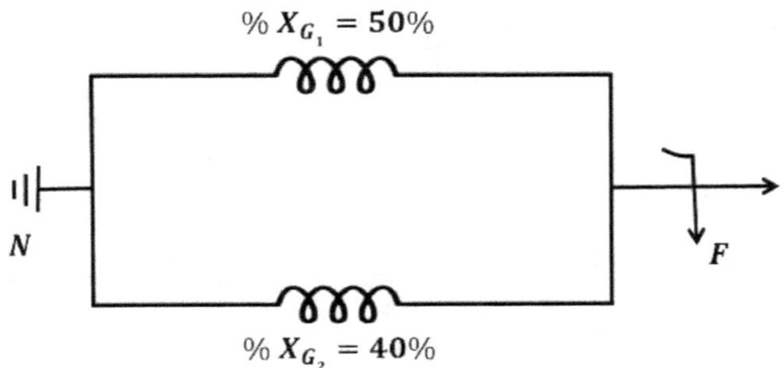

$$\% \, X_F = \% \, X_{G1} \parallel \% X_{G2}$$

$$= \frac{40*50}{40+50} = 22.22\%$$

$$\text{MVA sc} = \frac{100}{\% \, X_F} * \text{MVA}_b$$

$$= \frac{100}{22.22} * 40 = 90 \text{ MVA}$$

With feeder reactor, Short Circuit MVA = 60% * 90 = 54 MVA

$$\% \, X_{eq} = 22.2 + \% \, X_{Feeder} \qquad\qquad .....(7.1)$$

$$\text{Short Circuit MVA} = \frac{100}{\% \, X_{eq}} * \text{MVA}_b$$

$$54 = \frac{100}{\% \, X_{eq}} * 20$$

$$\% \, X_{eq} = 37.03\%$$

From (7.1) => 37.03% = 22.22% + % X_{Fee}

$$\% X_{Fee} = 14.81\%$$

$$X_{Feeder (p.u.)} = X_{Feeder}(\Omega) * \left(\frac{MVA_b}{KV_b{}^2}\right)$$

$$0.1481 = X_{Feeder}(\Omega) * \frac{20}{11^2}$$

$$X_{Feeder}(\Omega) = j \, 0.8964 \, \Omega$$

Q12. A power plant has 4 generators, two generators rating 15 MVA have reactance of 15 MVA and other two rating 20 MVA have 20% reactance these generators are connecting to the station bus bars. Power is drawn to the load centres to two power transformers rated 15 MVA each and having % X of 10%. It is required to design circuit breakers with rating that are to be providing at LT and HT side of transformer. The system is shown as below.

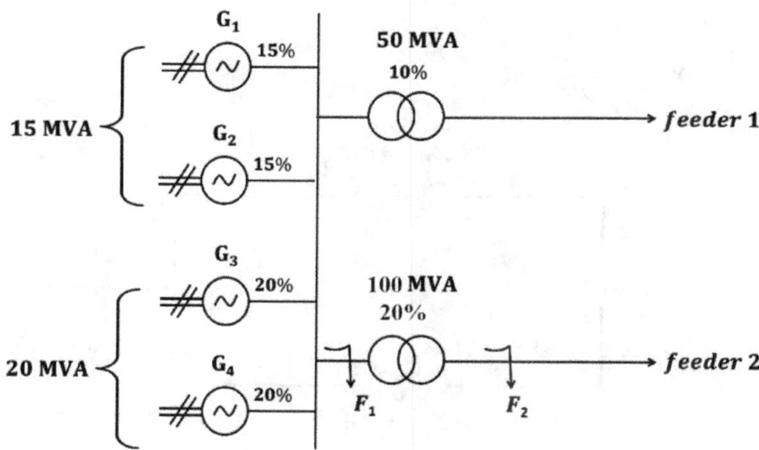

Solution:

Let base MVA be 20 MVA

Percentage reactance of generators G_1 and G_2,

$$\% \ X_{G1} \propto 15 \text{ MVA} \qquad \text{and } \% \ X_{G2} \propto 15 \text{ MVA}$$

$$15 \ \% \propto 15 \text{ MVA} \qquad \text{and } \% \ 15 \propto 15 \text{ MVA}$$

Then $\% \ X_{G1} \propto 20$ MVA

$$\% \ X_{G1} = \% \ X_{G2} = \frac{20*5}{15} = 20\%$$

Design of circuit breaker on HT side of transformer

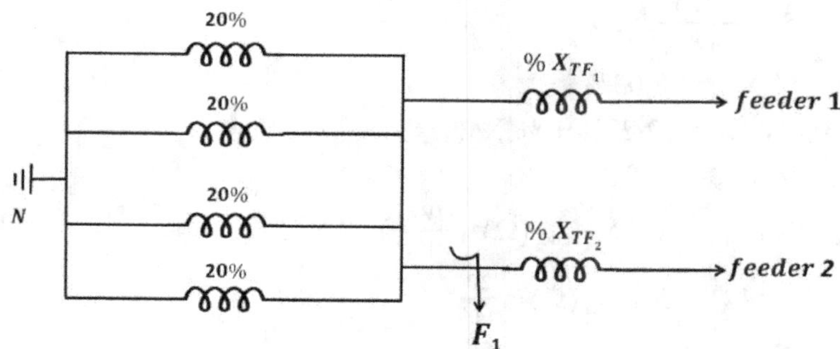

$$\% \ X_{F1} = \% \ X_{G1} \parallel \%X_{G2} \parallel \%X_{G3} \parallel \%X_{G4}$$

$$= \frac{20}{4} = 5 \ \%$$

Circuit breaking rating on LT side,

$$\text{SC MVA} = \frac{100}{\% \ X_{F1}} * \text{MVA}_b$$

$$= \frac{100}{5} * 20 = 400 \text{ MVA}$$

Design of circuit breaker on HT side of transformer

Percentage reactance of transmission line,

$$20\% \propto 100 \text{ MVA}$$

$$\% X_{TL} \propto 20 \text{ MVA}$$

$$= \frac{20*20}{100} = 4\%$$

$$\% X_{F2} = (\% X_{G1} \parallel \% X_{G2} \parallel \% X_{G3} \parallel \% X_{G4}) + \% X_{TF2}$$

$$= 5\% + 4\%$$

$$= 9\%$$

Circuit breaker rating on HV side

$$\text{SC MVA} = \frac{100}{9} * 20$$

$$= 222.22 \text{ MVA}$$

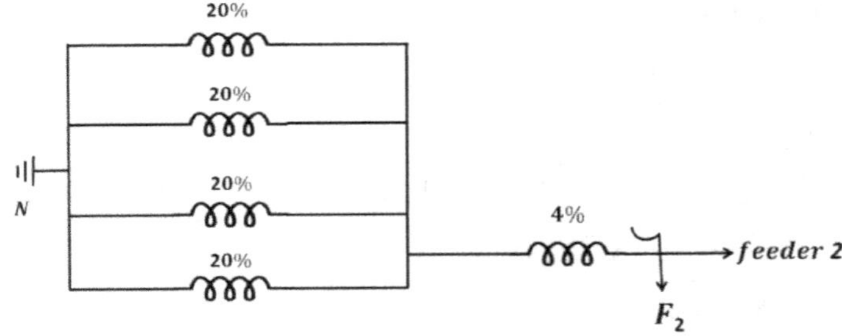

Q13. A generating station has 4generators; each rated 11 KV, 100 MVA and each having a sub transient reactance of 10%. A sectionalized bus bar reactor is placed as shown in figure. A 3–Φ fault occurs as shown in the figure. Determine fault level if %X = 0, ii) Determine reactance value in Ω to limit the fault level to 5000 MVA?

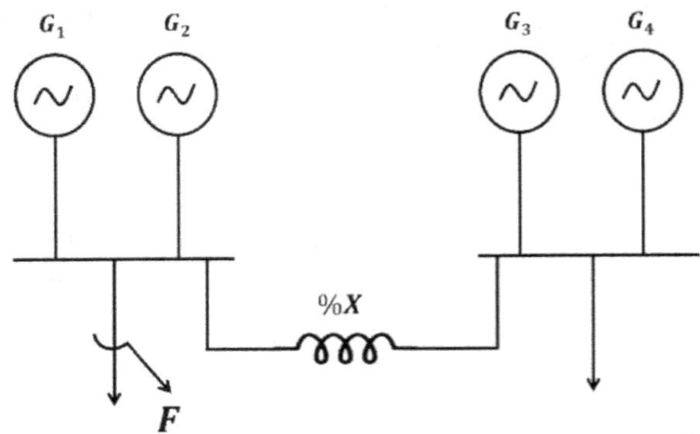

Solution:

$$\text{Base MVA} = 100$$

(i)   Base KV $= KV_b = 11$ KV

$$\% \, X = 0$$

$$\% \, X_F = (\% \, X_{G1} \parallel \% X_{G2} \parallel \% X_{G3} \parallel \% X_{G4})$$

$$= \frac{10}{4} = 2.5 \, \%$$

$$\text{Short circuit MVA} = \frac{100}{2.5} * 100$$

$$= 8000 \, \text{MVA}$$

(ii) $MVA_{SC} = 5000$ MVA

SC MVA has to be decreased from 4000 MVA to 5000 MVA

out of 5000 MVA, generators G_1 and G_2 supplies 4000 MVA to fault

Therefore, G_3 and G_4 have to supply (5000 – 4000)

$$= 1000 \, \text{MVA to the fault}$$

SC MVA supplied by G_3 and G_4

$$1000 = \frac{100}{\% X_{eq}} * 100$$

$$\% \, X_{eq} = 10\%$$

$$\% \, X_{eq} = \% \, (\% X_{G2} \parallel \% X_{G3} \parallel \% X_{G4}) + \% \, X$$

$$10 = \frac{10}{2} + \% \, X$$

$$\% \, X = 5\%$$

$$X_{p.u.} = 0.05 \, \text{p.u.}$$

$$X \, (\Omega) * \frac{MVA_b}{(KV_b)2} = 0.05$$

$$X \, (\Omega) = \frac{0.05 * 11^2}{100} = 0.0605 \, \text{ohm.}$$

Q14. An interconnecting generator reactor system is shown in the figure. The base values for the given % reactance are the ratings of the individual components. A 3–phase short–circuit occurs at fault point. Determine the fault current and fault KVA. The bus bar line–line voltage is 33 KV?

Solution:

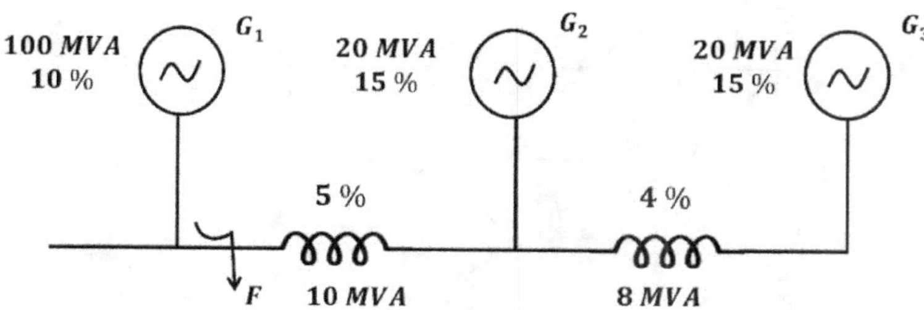

Solution:

Let base MVA, MVA $_b$ = 20 MVA

Percentage reactance of G1 with respect to base MVA

$$10\% \propto 10 \text{ MVA}$$

$$\% \text{ X}_{G1} \propto 20 \text{ MVA}$$

$$\% \text{ X}_{G1} = \frac{20*10}{10} = 20\%$$

Percentage reactance of transmission line with respect to MVA$_b$

1. $5\% \propto 10$ MVA

$$\% \text{ X}_{TL1} \propto 20 \text{ MVA}$$

$$\% \text{ X}_{TL1} = \frac{20*5}{10} = 10\%$$

2. $4\% \propto 8$ MVA

$$\% \text{ X}_{TL2} \propto 20 \text{ MVA}$$

$$\% \text{ X}_{TL2} = \frac{20*4}{8} = 10\%$$

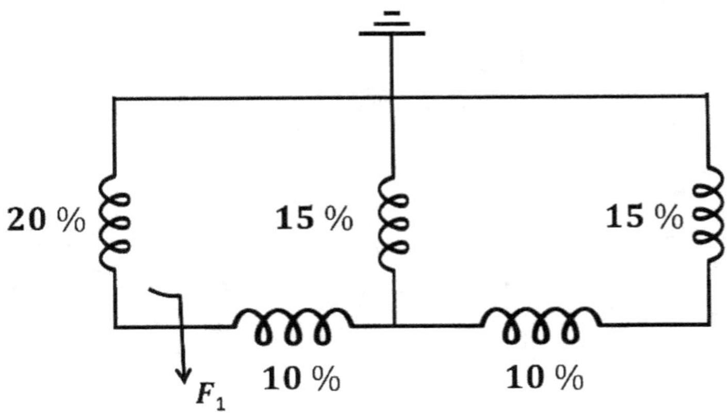

$$\% \text{ X}_{F1} = \% \text{ X}_{G1} \| ((\% \text{X}_{G3} + \%\text{X}_{2N}) \| (\% \text{ X}_{G2} + \% \text{ X}_{G2}))$$

$$= 20 \| [(15 + 10) \| (15 + 10)]$$

$$= 9.84\%$$

Short Circuit MVA $= \frac{100}{9.84} * 20 = 203.25$ MVA

Short Circuit VA $= \sqrt{3}$ V Isc

$$\rightarrow \frac{203.25 * 10^6}{\sqrt{3} * 33 * 10^3} = \text{Isc}$$

$$\text{I}_{SC} = 3551.07 \text{ A}$$

Q15.

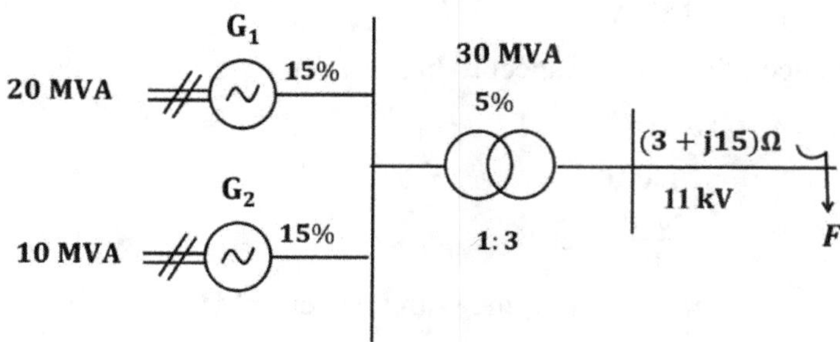

A three phase SC fault occurs at point F in the system shown. Calculate the fault current.

Solution:

Ans: $MVA_b = 20$ MVA

$$\% X_{G1\,(N)} = 15\%$$

$$\% X_{G2\,(N)} = \frac{20*10}{10} = 20\%$$

$$\% X_{TF1\,(N)} = \frac{5*20}{30} = 3.33\%$$

$$\% X_{TL\,(N)} = 100\,(3 + j5) * \frac{20}{33^2}$$

$$= (550 + j\,27.5)\,\%$$

Impedance diagram,

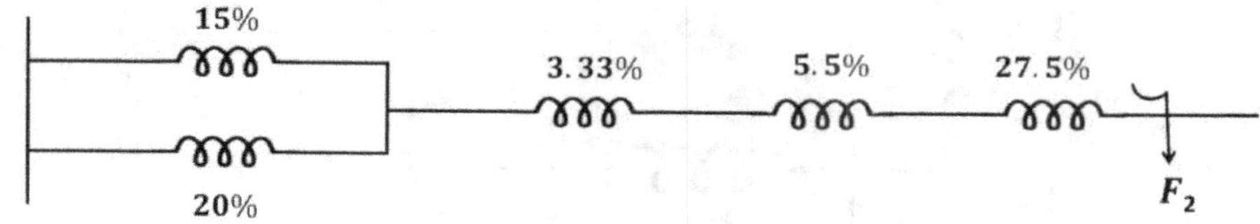

$$\% X_F = j\left[\left(\frac{20*15}{20+15}\right) + 3.33 + 27.5\right] + 5.50 = (j\,39.401 + 5.50)\%$$

Short circuit MVA, $MVA_{SC} = \dfrac{100*20}{j39.401 + 5.50} = 50.27$ MVA

Short circuit current,

$$10^6 * 50.27 = \sqrt{3} * 33 * 10^3\ ISC$$

$$ISC = 2638.632\ A$$

Q16. The estimating MVA at bus bar of a generating unit G_1 is 1000 MVA and short circuit MVA at the bus bar of another unit G_2 is 500 MVA. They are interconnecting through 3.6 Ω reactance. The operating voltage is 33 KV. Determine the maximum value of short circuit MVA.

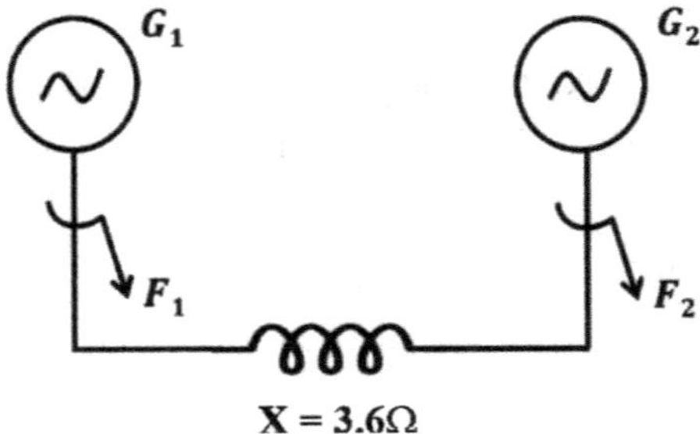

Solution:

Given, estimated SC MVA for G_1 = 1000 MVA

SC MVA for G_2 = 500 MVA

MVA = MVA_b = 100 MVA

$\% X_{G1} = \dfrac{100*100}{1000} = 100\%$

$\% X_{G2} = \dfrac{100*100}{500} = 20\%$

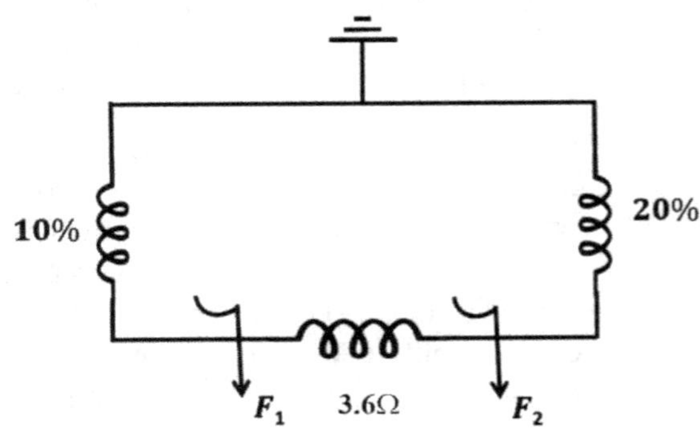

For transmission line $\% X = 100 * X_{pu}$

$$= 100 * X \text{ in ohm} * \frac{MVA_b}{(KV_b)^2} = 100 * \frac{100}{33^2} * 3.6$$

$$= 33\%$$

$\% X_{F1} = \dfrac{(33+20)*10}{33+30} = 8.41\ \%$

$\% X_{F2} = \dfrac{43*20}{43+20} = 13.65\ \%$

Short circuit MVA,

$$\text{SC MVA}_{F1} = \frac{100}{8.41} * 100 = 1189.06 \text{ MVA}$$

$$\text{SC MVA}_{F2} = \frac{100}{13.65} * 100 = 732.600 \text{ MVA}$$

Short circuit MVA is maximum at F1.

OBJECTIVE QUESTIONS

1. How the loads are classified in reactance or impedance diagram?

   Ans: The resistive and reactive loads can be represented by

   (a) Constant power representation

   (b) Constant current representation

   (c) Constant impedance representation

2. What is a single line diagram?

   Ans: A single line diagram is diagrammatic representation of power system in which the components are represented by their symbols and the interconnection between them are shown by a single line diagram (for three phase system also). The ratings and the impedances of the components are also marked on the single line diagram.

3. What are the components of a power system?

   Ans: The components of power system are generators, power transformers, transmission lines, substation transformers, distribution transformers and loads.

4. How an induction motor is represented in reactance diagram?

   Ans: For estimation of steady state fault current, the induction motor is neglected. But for estimation of fault current immediately after the fault i.e.to estimate sub–transient fault currents, the induction motor can be represented by a source in series with reactance.

5. What is impedance and reactance diagram?

   Ans: The impedance diagram is the equivalent circuit of power system in which power system components are represented by approximate equivalent circuits. The impedance diagram is used for load flow studies. The reactance diagram is the simplified equivalent circuit of power system in which various components are represented by their reactances. The reactance diagram can be obtained from impedance diagram with all resistive components neglected. The reactance diagram is used for fault calculations.

6. What are the approximations of impedance diagram?

   Ans:

   (a) The current limiting impedances connected between the generator neutral and ground are neglected as current do not flow through neutral under balanced conditions.

   (b) As the magnetizing current of a transformer is very low, shunt branches in equivalent circuit of induction motor are neglected.

   (c) If the inductive reactance of a component is very high when compared to resistance then the resistance can be omitted.

7. What are the approximations of reactance diagram?

   Ans:

   (a) The neutral reactances are neglected

   (b) Shunt branches in the equivalent circuits of transformer are neglected.

   (c) Resistances are neglected.

(d) All static loads and induction motors are neglected.

(e) The capacitance of the transmission lines are neglected.

8. What are Symmetrical Components?

Ans: An unbalanced system of N–related vectors can be resolved into N systems of balanced vectors. The N–set of balanced vectors are called symmetrical components. Each set of N–vectors are equal in magnitude and have equal phase displacements between adjacent vectors.

9. Define fault in a power system network?

Ans: Electrical fault is an abnormal condition, caused by equipment failures such as transformers, rotating machines, human errors and environmental conditions. The faults are associated with abnormal change in current, voltage and frequency of the system. These faults cause interruption to electric flows, equipment damages and even cause death of humans, birds and animals.

10. How the faults are classified?

Ans: Faults are classified into two types

(a) Shunt fault or Short circuit fault

(b) Series fault or Open circuit faults

Shunt type of faults involve power conductor or conductors to ground or short circuit between conductors.

Shunt faults are characterized by increase in current and fall in voltage and frequency.

Series faults occur with one or two broken conductors.

Series faults are characterized by increase in voltage and frequency and decrease in current.

Shunt faults are further classified into symmetrical and unsymmetrical shunt faults.

Unsymmetrical shunt faults are classified into a) LG b) LL c) LLG

Symmetrical shunt faults are classified into LLL and LLLG faults.

Series faults are classified into a) 1–open conductor fault b) 2–open conductor fault.

11. What is symmetrical and unsymmetrical fault?

Ans: A fault with equal fault current flowing in all the phases is known as symmetrical fault. A fault with unequal fault current flowing in all the phases is known as unsymmetrical fault.

12. Compare load flow studies and short circuit studies?

Ans:

(a) Resitance, R and reactance, X are considered in power flow studies and R is neglected in fault analysis.

(b) Y_{bus} is required for power flow studies and Z_{bus} is required for short circuit studies.

(c) Power flow study is performed to determine the exact voltages and currents whereas voltages are assumed as 1p.u. and pre–fault current is neglected in short circuit studies.

13. What are fault calculations?

Ans: The fault condition of a power system can be divided into sub–transient, transient and steady state periods. The currents in various parts of the system and during fault are different in these durations. The estimation of these currents for various types of faults at various locations in the system is referred as fault calculations.

14. What is the need for short circuit studies or fault analysis?

Ans: The short circuit studies are essential to design or develop the protective schemes for various parts of the system. The protective schemes consist of current and voltage sensing devices, protective relays and circuit breakers. The selection of these devices depends on various currents that flow during fault conditions.

15. What is the reason for transients during short circuits?

Ans: The faults or short circuits are associated with sudden changes in currents. Most of the components of the power system have inductive property which opposes any sudden changes and so the faults are associated with transients.

16. Define Synchronous reactance?

Ans: The reactance of a synchronous machine under steady state condition or the ratio of induced EMF and the steady state RMS current is known as synchronous reactance.

17. Define Sub–transient reactance?

Ans: The ratio of induced EMF on no–load and the sub–transient symmetrical RMS current is known as sub–transient reactance

18. Define transient reactance?

Ans: The ratio of induced EMF on no–load and the transient symmetrical RMS current is known as transient reactance

19. What is the significance of sub–transient reactance in short circuit studies?

Ans: The sub–transient reactance can be used to estimate the initial fault current immediately on the occurrence of fault. The maximum momentary short circuit rating of the circuit breaker used for the protection or fault clearing should be less than the initial fault current.

20. What is the significance of transient reactance in short circuit studies?

Ans: The transient reactance is used to estimate the transient state fault current. Most of the circuit breakers open their contacts only during this period. Therefore, a circuit breaker used for fault clearing, it's interrupting short circuit current rating should be less than the transient fault current.

21. What are the applications of sub–transient reactance?

Ans: Sub–transient reactance of generators and motors are used to determine the initial current on the occurrence of the short circuits.

22. When a synchronous machine is suddenly short circuited, what is the maximum possible instantaneous current in terms of symmetrical short circuit current?

Ans: The initial short circuit current will have its maximum possible value if the fault occurs when the voltage wave is going through zero. This maximum possible value will be double that of maximum symmetrical short circuit current.

23. How symmetrical faults are analyzed?

Ans: The symmetrical faults are analyzed using per unit reactance diagram of the power system. Once the reactance diagram of the power system is formed then the fault is simulated by short circuit or by connecting fault impedance at the fault point. The currents and voltages at various parts of the system can be estimated by (a) Kirchhoff's laws (b) Thevenin's Theorem c) Z_{bus} matrix.

24. What is the use of current limiting reactor?

Ans: Current limiting reactor limits the short circuit current to a safe value and protects the power system network. Current limiting reactors are large coils wound for high self–inductance and very low resistance.

25. Define infinite bus in short circuit studies?

Ans: A bus which maintains constant voltage before and after the fault is known as infinite bus.

26. Define Short Circuit Capacity of a bus?

Ans: The product of pre–fault voltage and post fault current is known as short circuit capacity of a bus.

27. How the current limiting reactors are classified?

Ans: Reactors are classified into three types (a) Generator reactor (b) Feeder Reactor (c) Busbar reactor. Generator reactors are connected in series with each generator. Bus bar reactors are classified into two types (a) Ring type (b) Tie–bar type.

8 Symmetrical Components

8.1 INTRODUCTION

Whenever fault occurs, power system network becomes unbalanced. An unbalanced power system is represented using symmetrical components.

Consider single line diagram of the power system network as shown in figure 8.1.

The network consists of an alternator connected to load through a step-up transformer and transmission line.

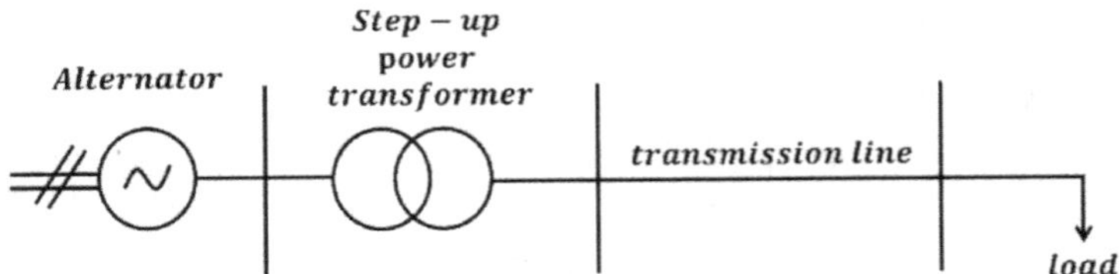

Figure 8.1 Power system network

Per unit diagram is drawn between phase and neutral between which the terminal voltage is calculated as in figure 8.2.

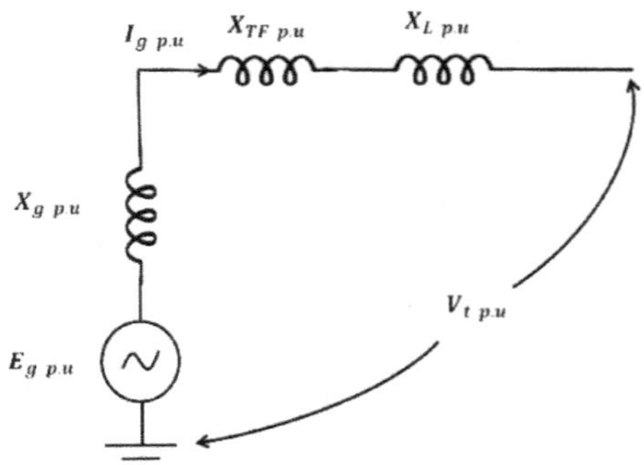

Figure 8.2 Per unit diagram

$$X_{tp.u.} = E_{gp.u.} - j\, I_{p.u.}\,(X_{gp.u.} + X_{TF\,p.u.} + X_{TL\,p.u.})$$
$$= E_{gp.u.} - j\, I_{p.u}\, X_{p.u.} \qquad\qquad .....(8.1)$$

Equation (8.1) represents per unit system analysis of a balanced power system network.

For an unbalanced power system network,

$$V_{t\,R\,p.u.} = E_{gR\,p.u.} - j\, I_{Rp.u.}\, X_{R\,p.u.}$$
$$V_{t\,Y\,p.u.} = E_{gY\,p.u.} - j\, I_{Yp.u.}\, X_{Y\,p.u.}$$
$$V_{t\,B\,p.u.} = E_{gB\,p.u.} - j\, I_{Bp.u.}\, X_{B\,p.u.}$$

8.2 CLASSIFICATION OF UNBALANCED SYSTEMS

S.No.	Internal Unbalance	External Unbalance
1	Occurs due to changes in load demand	Occurs due to lightning storms, faults, trees falling across transmission lines
2	Neutral current, $I_n \leq 10\%\ I_{FL}$	Neutral current, $I_n \geq 90\%\ I_{FL}$
3	$V_R \cong V_Y \cong V_B$	$V_R \neq V_Y \neq V_B$
4	Voltages are within regulation	Voltages are not within regulation

8.3 SIGNIFICANCE OF OPERATOR 'k' OR 'a' OR 'α'

Operator 'k' rotates a vector in anti clock wise direction by 120^0

$$k^3 = 1 \Rightarrow k = 1 = 1 + j\,0$$
$$|k| = \sqrt{1^2 + 0^2} = 1$$
$$\text{Polar form, } k = 1\,\underline{|120^0}$$
$$= -0.5 + j\,0.866$$
$$k^2 = 1\,\underline{|240^0} = -0.5 - j0.866$$
$$k + k^2 = -1$$
$$k + k^2 + 1 = 0$$

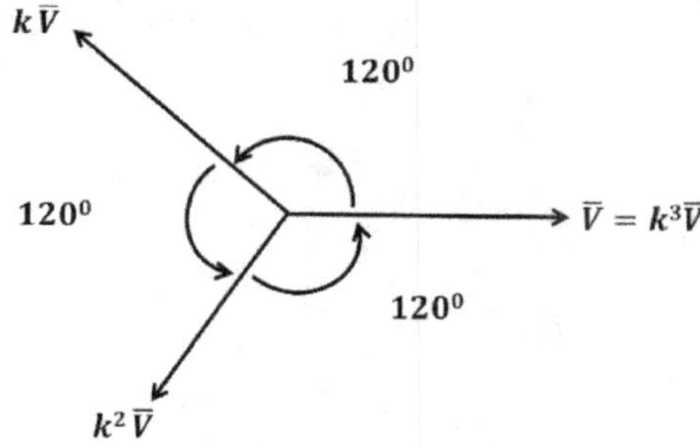

Figure 8.3 Operator–k

The magnitude of each vector is same.

A three phase power system network consists of two possible phase sequences i.e. based on clockwise & anti–clockwise directions (RBY and RYB)

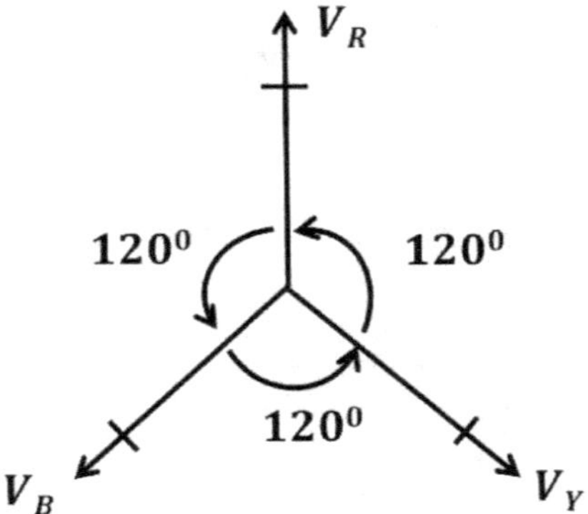

Figure 8.4 Three phase network

The unknowns can be represented by symmetrical components.

$$V_{tR} = V_{R0} + V_{R1} + V_{R2}$$
$$V_{tY} = V_{Y0} + V_{Y1} + V_{Y2}$$
$$V_{tB} = V_{B0} + V_{B1} + V_{B2}$$

Positive sequence components

These components (V_{R1}, V_{Y1} and V_{B1})are equal in magnitude and displaced from each other by 120^0 with phase sequence same as phase sequence of three phase power system as shown in figure 8.5.

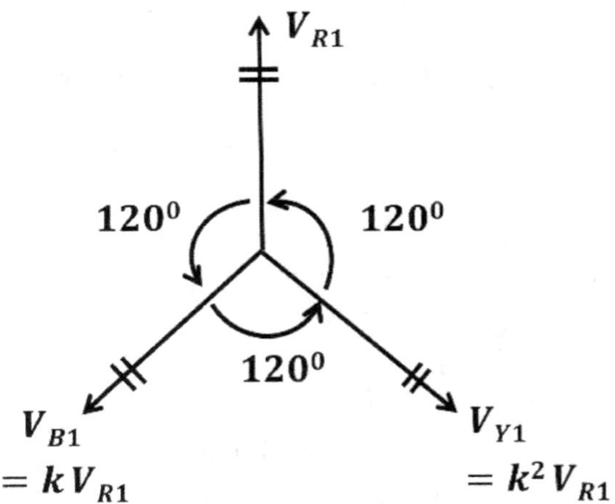

Figure 8.5 Positive sequence components

Negative sequence components

These components (V_{R2}, V_{Y2} and V_{B2}) are equal in magnitude and displaced from each other by 120^0 with phase sequence opposite to phase sequence of three phase power system as shown in figure 8.6.

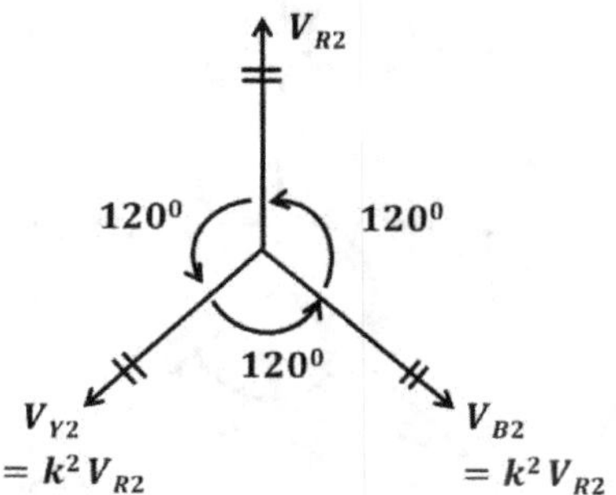

Figure 8.6 Negative sequence components

Zero sequence components

Zero sequence components V_{R0}, V_{YO}, V_{BO} are equal in magnitude without any phase sequence as shown in figure 8.7.

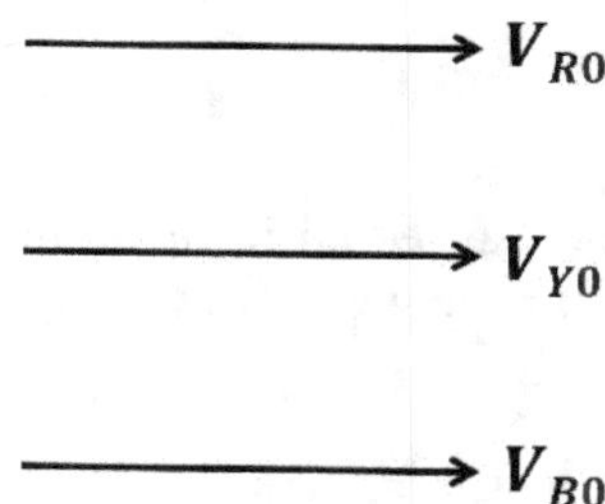

Figure 8.7 Zero sequence components

Operator–k minimises nine symmetrical components into 3 symmetrical components

$$V_{tR} = V_{R0} + V_{R1} + V_{R2} = V_{R0} + V_{R1} + V_{R2}$$
$$V_{tY} = V_{Y0} + V_{Y1} + V_{Y2} = V_{R0} + K^2 V_{R1} + K V_{R2}$$
$$V_{tB} = V_{B0} + V_{B1} + V_{B2} = V_{R0} + K V_{R1} + K^2 V_{R2}$$

$$\begin{bmatrix} V_{tR} \\ V_{tY} \\ V_{tB} \end{bmatrix} = \begin{bmatrix} 1 & 1 & 1 \\ 1 & K^2 & K \\ 1 & K & K^2 \end{bmatrix} \begin{bmatrix} V_{R0} \\ V_{R1} \\ V_{R2} \end{bmatrix}$$

Expressing symmetrical components in terms of unbalance voltages

$$\begin{bmatrix} V_{R0} \\ V_{R1} \\ V_{R2} \end{bmatrix} = \begin{bmatrix} 1 & 1 & 1 \\ 1 & K^2 & K \\ 1 & K & K^2 \end{bmatrix} \begin{bmatrix} V_{tR} \\ V_{tY} \\ V_{tB} \end{bmatrix} = [A] \begin{bmatrix} V_{tR} \\ V_{tY} \\ V_{tB} \end{bmatrix}$$

$$A^{-1} = \frac{1}{3(K-K^2)} \begin{bmatrix} K^2(K^2-1) & (K-K^2) & (K-K^2) \\ (K-K^2) & (K^2-1) & (1-K) \\ (K-K^2) & (1-K) & (K^2-1) \end{bmatrix}$$

$$= \frac{1}{3(K-K^2)} \begin{bmatrix} (K-K^2) & (K-K^2) & (K-K^2) \\ (K-K^2) & (K^2-1) & (1-K) \\ (K-K^2) & (1-K) & (K^2-1) \end{bmatrix}$$

$$= \frac{1}{3} * \begin{bmatrix} 1 & 1 & 1 \\ 1 & \dfrac{K^2-1}{K-K^2} & \dfrac{1-K}{K-K^2} \\ 1 & \dfrac{1-K}{K-K^2} & \dfrac{K^2-1}{K-K^2} \end{bmatrix}$$

$$\frac{K^2-1}{K-K^2} = \frac{K^2-K^3}{K(1-K)} = K \ (\because K^3 = 1 \ \& \ K^4 = K)$$

$$\frac{1-K}{K-K^2} = \frac{K^3-K^4}{(K-K^2)} = K^2$$

$$A^{-1} = \frac{1}{3} * \begin{bmatrix} 1 & 1 & 1 \\ 1 & K & K^2 \\ 1 & K^2 & K \end{bmatrix}$$

$$\begin{bmatrix} V_{R0} \\ V_{R1} \\ V_{R2} \end{bmatrix} = \frac{1}{3} * \begin{bmatrix} 1 & 1 & 1 \\ 1 & K & K^2 \\ 1 & K^2 & K \end{bmatrix} \begin{bmatrix} V_{tR} \\ V_{tY} \\ V_{tB} \end{bmatrix}$$

8.4 PER UNIT NEUTRAL CURRENT

Case (i): Balanced Condition

$$I_{np.u.} = I_{R\,p.u.} + I_{Yp.u.} + I_{Bp.u.} = 0$$

Case (i): Unbalanced Condition

$$I_{np.u.} = (I_{R0p.u.} + I_{R1p.u.} + I_{R2p.u.}) + (I_{Y0p.u.} + I_{Y1p.u.} + I_{Y2p.u.}) + (I_{B0p.u.} + I_{B1p.u.} + I_{B2p.u.})$$

$$= (I_{R0p.u.} + I_{R1p.u.} + I_{R2p.u.}) + (I_{R0p.u.} + K^2 I_{R1p.u.} + K I_{R2p.u.}) + (I_{R0p.u.} + K I_{R1p.u.} + K^2 I_{R2p.u.})$$

$$= 3 I_{R0p.u.} + I_{R1p.u.}(1+K+K^2) + I_{R2p.u.}(1+K+K^2)$$

$$= 3 I_{R0p.u.}$$

Only zero sequence current flows from ground to neutral of a star connected three phase alternator.

Thus ground is the reference to the flow of zero sequence current and neutral is the reference to the flow of positive and negative sequence components.

8.5 METHOD OF CONNECTIONS BETWEEN NEUTRAL AND GROUND OF A THREE PHASE STAR CONNECTED ALTERNATOR

The neutral and ground of a three-phase star connected alternator can be connected by three methods.

Method 1: Neutral Grounding

$$V_{np.u.} = I_{np.u.} X_{np.u.} = I_{np.u.} * 0$$
$$= 0$$

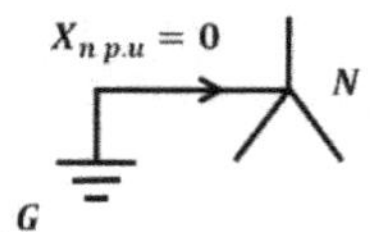

Method 2: Reactance neutral / Grounding

$$V_{np.u.} = I_{np.u.} X_{np.u.}$$
$$V_{np.u.} = (3I_{R0p.u.}) X_{np.u.} = I_{R0} (3X_{np.u.})$$

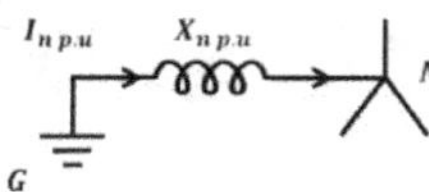

Method 3: Isolated neutral / grounding

$$V_{np.u.} = I_{np.u.} X_{np.u.}$$
$$= 0$$

8.6 ZERO SEQUENCE NETWORKS OF TRANSFORMERS

The transformer windings can be connected either in star or delta

Star connection of transformer windings:

(a) Solid neutral (b) Reactance neutral (c) Isolated neutral.

Representation of a zero sequence network is shown in the figure 8.8.

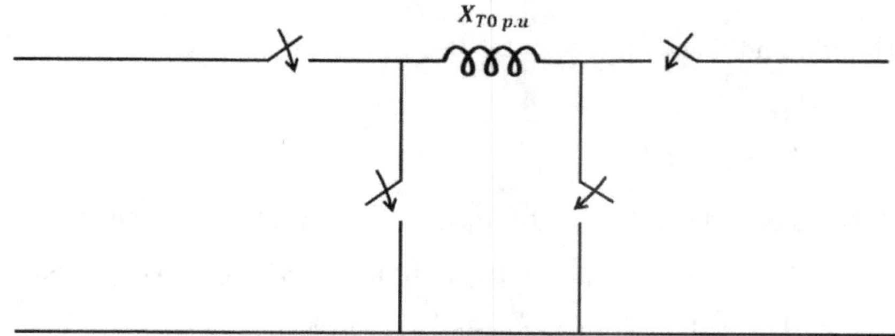

Figure 8.8 Equivalent circuit of Zero sequence network of a transformer

$X_{T0p.u}$ is the per unit zero sequence reactance of transformer.

For star connections of transformer windings operate series switch

➤ For solid neutral or solid grounding, close series switch with $X_{np.u.} = 0$

➤ For reactance neutral close the series switch with $3 X_{np.u.}$

➤ For Isolated neutral open series switch or close series switch $X_{np.u.} = \infty$

For Delta connection of transformer windings, close shunt switch with $X_{np.u.} = 0$

Determination of equivalent per unit zero sequence reactances.

1.

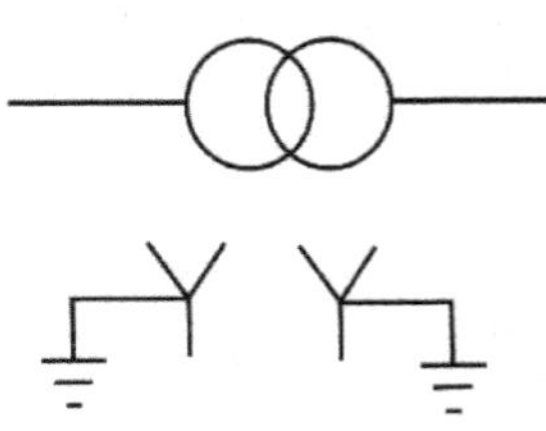

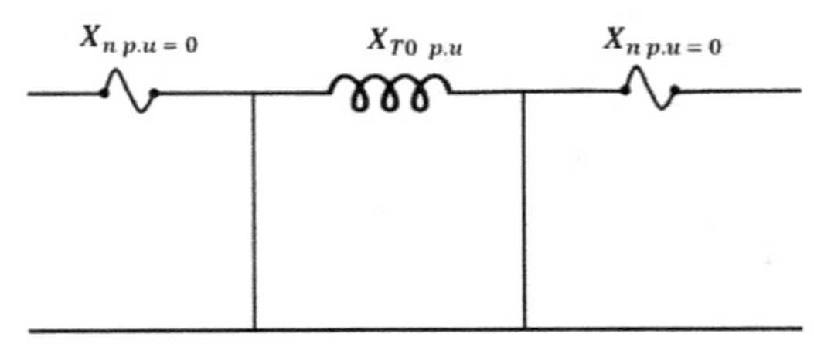

$$X_{0eq\,p.u} = 0 + X_{T0\,p.u} + 0$$
$$= X_{T0\,p.u}$$

2.

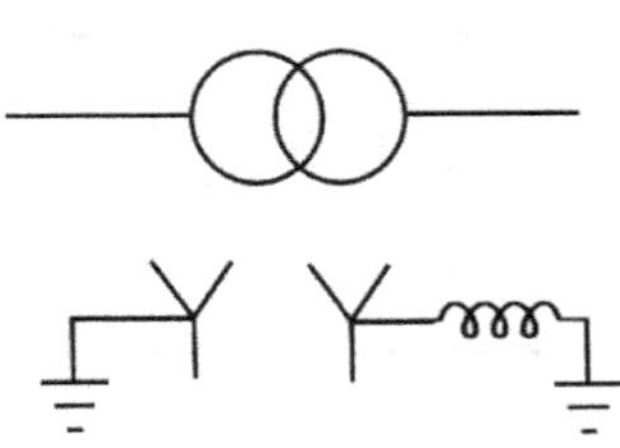

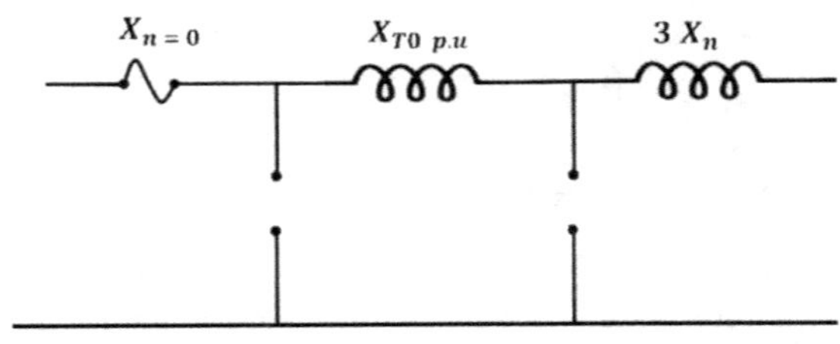

$$X_{0eq\,p.u} = 0 + X_{T0\,p.u} + 3X_{n\,p.u}$$
$$= X_{T0\,p.u} + 3X_{n\,p.u}$$

3.

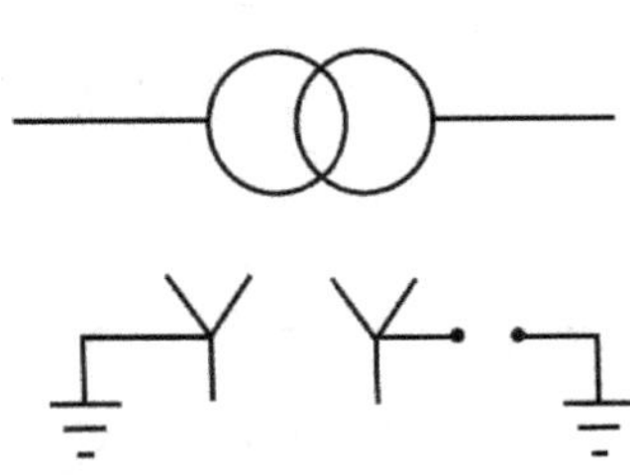

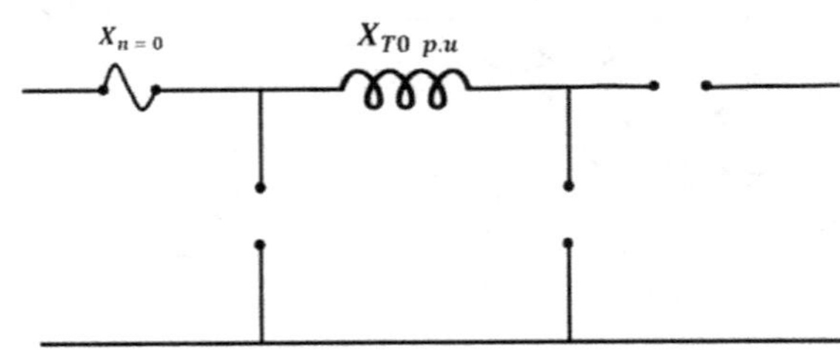

$$X_{0eq\,p.u} = 0 + X_{T0\,p.u} + \infty = \infty$$

4.

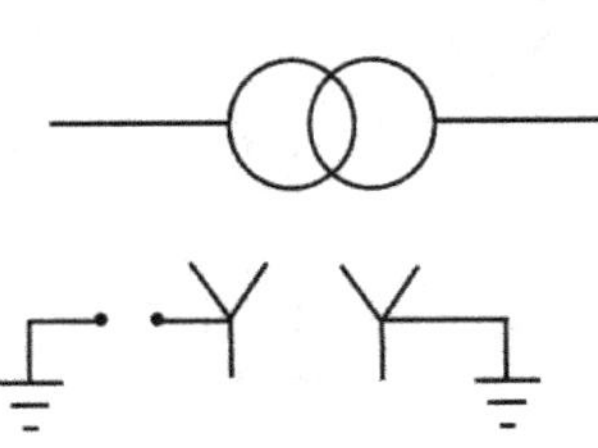

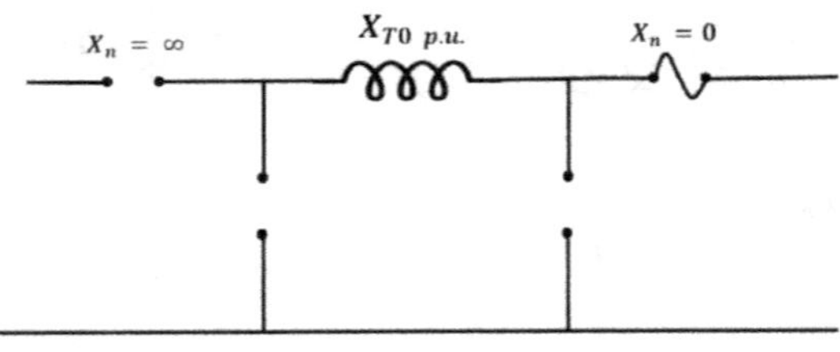

$$X_{0eq\,p.u.} = \infty$$

5.

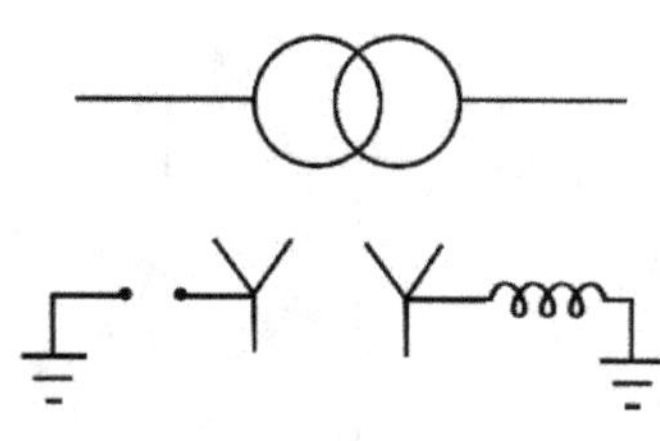

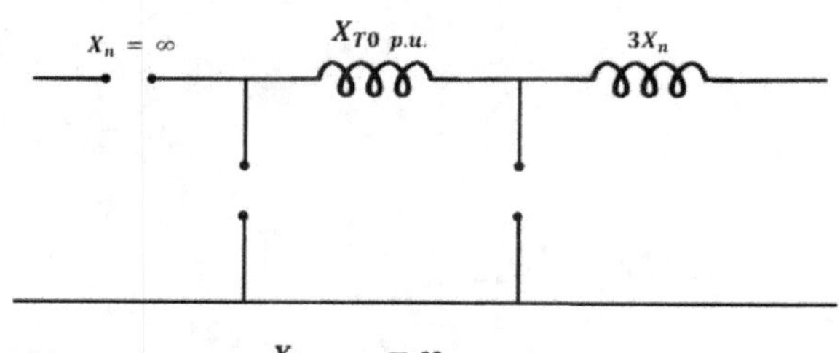

$$X_{0eq\ p.u.} = \infty$$

6.

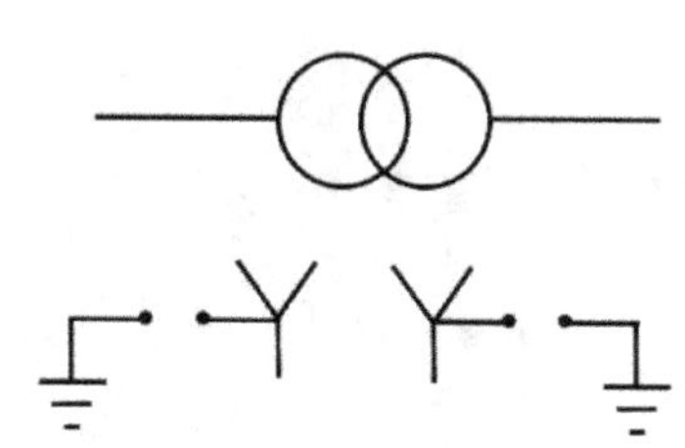

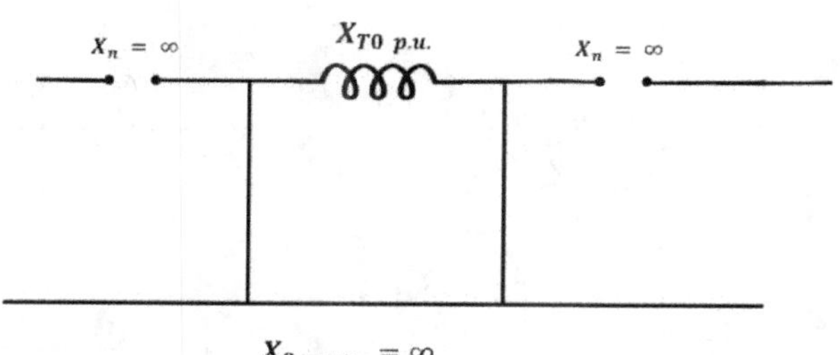

$$X_{0eq\ p.u.} = \infty$$

7.

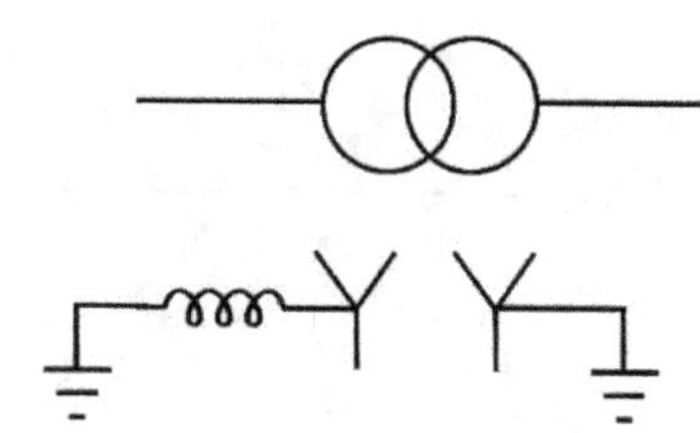

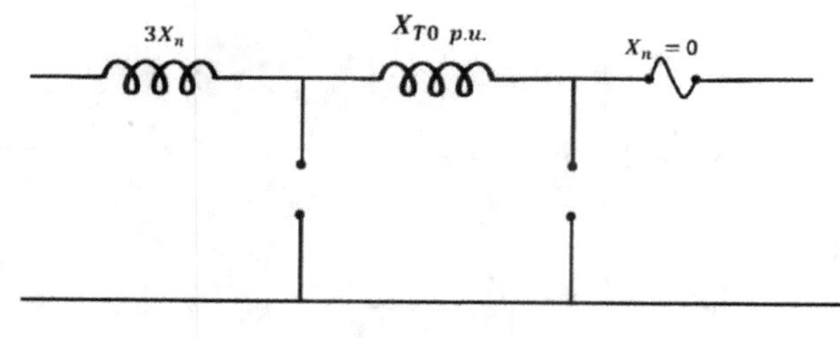

$$X_{0eq\ p.u} = X_{T0\ p.u.} + 3X_{n\ p.u.}$$

8.

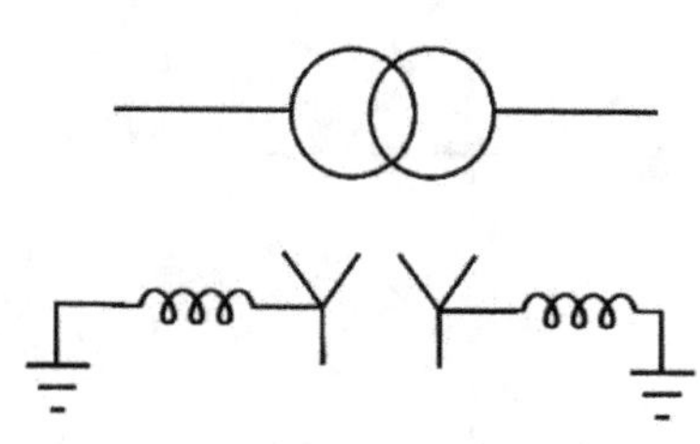

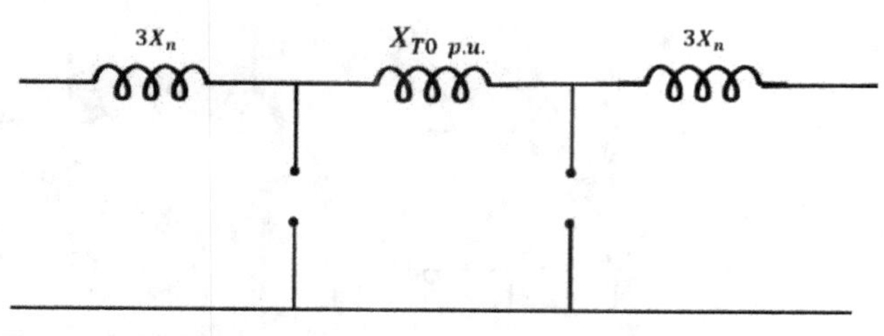

$$X_{0eq\ p.u} = X_{T0\ p.u.} + 3X_{n\ p.u.}$$

9.

$3X_n$
$X_{T0\ p.u.}$
$X_n = \infty$
$X_{0eq\ p.u} = \infty$

10.

$X_{T0\ p.u.}$
$X_n = 0$
$X_n = 0$
$X_{0eq\ p.u} = X_{T0\ p.u}$

11.

$X_{T0\ p.u.}$
$3X_n$
$X_n = 0$
$X_{0eq\ p.u} = X_{T0\ p.u} + 3X_{n\ p.u}$

12.

$X_{T0\ p.u.}$
$X_n = \infty$
$X_n = 0$
$X_{0eq\ p.u} = \infty$

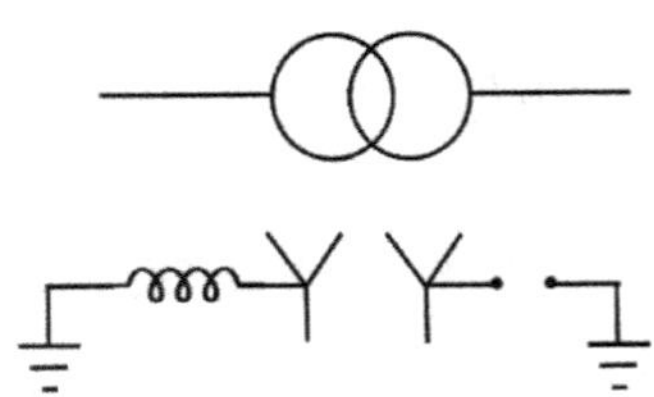

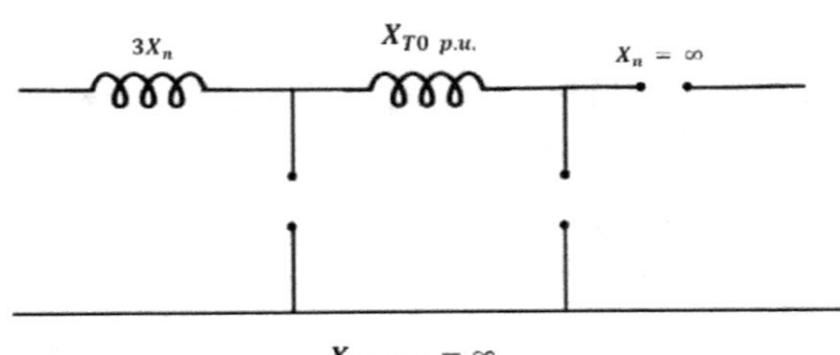

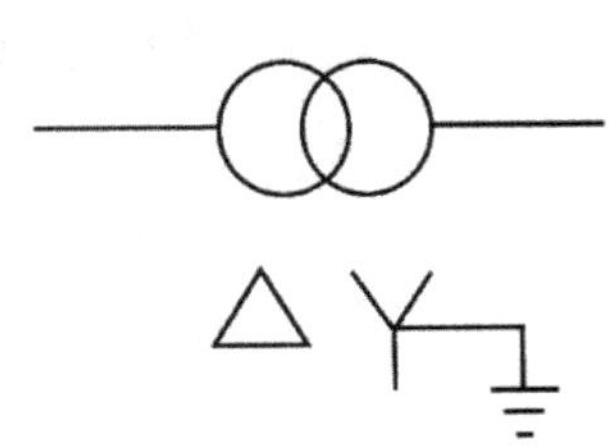

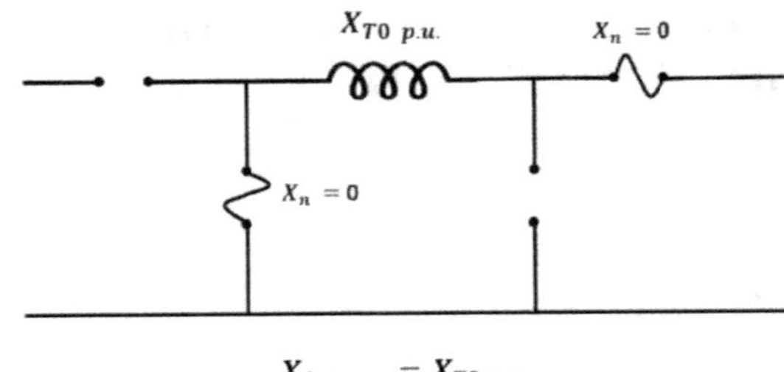

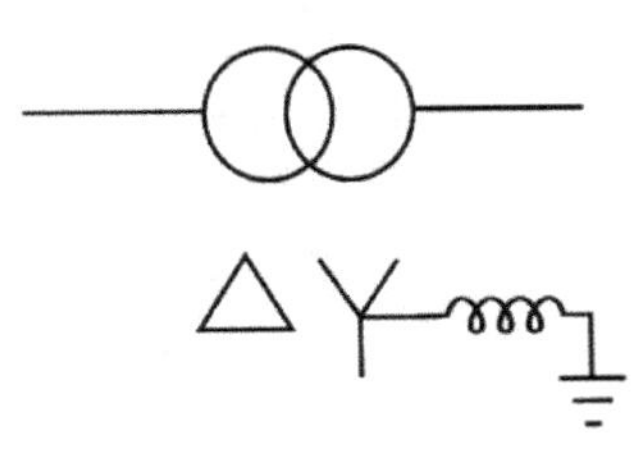

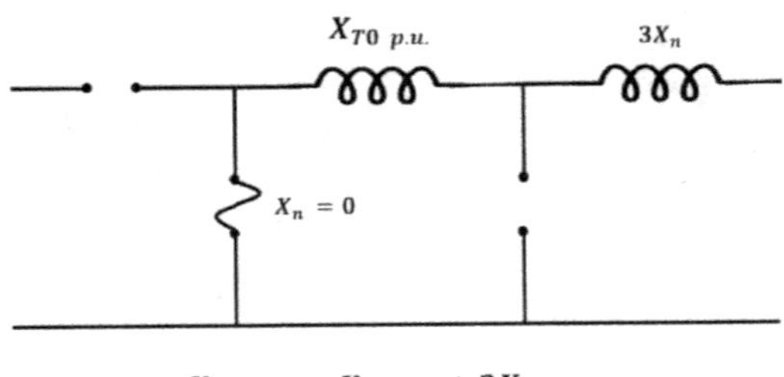

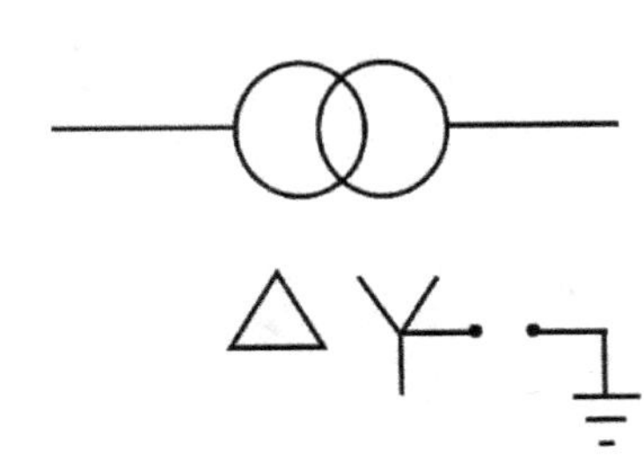

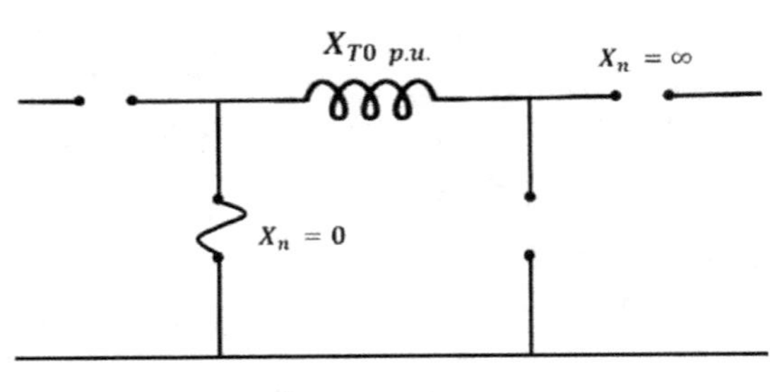

13.

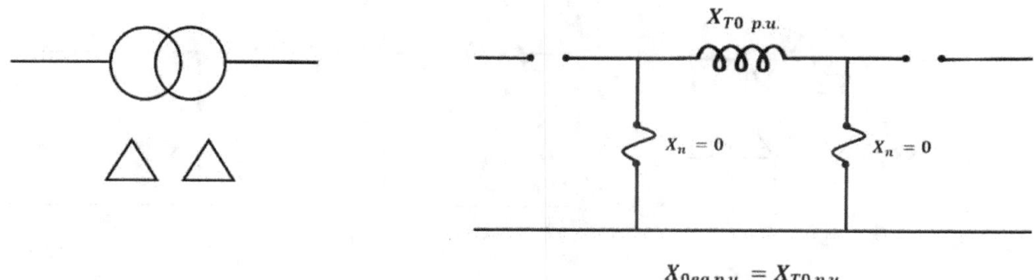

8.7 SEQUENCE NETWORKS OF AN ALTERNATOR OPERATING UNDER NO LOAD CONDITION

Consider a three-phase alternator operating at no-load condition as in figure 8.9.

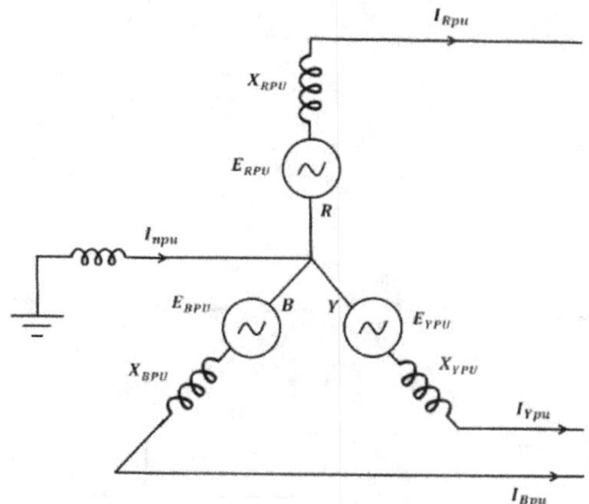

Figure 8.9 A three phase alternator operating at no–load

Positive Sequence Network

$$V_{R1\ p.u.} = E_{R1\ p.u.} - j\ I_{R1p.u.}\ X_{R1p.u.}$$
$$= E_{R1\ p.u.} - j\ I_{R1p.u.}\ X_{1eqp.u.}$$

Where $X_{1eqp.u.}$ is the equivalent p.u. positive sequence Reactance and is equal to $X_{R1p.u.}$.

Negative Sequence Network

$$V_{R2p.u.} = E_{R2p.u.} - j\ I_{R2p.u.}\ X_{R2p.u.}$$
$$V_{R2p.u.} = 0 - j\ I_{R2p.u.}\ X_{R2ep.u.}$$
$$= -j\ I_{R2p.u.}\ X_{2eq\ p.u.}$$

Where $X_{2eqp.u.}$ is the equivalent p.u. negative sequence

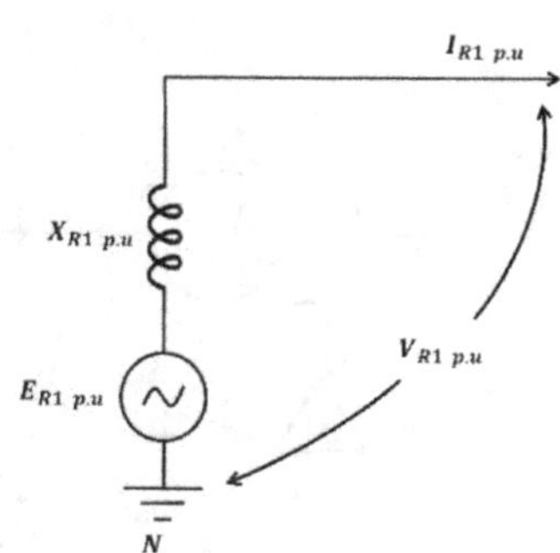

Positive Sequence Network

Reactance and is equal to $X_{R2p.u.}$

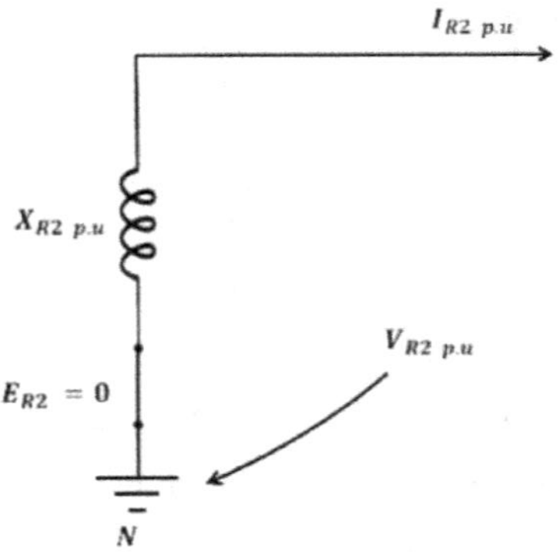

Negative Sequence Network

Zero Sequence Network

$$V_{R0p.u.} = E_{R0p.u.} - j\ I_{R0p.u.}\ (X_{R0p.u.} + 3X_n)$$

$$V_{R0p.u.} = -j\ I_{R0}(X_{0eqp.u})$$

Where $X_{0eqp.u.}$ is the equivalent p.u. zero sequence

Reactance and is equal to $(X_{R0p.u.} + 3X_n)$

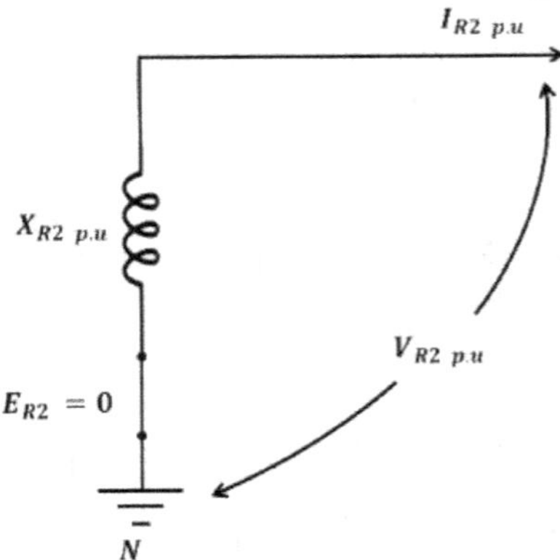

Zero Sequence Network

8.8 REACTANCE OFFERED BY ALTERNATOR DURING FAULT

An alternator consists of armature winding, field winding and damper winding connected in parallel as shown in the figure 8.10

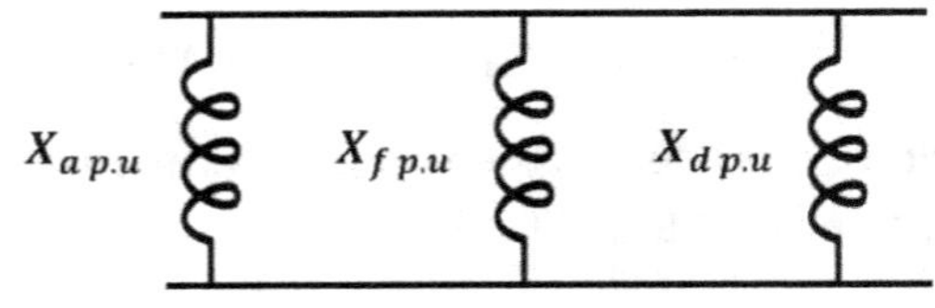

Figure 8.10 An alternator with three windings

1. **Sub–transient reactance:** The reactance offered by alternator immediately after occurrence of faults known as sub–transient reactance.

$$X''_{dp.u.} = (X_{ap.u.} \,/\!/\, X_{fp.u.} \,/\!/\, X_{dampp.u.})$$

Sub transient duration exists for 2 cycles during which the damper winding is disconnected.

2. **Transient reactance:** The reactance offered by alternator after two cycles of occurrence of fault.

$$X'_{dp.u.} = (X_{ap.u.} \,/\!/\, X_{fp.u.})$$

Transient duration exist from 2 to 4 cycles during which field winding is disconnected.

3. **Direct axis / steady state reactance:** The reactance offered by alternator after clearance of fault after 4cycles of Fault occurence.

$$X''_{dp.u.} = X_{ap.u.}$$

If fault is not cleared within 5 cycles then power system becomes unstable.

Q1. Draw an impedance diagram for the system shown by expressing all values as p.u values.

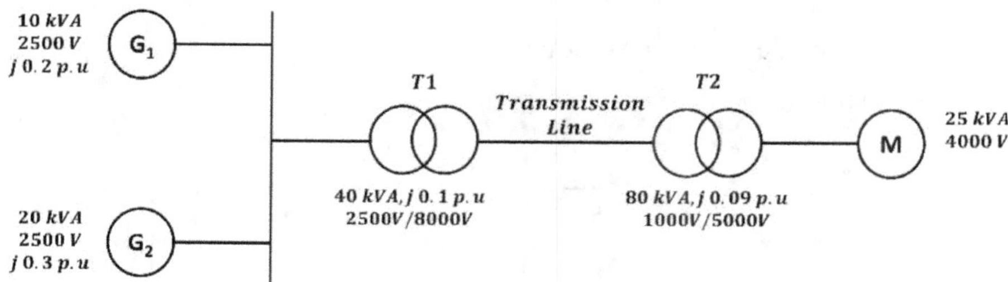

Select the base KVA as 50 KVA & 2500 V as base voltage

Solution:

From the formula,

$$X_{p.u.\ new} = X_{p.u old} * [(\frac{MVAb,New}{MVA,old}) * (\frac{KVb,old}{KVb,new})^2]$$

For generator G_1, $X_{p.u new} = j\,0.2 * [\frac{50}{10}] * [(\frac{2500}{2500})^2] = 1$p.u.

For generator G_2, $X_{p.u.\ new} = j\,0.3 * [\frac{50}{20}] * [(\frac{2500}{2500})^2] = 0.75$p.u.

For transformer T_1, $X_{p.u.\ new} = j\,0.1 * [\frac{50}{40}] * [(\frac{2500}{2500})^2] = 0.125$ p.u.

For transformer T_2, $X_{p.u.\ new} = j\,0.09 * [\frac{50}{80}] * [(\frac{5000}{2500})^2] = 0.2$ p.u.

Note: Voltages for step up transformer with respect to alternator side & for step down transformer with respect to load side.

For Transmission line,

$$X_{TL}\ (p.u.) = X_{TL}(\Omega)\ [\frac{MVAb}{(KVb)2}]$$

$$= (50 + j\ 200)\ [\frac{50*1000}{(8000)2}] = 0.039 + j\ 0.150\ \text{p.u. on generator side}$$

For Motor (load),

$$X_m\ (p.u.) = \frac{Actual\ value\ MVA}{Base\ value\ MVA} = \frac{25}{50} = 0.5\ \text{p.u.}$$

The p.u. impedance diagram is

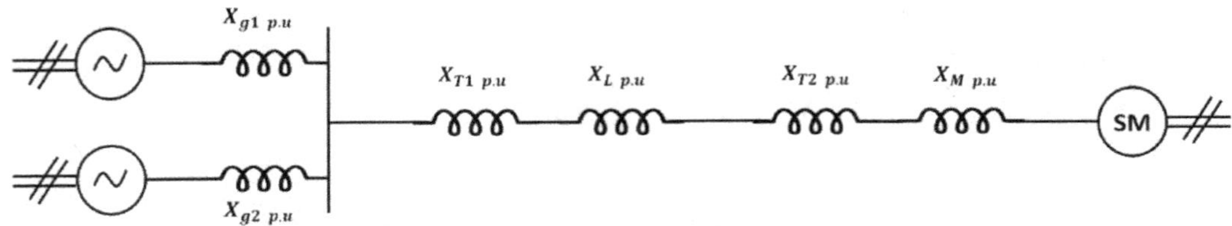

Q2. For the power system shown in the figure below the specifications of the components are as follows:

G1: 25 kV, 100 MVA, 9%

G2: 25 kV. 100 MVA, 9%

T1: 25/220 kV, 90 MVA, X_1= 12%

T2: 220/25 kV, 90 MVA, X_1= 12%

Transmission Line: 220 kV, X=150 Ω

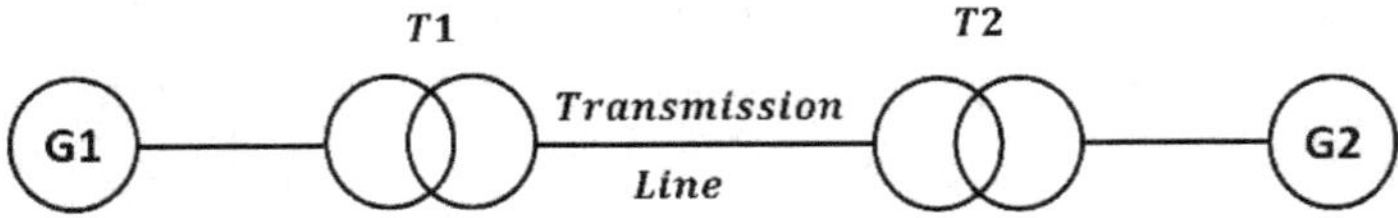

Select 25 KV as base voltage at the generator G1 and 200 MVA as base MVA and draw the p.u. impedance diagram?

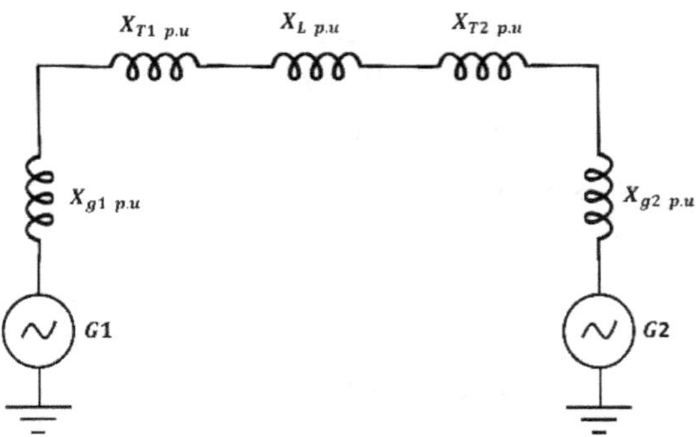

Solution:

$$MVA_b = 200 \text{ MVA}$$

$$kV_b = 25 \text{ kV}$$

$$Z_{p.u. new} = Z_{p.u. old} * [(\frac{MVAb,New}{MVAb,old}) * (\frac{KVb,old}{KVb,new})^2]$$

For G1, $X_{G1p.u. new} = j\, 0.09 * [\frac{200}{100}] * [(\frac{25}{25})^2] = j\, 0.18$

For G2, $X_{G2p.u. new} = j\, 0.1$ (same as for G1)

T1: $X_{p.u. new} = 0.12\, \Omega * [\frac{200}{90}] * [(\frac{25}{25})^2] = j\, 0.26$

T2: $X_{p.u. new} = 0.12 * [\frac{200}{90}] * [(\frac{25}{25})^2] = j\, 0.26$

Transmission line:

$$X_{T1\, p.u\, new} = 150 * \frac{200}{(220)2} = j\, 0.61983$$

OBJECTIVE QUESTIONS

1. _______________are used to represent an unbalanced power system network.

2. A three phase power system network consists of _______________ possible phase sequences.

3. A power system network is subjected to _______________ and _______________ unbalanced conditions during fault.

4. _______________ sequence components are equal in magnitude and displaced from each other by 120° with phase sequence same as three phase power system network.

5. _______________ sequence components are equal in magnitude and displaced from each other by 120° with phase sequence opposite to three phase power system network.

6. _______________ sequence components are equal in magnitude without any phase sequence.

7. _______________ rotates a vector in anti–clock wise direction by 120°

8. The relation between 1,k and k^2 is _______________

9. The positive sequence generated voltage during fault is _______________ and negative, zero sequence generated voltages during fault are _______________ and _______________

10. Neutral current during fault is _______________

11. _______________ is the reference to the flow of zero sequence current.

12. _______________ is the reference to the flow of positive and negative sequence currents.

13. Voltage to neutral is measured between _______________ .

14. Voltage to neutral, Vn = _______________

15. The neutral and ground of a three-phase transformer or alternator can be connected by three methods _______________

16. _______________ switches are operated for star connection of three phase power system components.

17. ______________ switches are operated for delta connection of three phase power system components.
18. The series switch is closed by reactance, $X_n = 0$ for ______________ neutral
19. The series switch is closed by reactance, $3X_n$ for ______________ neutral
20. The series switch is closed by reactance $X_n = $ infinity for ______________ neutral
21. The shunt switch is closed by reactance, $X_n = 0$ for ______________ connection.
22. The zero-sequence reactance offered by alternator during fault, $X_{oeq} = $ ______________
23. The positive sequence reactance offered by alternator during fault, $X_{1eq} = $ ______________
24. The negative sequence reactance offered by alternator during fault, $X_{2eq} = $ ______________
25. The reactance offered by alternator immediately after occurrence of fault is known as ______________
26. The reactance offered by alternator after two cycles of occurrence of fault is known as ______________
27. The reactance offered by alternator after four cycles of occurrence of fault is known as ______________
28. The relation between the reactances after fault is ______________
29. For a balanced three phase system, the ______________ is same as positive sequence reactance diagram.
30. In a transmission line, zero sequence reactance is ______________ times the positive sequence reactance.
31. The ______________ networks of alternator will not have any sources.
32. In a three winding transformer, the ______________ windings may have different KVA ratings.
33. The impedance per phase of a transmission line for ______________ currents is independent of phase sequence.
34. The series impedances of all sequences are ______________ for all types of transformers.
35. ______________ current can flow in one winding of a transformer only when there is a path for it in the other winding.
36. In loads, the ______________ currents will flow in the network only if a return path exist for it.
37. An ______________ in a circuit is defined as any failure which interferes with the normal current.
38. In ______________ faults, fault currents are same in all the phases.
39. The presence of ______________ current makes the short circuit current asymmetrical.
40. ______________ determines the capacity of circuit breaker.
41. The estimation of various fault currents are commonly referred as ______________ .

9 Unsymmetrical Fault Analysis

9.1 INTRODUCTION

As most of the electrical power system network consisting of generators, transformers, transmission lines and distribution circuits is transmission lines, faults occur mostly on the transmission circuits. In case of over head transmission lines, insulation is air or a high resistivity material. The overhead lines may be short circuited by birds, kite strings, branches of trees etc. Adequate protection has been provided to the power system components with appropriate relays and circuit breakers.

The features of faults are:

1. Interruption of power supply to consumers.

2. Substantial decrease in voltage and frequency.

3. Decrease in stability of parallel operation.

4. Possibility of drop out of generators and separate generating stations.

9.2 CLASSIFICATION OF FAULTS

Faults on power systems are classified into two types.

1. Open circuit or series faults.

2. Short circuit or shunt faults.

Open circuit or series faults

1. Open circuit faults are characterized by decrease in current and increase in voltage and frequency.

2. Open circuit faults occur due to sudden increase in consumer load demand or mis-operation of protective devices like fuses and circuit breakers.

Classification of open circuit or series faults

Open circuit or series faults on power systems are classified into two types as follows.

1. Symmetrical open circuit faults.
2. Unsymmetrical open circuit faults.

Symmetrical open circuit fault

1. Fault is in R,Y and B-phases, as fuse is blown out in all phases.

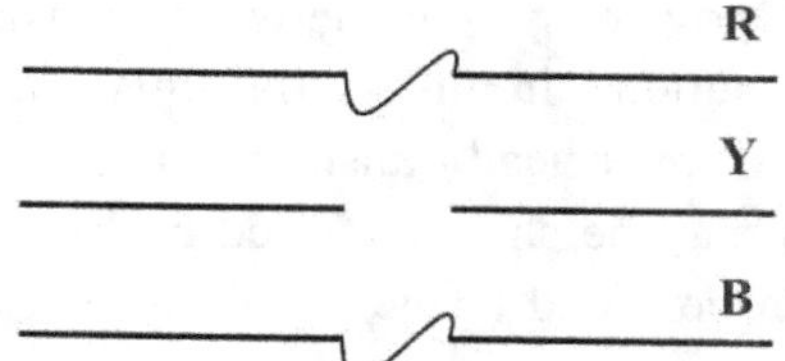

Unsymmetrical open circuit fault

1. Fault is in Y-phase as fuse is blown out in Y-phase

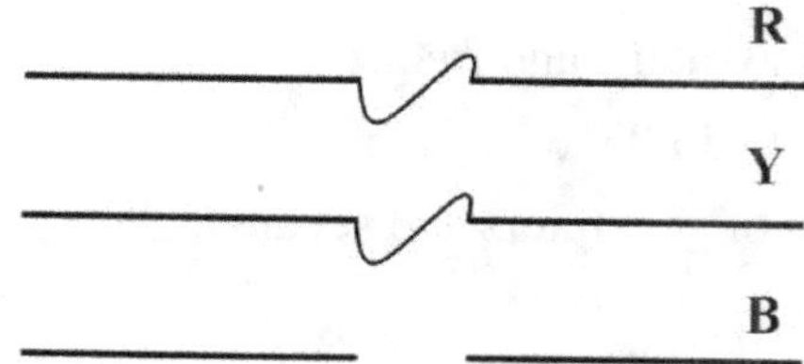

2. Fault is in B-phase as fuse is blown out in B-phase

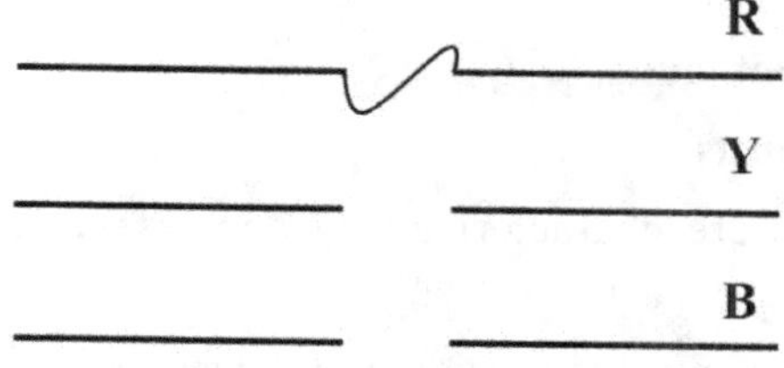

3. Fault is in Yand B-phases as fuse is blown out in Y and B-phases

No fault Condition

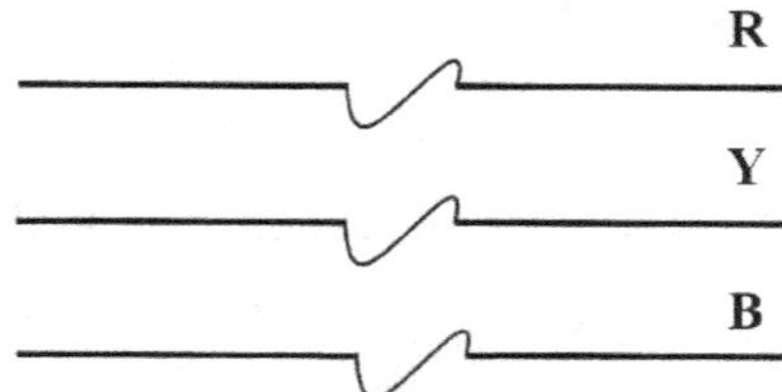

Short circuit or shunt faults

1. Short circuit faults or shunt faults are characterized by increase in current and decrease in voltage and frequency.
2. Short circuit faults or shunt faults occur due to trees falling across transmission lines, lightning surges etc.

Classification of short circuit or shunt faults

Short circuit or shunt faults on power systems are classified into two types as follows.

1. Symmetrical short circuit faults.
2. Unsymmetrical short circuit faults.

Unsymmetrical short circuit faults

Unsymmetrical short circuit faults are classified into three types.

(a) Line to Ground fault(L-G fault)
(b) Line to Line fault(L-L fault)
(c) Line to Line to Ground fault(L-L-G fault)

Symmetrical short circuit faults

Symmetrical short circuit faults are classified into two types.

(a) Line to Line to Line fault(L-L-L fault)
(b) Line to Line to Line to Ground fault(L-L-L-G fault)

9.3   LINE TO GROUND FAULT ANALYSIS (L-G FAULT)

Representation

Consider an alternator operating under no load condition as shown in figure 9.1.

Let the ground and neutral are connected solidly i.e., solid neutral or solid grounding, $X_{n\ p.u}=0$.

Assume line to ground fault occur between R phase and ground.

Assume solid fault i.e., $X_{f\ p.u}= 0$.

During fault, all unbalanced electrical quantities are represented using symmetrical components.

$$E_{R\ p.u} \text{ is represented as } E_{R0\ p.u}, E_{R1\ p.u} \text{ and } E_{R2\ p.u}$$

E_Y p.u is represented as E_{Y0} p.u, E_{Y1} p.u and E_{Y2} p.u

E_B p.u is represented as E_{B0} p.u, E_{B1} p.u and E_{B2} p.u

X_R p.u is represented as X_{R0} p.u, X_{R1} p.u and X_{R2} p.u

X_Y p.u is represented as X_{Y0} p.u, X_{Y1} p.u and X_{Y2} p.u

X_B p.u is represented as X_{B0} p.u, X_{B1} p.u and X_{B2} p.u

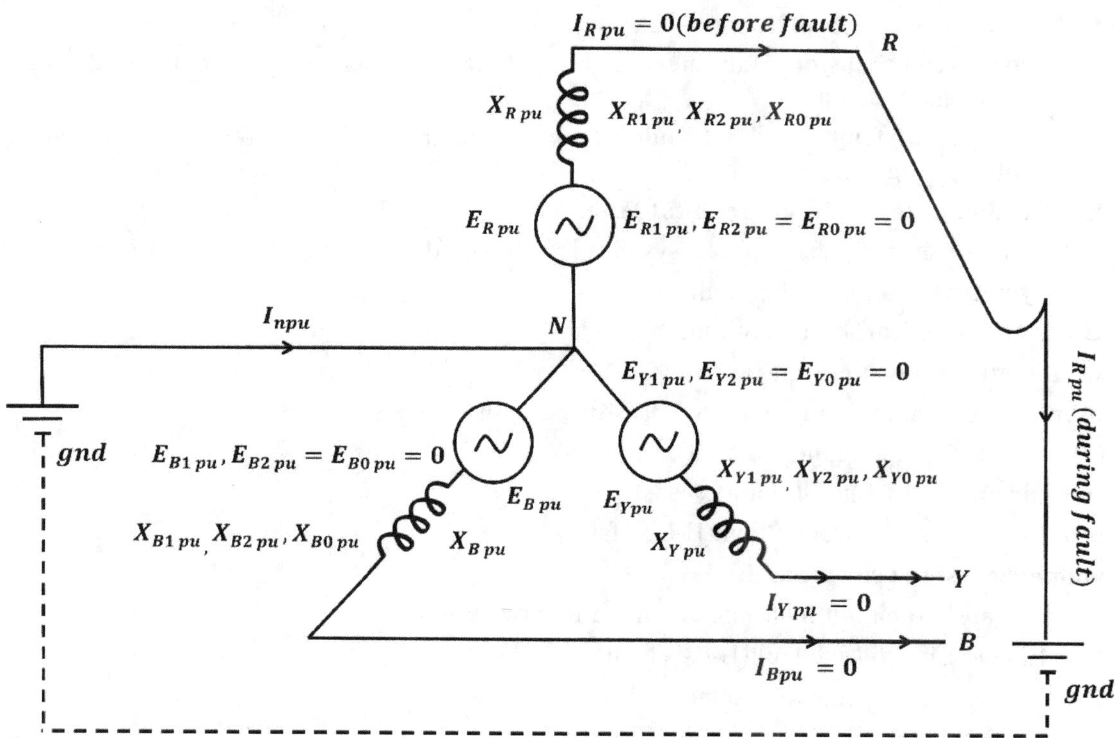

Figure 9.1 LG fault analysis

Boundary Conditions

Number of boundary conditions = number of sequence networks = 3.

The boundary conditions are:

1. I_Y p.u = 0 and I_B p.u = 0

2. V_R p.u. = 0

3. I_f p.u = I_R p.u

Sequence current components of R-Phase

$$I_{R0\ p.u} = \frac{1}{3}\left(I_{R\ p.u} + I_{Y\ p.u} + I_{B\ p.u}\right) = \frac{I_R}{3}\ p.u$$

$$I_{R1\ p.u} = \frac{1}{3}\left(I_{R\ p.u} + K\ I_{Y\ p.u} + K^2 I_{B\ p.u}\right) = \frac{I_R}{3}\ p.u$$

$$I_{R2\ p.u} = \frac{1}{3}\left(I_{R\ p.u} + K^2 I_{Y\ p.u} + K I_{B\ p.u}\right) = \frac{I_R}{3}\ p.u$$

$$I_{R0\ p.u} = I_{R1\ p.u} = I_{R2\ p.u} = \frac{I_R}{3}\ p.u \qquad\qquad .....(9.1)$$

Zero sequence current, $I_{R0\ p.u}$ flows for faults involved with ground.

Positive sequence current, $I_{R1\ p.u}$ flows for every fault because $I_{R1\ p.u}$ is driven by $E_{R1\ p.u}$.

Negative sequence current, $I_{R2\ p.u}$ flows for unsymmetrical faults.

Per unit Positive sequence current in terms of per unit positive sequence generated Voltage

As the boundary condition, $V_{R0\ p.u} = 0$

$$\Rightarrow V_{R0\ p.u} + V_{R1\ p.u} + V_{R2\ p.u} = 0$$

$$\Rightarrow \left(-jI_{R0\ p.u}X_{0\ eq\ p.u}\right) + \left(E_{R1\ p.u} - jI_{R1\ p.u}X_{1\ eq\ p.u}\right) + \left(-jI_{R2\ p.u}X_{2\ eq\ p.u}\right) = 0$$

$$\Rightarrow I_{R1\ p.u}j\{X_{0\ eq\ p.u} + X_{1\ eq\ p.u} + X_{2\ eq\ p.u}\} = E_{R1\ p.u}$$

$$\Rightarrow I_{R1\ p.u} = \frac{E_{R1\ p.u}}{j\{X_{0\ eq\ p.u} + X_{1\ eq\ p.u} + X_{2\ eq\ p.u}\}} \qquad .....(9.2)$$

Per unit fault current

As fault involves $R-$ phase, $I_{f\ p.u} = I_{R\ p.u} = 3I_{R1\ p.u}$

$$= \frac{3E_{R1\ p.u}}{j\{X_{0\ eq\ p.u} + X_{1\ eq\ p.u} + X_{2\ eq\ p.u}\}}$$

Fault Current in Amperes

$$\text{Base voltampere, } VA_b = \sqrt{3}\cdot V_b\cdot I_b \Rightarrow I_b = \frac{(VA)b}{\sqrt{3}V_b}$$

and $\qquad I_{f\ p.u} = \dfrac{I_{f\ amp}}{I_{b\ amp}} \Rightarrow I_{f\ amp} = I_{f\ pu}I_{b\ amp}$

Perunit Sub transient reactance

The Sub transient reactance offered by alternator for the fault is

$$\Rightarrow X_{d\ p.u}^{''} = j\{X_{0\ eq\ p.u} + X_{1\ eq\ p.u} + X_{2\ eq\ p.u}\} \qquad .....(9.3)$$

Sequence Networks

Positive, negative & zero sequence networks are connected in series as shown in figure 9.2.

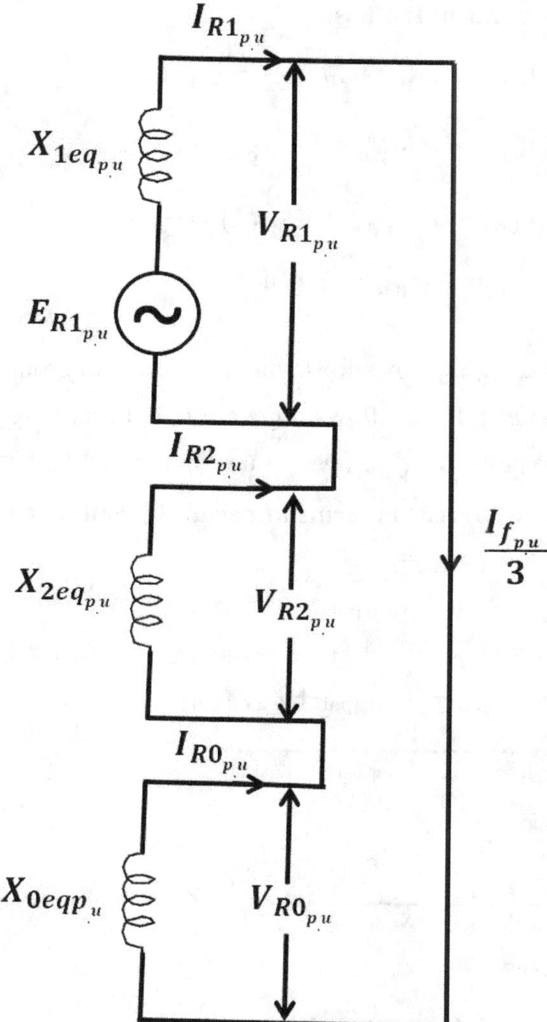

Figure 9.2 Sequence network for LG fault

Neutral current

Case 1: Balanced condition

$$\Rightarrow I_{n\ p.u} = I_{R\ p.u} + I_{Y\ p.u} + I_{B\ p.u} = 0$$

Case 2: Unbalanced condition

$$\Rightarrow I_{n\ p.u} = I_{R\ p.u} + I_{Y\ p.u} + I_{B\ p.u} = 3I_{R0\ p.u}$$

Per unit voltage to neutral

Case 1: Balanced condition

$$\Rightarrow V_{n\ p.u} = I_{n\ p.u} \cdot X_{n\ p.u} = 0$$

Case 2: Unbalanced condition

$$\Rightarrow V_{n\ p.u} = I_{n\ p.u} \cdot X_{n\ p.u} = 3I_{R0\ p.u} \cdot X_{n\ p.u} = I_{R0\ p.u}\left(3X_{n\ p.u}\right)$$

$$\Rightarrow \qquad\qquad\qquad\qquad = I_{n\ p.u} \cdot X_{n\ p.u}$$

$$\Rightarrow \qquad\qquad\qquad\qquad = I_{f\ p.u} \cdot X_{n\ p.u}$$

Fault current in amperes expressed in terms of $X_{d\ p.u}^{''}$:-

Per unit subtransient reactance, $X_{d\ p.u}^{''} = X_{d}^{''}\ ohms \cdot \dfrac{I_b}{V_b}$

$$= \frac{I_b}{\left(\dfrac{V_b}{X_d^{''}}\right)} = \frac{I_{b\ amp}}{I_{f\ amp}}$$

$$\Rightarrow I_{f\ amp} = \frac{I_{b\ amp}}{X_{d\ p.u}^{''}} \qquad\qquad\qquad\qquad .....(9.4)$$

Fault MVA expressed in terms of $X_{d\ p.u}^{''}$

$$V_b I_{f\ amp} = \frac{V_b I_{b\ amp}}{X_{d\ p.u}^{''}}$$

$$(VA)_{sc} = \frac{(VA_b)}{X_{d\ p.u}^{''}}$$

$$MVA_{SC} = \frac{MVA_b}{X_{d\ p.u}^{''}}$$

Considering neutral reactance and fault reactance, per unit sub-transient Reactance

$$X_{d\ p.u}^{''} = j\left\{X_{1\ eq\ p.u} + X_{2\ eq\ p.u} + X_{0\ eq\ p.u}\right\}$$

Where, $\qquad X_{0\ eq\ p.u} = X_{R0\ p.u} + 3X_{n\ p.u} + 3X_{f\ p.u}$

9.4 LINE TO LINE FAULT ANALYSIS (L-L FAULT)

Representation

Consider an alternator operating under no load condition as shown in figure 9.3.

Let the ground and neutral are connected solidly i.e., solid neutral or solid grounding,

$$X_{n\ p.u} = 0.$$

Assume line to line fault occur between R phase and Y phase.

Assume solid fault i.e., $X_{f\ p.u} = 0$.

During fault, all unbalanced electrical quantities are represented using symmetrical components.

$$E_{R\ p.u} \text{ is represented as } E_{R0\ p.u}, E_{R1\ p.u} \text{ and } E_{R2\ p.u}$$

$$E_{Y\ p.u} \text{ is represented as } E_{Y0\ p.u}, E_{Y1\ p.u} \text{ and } E_{Y2\ p.u}$$

$$E_{B\ p.u} \text{ is represented as } E_{B0\ p.u}, E_{B1\ p.u} \text{ and } E_{B2\ p.u}$$

$$X_{R\ p.u} \text{ is represented as } X_{R0\ p.u}, X_{R1\ p.u} \text{ and } X_{R2\ p.u}$$

$$X_{Y\ p.u} \text{ is represented as } X_{Y0\ p.u}, X_{Y1\ p.u} \text{ and } X_{Y2\ p.u}$$

$$X_{B\ p.u} \text{ is represented as } X_{B0\ p.u}, X_{B1\ p.u} \text{ and } X_{B2\ p.u}$$

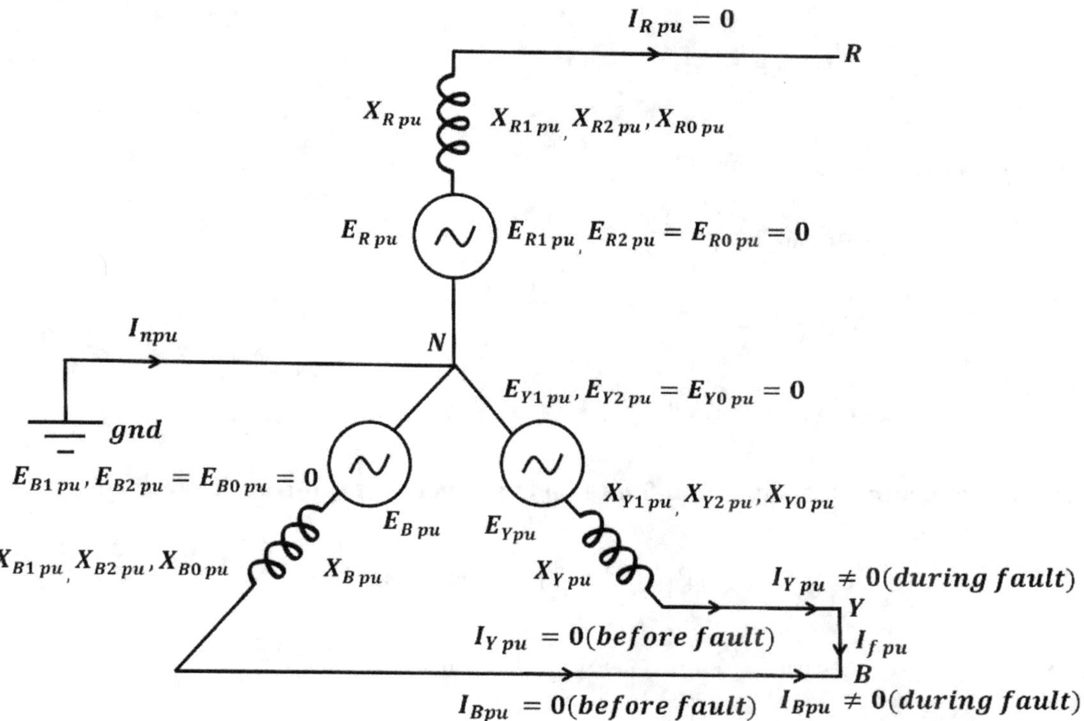

Figure 9.3 LL fault analysis

Boundary Conditions

Number of boundary conditions = number of sequence networks = 3.

The boundary conditions are:

1. $I_{R\ p.u} = 0$

2. $V_{Y\ p.u} = V_{B\ p.u}$

3. $I_{f\ p.u} = I_{Y\ p.u} = I_{B\ p.u}$

Per unit sequence current components of R-Phase

$$I_{R0\ p.u} = \frac{1}{3}\left(I_{R\ p.u} + I_{Y\ p.u} + I_{B\ p.u}\right) = 0$$

$I_{R0\ p.u}$ do not flow for faults not involved with ground.

$$I_{R1\ p.u} = \frac{1}{3}\left(I_{R\ p.u} + K\,I_{Y\ p.u} + K^2 I_{B\ p.u}\right)$$

$$= \frac{1}{3}\left(0 + K\,I_{Y\ p.u} - K^2 I_{Y\ p.u}\right)$$

$$= \frac{I_{Y\ p.u}}{3}(K - K^2)$$

$I_{R1\ p.u}$ flows for any fault because $E_{R1\ p.u}$ drives $I_{R1\ p.u}$

$$I_{R2\ p.u} = \frac{1}{3}\left(I_{R\ p.u} + K^2\,I_{Y\ p.u} + KI_{B\ p.u}\right)$$

$$= \frac{I_{Y\ p.u}}{3}(K^2 - K)$$

$I_{R2\ p.u}$ flows for unsymmetrical faults.

The sequence current components of R-phase are,

$$I_{R0\ p.u} = 0;\ I_{R1\ p.u} = -I_{R2\ p.u}$$

Sequence voltage components of R-phase

As per the boundary condition, $V_{Y\ p.u} = V_{B\ p.u}$

$$\Rightarrow V_{Y0\ p.u} + V_{Y1\ p.u} + V_{Y2\ p.u} = V_{B0\ p.u} + V_{B1\ p.u} + V_{B2\ p.u}$$

$$\Rightarrow V_{R0\ p.u} + K^2 V_{R1\ p.u} - V_{R0\ p.u} - KV_{R1\ p.u} + K^2 V_{R2\ p.u}$$

$$\Rightarrow (K^2 - K)V_{R1\ p.u} = (K^2 - K)V_{R2\ p.u}$$

$$\Rightarrow V_{R1\ p.u} = V_{R2\ p.u} \qquad\qquad .....(9.5)$$

Per unit positive sequence current in terms of per unit positive sequence generated Voltage

$$\text{As } V_{R1\ p.u} = V_{R2\ p.u}$$

$$\Rightarrow E_{R1\ p.u} - j\,I_{R1\ p.u}X_{1eq\ p.u} = -jI_{R2\ p.u}X_{2eq\ p.u}$$

$$= j\,I_{R1\ p.u}X_{1eq\ p.u}$$

$$\Rightarrow I_{R1\ p.u}j\{X_{1eq\ p.u} + X_{2eq\ p.u}\} = E_{R1\ p.u}$$

$$\Rightarrow I_{R1\ p.u} = \frac{E_{R1\ p.u}}{j\{X_{1eq\ p.u}+X_{2eq\ p.u}\}} \qquad\qquad .....(9.6)$$

Per unit fault current

Fault current in per unit is,

$$I_{f\ p.u} = I_{Y\ p.u}$$

$$= I_{Y0\ p.u} + I_{Y1\ p.u} + I_{Y2\ p.u}$$

$$= I_{R0\ p.u} + K^2 I_{R1\ p.u} + KI_{R2\ p.u}$$

$$= 0 + K^2 I_{R1\ p.u} - K I_{R1\ p.u}$$

$$= (K^2 - K) I_{R1\ p.u}$$

$$I_{f\ p.u} = -\sqrt{3} I_{R1\ p.u}$$

$$I_{f\ p.u} = \frac{|\sqrt{3}| I_{R1\ p.u}}{j\{X_{1eq\ p.u} + X_{2eq\ p.u}\}} \qquad \dots(9.7)$$

Fault Current in Amperes

$$\text{Base voltampere, } VA_b = \sqrt{3} \cdot V_b \cdot I_b \Rightarrow I_b = \frac{(VA)b}{\sqrt{3} V_b}$$

$$\text{and } I_{f\ p.u} = \frac{I_{f\ amp}}{I_{b\ amp}} \Rightarrow I_{f\ amp} = I_{f\ pu} I_{b\ amp}$$

Perunit Sub transient reactance

The Sub transient reactance offered by alternator for the fault is

$$X_{d\ p.u}^{''} = j\{X_{1\ eq\ p.u} + X_{2\ eq\ p.u}\} \qquad \dots(9.8)$$

Sequence Networks

As $I_{R1\ p.u} = -I_{R2\ p.u}$, negative & zero sequence networks are connected in anti parallel

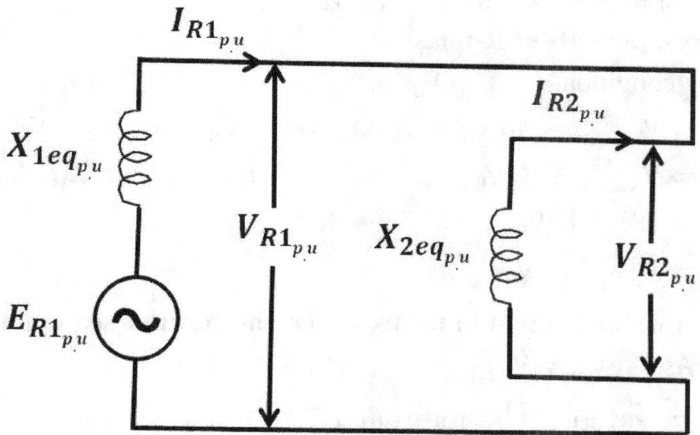

Figure 9.4 Sequence network for LL fault

Neutral current

Case 1: Balanced condition

$$\Rightarrow I_{n\ p.u} = I_{R\ p.u} + I_{Y\ p.u} + I_{B\ p.u} = 0$$

Case 2: Unbalanced condition

$$\Rightarrow I_{n\ p.u} = I_{R\ p.u} + I_{Y\ p.u} + I_{B\ p.u} = 3 I_{R0\ p.u} = 0$$

Per unit voltage to neutral

Case 1: Balanced condition

$$V_{n\ p.u} = I_{n\ p.u} \cdot X_{n\ p.u} = 0$$

Case 2: Unbalanced condition

$$V_{n\ p.u} = I_{n\ p.u} \cdot X_{n\ p.u} = 0$$

Fault current in amperes expressed in terms of $X_{d\ p.u}^{''}$

Per unit subtransient reactance, $X_{d\ p.u}^{''} = X_d^{''}\ ohms \cdot \dfrac{I_b}{V_b}$

$$\Rightarrow X_{dp.u} = \frac{I_b}{\left(\dfrac{V_b}{X_d^{''}}\right)} = \frac{I_{b\ amp}}{I_{f\ amp}}$$

$$\Rightarrow I_{f\ amp} = \frac{I_{b\ amp}}{X_{d\ p.u}^{''}} \qquad\qquad\qquad .....(9.9)$$

Fault MVA expressed in terms of $X_{d\ p.u}^{''}$

$$V_b I_{f\ amp} = \frac{V_b I_{b\ amp}}{X_{d\ p.u}^{''}}$$

$$(VA)_{sc} = \frac{(VA_b)}{X_{d\ p.u}^{''}}$$

$$MVA_{SC} = \frac{MVA_b}{X_{d\ p.u}^{''}}$$

Considering neutral reactance and fault reactance fault, Per unit sub-transient Reactance

$$X_{d\ p.u}^{''} = j\{X_{1\ eq\ p.u} + X_{2\ eq\ p.u} + X_{f\ p.u}\}$$

9.5 LINE TO LINE TO GROUND FAULT ANALYSIS (L-L-G FAULT)

Representation

Consider an alternator operating under no load condition as shown in figure 9.5.

Let the ground and neutral are connected solidly i.e., solid neutral or solid grounding

$$X_{n\ p.u} = 0.$$

Assume line to line to ground fault occur between R phase, Y phase and ground.

Assume solid fault i.e.,, $X_{f\ p.u} = 0$.

During fault, all unbalanced electrical quantities are represented using symmetrical components.

E_R p.u is represented as E_{R0} p.u, E_{R1} p.u and E_{R2} p.u

E_Y p.u is represented as E_{Y0} p.u, E_{Y1} p.u and E_{Y2} p.u

E_B p.u is represented as E_{B0} p.u, E_{B1} p.u and E_{B2} p.u

X_R p.u is represented as X_{R0} p.u, X_{R1} p.u and X_{R2} p.u

X_Y p.u is represented as X_{Y0} p.u, X_{Y1} p.u and X_{Y2} p.u

X_B p.u is represented as X_{B0} p.u, X_{B1} p.u and X_{B2} p.u

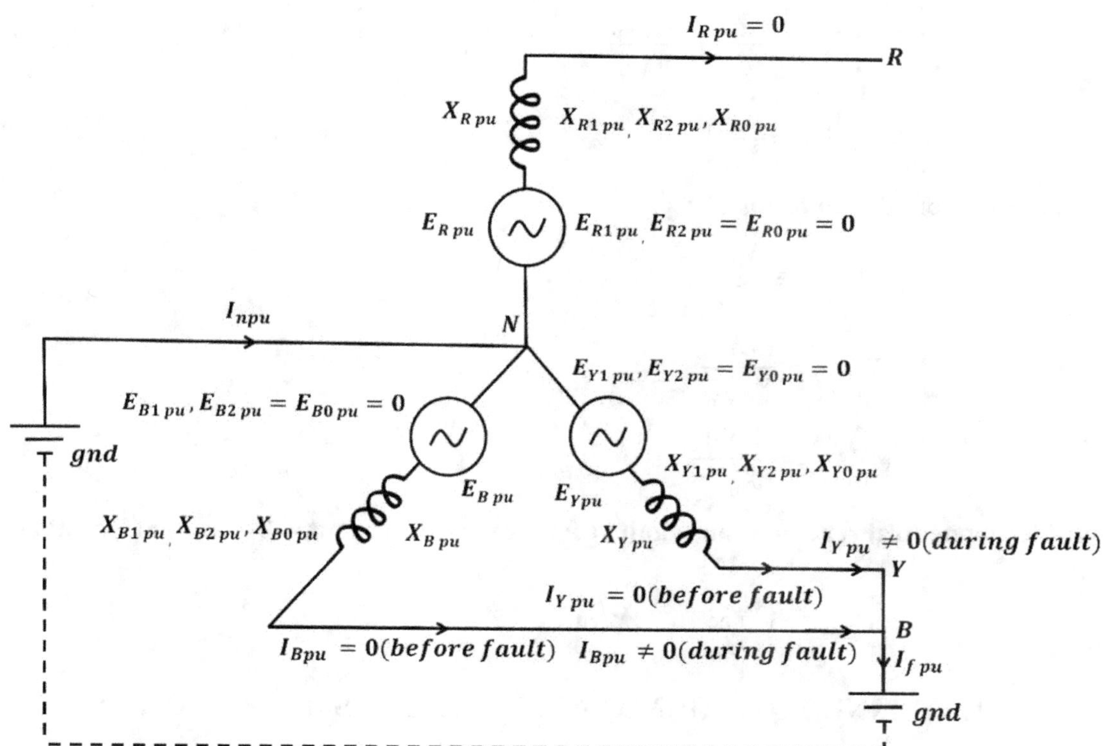

Figure 9.5 LLG fault analysis

Boundary Conditions

Number of boundary conditions = number of sequence networks = 3.

The boundary conditions are:

1. I_R p.u $= 0$

2. V_Y p.u $= V_B$ p.u $= 0$

3. I_f p.u $= I_Y$ p.u $+ I_B$ p.u

Per unit sequence voltage components of R-Phase

$$V_{R0\ p.u} = \frac{1}{3}\left(V_{R\ p.u} + V_{Y\ p.u} + V_{B\ p.u}\right) = \frac{V_R}{3}\ p.u.$$

$$V_{R1\ p.u} = \frac{1}{3}\left(V_{R\ p.u} + KV_{Y\ p.u} + K^2V_{B\ p.u}\right) = \frac{V_R}{3}\ p.u.$$

$$V_{R2\ p.u} = \frac{1}{3}\left(V_{R\ p.u} + K^2V_{Y\ p.u} + KV_{B\ p.u}\right) = \frac{V_R}{3}\ p.u.$$

$$V_{R0\ p.u} = V_{R1\ p.u} = V_{R2\ p.u} = \frac{V_R}{3}\ p.u. \qquad \qquad .....(9.10)$$

Expressing $I_{R0\ p.u}$ and $I_{R2\ p.u}$ in terms of $I_{R1\ p.u}$

From 9.10,

$$V_{R0\ p.u} = V_{R1\ p.u}$$

$$j\,I_{R1\ p.u}X_{0eq\ p.u} = E_{R1\ p.u.} - jI_{R1\ p.u}X_{1eq\ p.u}$$

$$I_{R0\ p.u} = \frac{-\{E_{R1\ p.u.} - jI_{R1\ p.u}X_{1eq\ p.u}\}}{j\,X_{0eq\ p.u}} \qquad \qquad .....(9.11)$$

From 9.10,

$$V_{R2\ p.u} = V_{R1\ p.u}$$

$$j\,I_{R2\ p.u}X_{2eq\ p.u} = E_{R1\ p.u.} - jI_{R1\ p.u}X_{1eq\ p.u}$$

$$I_{R2\ p.u} = \frac{-\{E_{R1\ p.u.} - jI_{R1\ p.u}X_{1eq\ p.u}\}}{j\,X_{2eq\ p.u}} \qquad \qquad .....(9.12)$$

Per unit positive sequence current in terms of per unit positive sequence generated Voltage

$$I_{R\ p.u} = 0 \Rightarrow I_{R0\ p.u} + I_{R1\ p.u} + I_{R2\ p.u} = 0$$

$$\frac{-\{E_{R1\ p.u.} - jI_{R1\ p.u}X_{1eq\ p.u}\}}{j\,X_{0eq\ p.u}} + I_{R1\ p.u} + \frac{-\{E_{R1\ p.u.} - jI_{R1\ p.u}X_{1eq\ p.u}\}}{j\,X_{2eq\ p.u}} = 0$$

$$I_{R1\ p.u} = \frac{E_{R1\ p.u.} - jI_{R1\ p.u}X_{1eq\ p.u}}{j\,X_{0eq\ p.u}} + \frac{E_{R1\ p.u.} - jI_{R1\ p.u}X_{1eq\ p.u}}{j\,X_{2eq\ p.u}}$$

$$I_{R1\ p.u} = \frac{E_{R1\ p.u.}}{j\{X_{1\ eq\ p.u} + (X_{2\ eq\ p.u} \| X_{0\ eq\ p.u})\}} \qquad \qquad .....(9.13)$$

Per unit fault current

$$I_{f\ p.u} = I_{Y\ p.u} + I_{B\ p.u}$$

$$= \left(I_{Y1\ p.u} + I_{Y2\ p.u} + I_{Y0\ p.u}\right) + \left(I_{B0\ p.u} + I_{B1\ p.u} + I_{B2\ p.u}\right)$$

$$= \left(K^2I_{R1\ p.u} + K\,I_{R2\ p.u} + I_{R0\ p.u}\right) + \left(I_{R0\ p.u} + K\,I_{R1\ p.u} + K^2I_{R2\ p.u}\right)$$

$$= 2\,I_{R0\ p.u} + (K^2 + K)\left(I_{R1\ p.u} + I_{R2\ p.u}\right)$$

$$= 2\,I_{R0\ p.u} + (-1)\left(-I_{R0\ p.u}\right) = 3I_{R0\ p.u}$$

Fault Current in Amperes

$$\text{Base voltampere, } VA_b = \sqrt{3} \cdot V_b \cdot I_b \Rightarrow I_b = \frac{(VA)b}{\sqrt{3}V_b}$$

$$\text{and } I_{f\ p.u} = \frac{I_{f\ amp}}{I_{b\ amp}} \Longrightarrow I_{f\ amp} = I_{f\ pu}I_{b\ amp}$$

Perunit Sub transient reactance

The Sub transient reactance offered by alternator for the fault is

$$X_{d\ p.u}'' = j\{X_{1\ eq\ p.u} + (X_{2\ eq\ p.u} \parallel X_{0\ eq\ p.u})\} \qquad \qquad .....(9.14)$$

Sequence Networks

Positive sequence network is connected in series with parallel combination of negative& zero sequence networks as shown in figure 9.6.

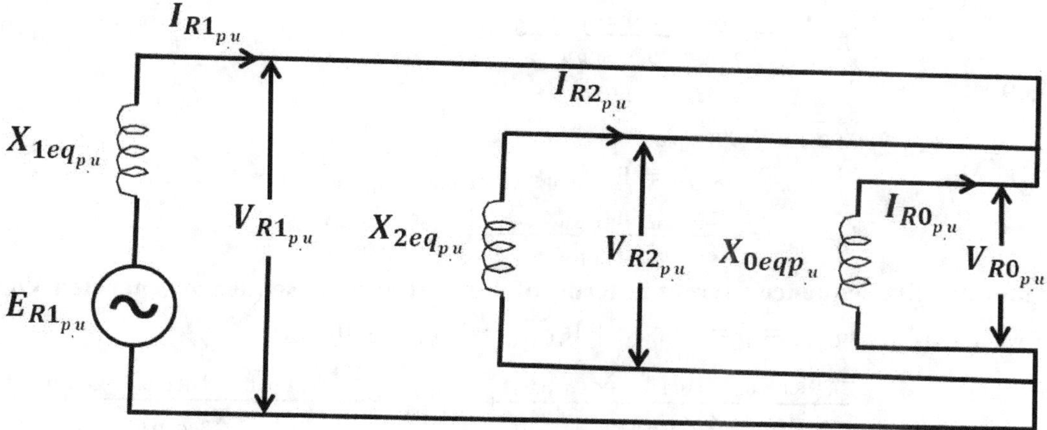

Figure 9.6 Sequence network for LLG fault

From current division rule,

$$I_{R2\ p.u} = \frac{-I_{R1\ p.u}\ X_{0\ eq\ p.u}}{X_{2\ eq\ p.u} + X_{0\ eq\ p.u}}$$

$$I_{R0\ p.u} = \frac{-I_{R1\ p.u}\ X_{2\ eq\ p.u}}{X_{0\ eq\ p.u} + X_{2\ eq\ p.u}}$$

Neutral current

Case 1: Balanced condition

$$\Rightarrow I_{n\ p.u} = I_{R\ p.u} + I_{Y\ p.u} + I_{B\ p.u} = 0$$

Case 2: Unbalanced condition

$$\Rightarrow I_{n \ p.u} = I_{R \ p.u} + I_{Y \ p.u} + I_{B \ p.u} = 3I_{R0 \ p.u}$$

Per unit voltage to neutral

Case 1: Balanced condition

$$V_{n \ p.u} = I_{n \ p.u} \cdot X_{n \ p.u} = 0$$

Case 2: Unbalanced condition

$$V_{n \ p.u} = I_{n \ p.u} \cdot X_{n \ p.u} = 3I_{R0 \ p.u} \cdot X_{n \ p.u} = I_{R0 \ p.u}\left(3X_{n \ p.u}\right)$$
$$= I_{n \ p.u} \cdot X_{n \ p.u}$$
$$= I_{f \ p.u} \cdot X_{n \ p.u}$$

Fault current in amperes expressed in terms of $X_{d \ p.u}^{''}$

Per unit subtransient reactance, $X_{d \ p.u}^{''} = X_{d}^{''} \ ohms \cdot \dfrac{I_b}{V_b}$

$$\Rightarrow X_{d \ p.u}^{''} = \dfrac{I_b}{\left(\dfrac{V_b}{X_{d}^{''}}\right)} = \dfrac{I_{b \ amp}}{I_{f \ amp}}$$

$$\Rightarrow I_{f \ amp} = \dfrac{I_{b \ amp}}{X_{d \ p.u}^{''}} \qquad\qquad .....(9.15)$$

Fault MVA expressed in terms of $X_{d \ p.u}^{''}$

$$V_b I_{f \ amp} = \dfrac{V_b I_{b \ amp}}{X_{d \ p.u}^{''}}$$

$$(VA)_{sc} = \dfrac{(VA_b)}{X_{d \ p.u}^{''}}$$

$$MVA_{SC} = \dfrac{MVA_b}{X_{d \ p.u}^{''}}$$

Considering neutral reactance and fault reactance fault, per unit sub-transient Reactance

$$X_{d \ p.u}^{''} = j\left\{X_{1 \ eq \ p.u} + \left(X_{2 \ eq \ p.u} \ \| \ X_{0 \ eq \ p.u}\right)\right\}$$

where, $\qquad X_{0 \ eq \ p.u} = X_{R0 \ p.u} + 3 \ X_{n \ pu} + 3 \ X_{f \ p.u.}$

Q1. A 110 MVA, 11kV synchronous generator is connected through a 5-cycle circuit breaker to a transformer. It has direct axis sub-transient reactances, $X_d'' = 19\%$, transient reactance, $X_d' = 26 \%$ and synchronous reactance, $X_d = 130\%$. It is operating at no load and rated voltage when a three-phase short circuit occurs between the breaker and the transformer. Determine the maximum possible dc component of the short-circuit current in the breaker?

Solution:

Given data:

Sub transient reactance, $\mathbf{X_d''}$=19%

Transient reactance, $\mathbf{X_d'}$ =26%

Synchronous reactance, $\mathbf{X_d}$=130%

$$\text{Per unit fault current, } I_{f\ p.u} = \frac{E_{R1\ p.u}}{\{X_{1\ eq\ p.u}\}} = \frac{1.0}{0.19} = 5.26 \text{ p. u.}$$

$$\text{Base current, } I_b = \frac{(VA)b}{\sqrt{3}V_b} = \frac{110 \times 10^6}{\sqrt{3} \times 11 \times 10^3} = 577.338 \text{ KA}$$

Fault current in amp, $I_{f\ amp} = I_{f\ pu} \times I_{base\ amp} = 5.26 \times 577.388 = 3036.8$ KA

DC component $= \sqrt{2}I_f = 42.96$ KA

Q2. A 100 MVA, 18 kV synchronous generator having reactances $X_d'' = 19\%$, $X_d' = 26\%$ and $X_d = 130\%$ is connected to a transformer through a 5-cycle circuit breaker. It is operating at no load and rated voltage when a three-phase short circuit occurs between the breaker and the transformer.

(a) What is the sustained short-circuit current in the breaker?

(b) What is the initial symmetrical rms current in the breaker?

Solution:

Given data:

Sub transient reactance, X_d''=19%

Transient reactance, X_d'=26%

Synchronous reactance, X_d=130%

(a) Per unit fault current

$$I_{f\ p.u} = \frac{E_{R1\ p.u}}{X_{d\ p.u}''} = \frac{1.0}{1.30} = 0.769 \text{ p. u.}$$

$$\text{Base current, } I_b = \frac{(VA)b}{\sqrt{3}V_b} = \frac{100 \times 10^6}{\sqrt{3} \times 18 \times 10^3} = 3208.1 \text{ A}$$

Fault current in amp, $I_{f\ amp} = I_{f\ pu} \times I_{base\ amp} = 0.769 \times 3208.1 = 2467$ A

(b) Per unit fault current

$$I_{f\ p.u} = \frac{E_{R1\ p.u}}{X_{d\ p.u}'} = \frac{1.0}{0.19} = 5.26 \text{ p. u.}$$

$$\text{Base current, } I_b = \frac{(VA)b}{\sqrt{3}V_b} = \frac{100 \times 10^6}{\sqrt{3} \times 18 \times 10^3} = 3208.1 \text{ A}$$

Fault current in amp, $I_{f\ amp} = I_{f\ pu} \times I_{base\ amp} = 5.26 \times 3208.1 = 16882$ A

Q3. A line to line fault and L-G fault has occurred at the load side of the transmission line. Determine the neutral to ground connection for which the intensity of L-L fault and L-L-G fault are equal.

Solution:

$$\text{LL fault, } I_{R1\ p.u.} = \frac{E_{R1\ p.u.}}{j\{X_{1\ eq\ p.u.} + X_{2\ eq\ p.u.}\}}$$

$$\text{LLG fault, } I_{R1\ p.u.} = \frac{E_{R1\ p.u.}}{j\{X_{1\ eq\ p.u.} + (X_{2\ eq\ p.u.} \parallel X_{0\ eq\ p.u.})\}}$$

For equal severity of faults,

$$\Rightarrow X_{1\ eq\ p.u.} + X_{2\ eq\ p.u.} = X_{1\ eq\ p.u.} + (X_{2\ eq\ p.u.} \parallel X_{0\ eq\ p.u.})$$

$$\Rightarrow X_{2\ eq\ p.u.} = (X_{2\ eq\ p.u.} \parallel X_{0\ eq\ p.u.})$$

$$\Rightarrow X_{0\ eq\ p.u.} = \infty$$

$$\Rightarrow X_{0\ eq\ p.u.} = X_{R0\ p.u.} + 3X_{n\ p.u.} = \infty$$

$$\Rightarrow X_{n\ p.u.} = \infty \text{ (isolated neutral)}$$

Therfore, the severity of LL fault and LLG faults are equal for isolated neutral.

Q4. A 25 MVA, 11 kV Star-connected synchronous generator has its neutral point solidly grounded. The generator is operating at no load at rated voltage. Their sequence reactances are $X_1 = X_2 = 0.20$ p. u. and $X_0 = 0.08$ p. u.

(a) What is the symmetrical sub-transient line current for the single line to ground fault?

(b) What is the symmetrical sub-transient line current for double line fault?

Solution:

Given data:

Positive sequence reactance, $X_1 = 0.2$ p.u.

Negative sequence reactance, $X_2 = 0.2$ p.u.

Zero sequence reactance, $X_0 = 0.2$ p.u.

The symmetrical sub-transient line current for the single line to ground fault is,

$$I_{f\ p.u} = \frac{3E_{R1\ p.u}}{j\{X_{0\ eq\ p.u} + X_{1\ eq\ p.u} + X_{2\ eq\ p.u}\}} = \frac{3 \times 1.0}{j\{0.2 + 0.2 + 0.08\}}$$

$$= -j2.08 \text{p. u.}$$

The symmetrical sub-transient line current for the double line fault is,

$$I_{f\ p.u} = \frac{\left|\sqrt{3}\right| E_{R1\ p.u}}{j\{X_{1eq\ p.u} + X_{2eq\ p.u}\}} = \frac{\left|\sqrt{3}\right| E_{R1\ p.u}}{j\{0.2 + 0.2\}} = -j4.33 \text{p. u}$$

Q5. The Z_{Bus} of a system is

$$Z_{Bus} = \begin{bmatrix} 0.1 & 0.1 & 0.1 \\ 0.1 & 0.2 & 0.1 \\ 0.1 & 0.1 & 0.3 \end{bmatrix} \text{p.u.}$$

If a 3-phase fault occurs at Bus-2, then determine the p.u. fault current in each phase?

Solution:

Given data:

A three phase fault occurs at bus-2.

Per unit fault current at bus2 is, $I_{f\ p.u} = \dfrac{E_{R1\ p.u}}{\{X_{22\ eq\ p.u}\}} = \dfrac{1.0}{j0.2} = 5\text{p.u.}$

Q6. The positive sequence bus impedance matrix of a 4-Bus power system network is given below:

$$Z_{1,\ Bus} = \begin{bmatrix} j0.724 & j0.620 & j0.656 & j0.644 \\ j0.620 & j0.738 & j0.642 & j0.660 \\ j0.656 & j0.642 & j0.0702 & j0.676 \\ j0.664 & j0.660 & j0.676 & j0.719 \end{bmatrix}$$

If a symmetrical fault occurs on Bus-2, then determine the positive sequence component of fault current?

Solution:

Given data:

A three phase fault occurs at bus-2.

Per unit positive sequence fault current at bus2 is, $I_{f\ p.u} = \dfrac{E_{R1\ p.u}}{\{X_{22\ eq\ p.u}\}} = \dfrac{1.0}{j0.738} = -j1.35\text{p.u}$

Q7. A 100 MVA, 20 kV generator has $X_d'' = X_1 = X_2 = 20\%$ and $X_0 = 5\%$. Its neutral is grounded through a reactance of 0.32 ohms. The generator is operating at rated voltage at no-load and is disconnected from the system when a single line-to-ground fault occurs at its terminals. Determine the sub-transient current in the fault phase?

Solution:

Given data:

Sub transient reactance, $X_d'' = X_1 = X_2 = 20\%$

Transient reactance, $X_0 = 5\%$

Neutral reactance $= 0.32\Omega$

Per unit neutral reactance, $X_{n\ p.u.} = X_\Omega \dfrac{MVA_{base}}{KV_{base}^2} = 0.32 \times \dfrac{100}{20^2} = 0.08 \text{ p.u.}$

For L-G fault, $I_{f\ p.u.} = I_{n\ p.u.} = 3\,I_{R1\ p.u.} = \dfrac{3E_{R1\ p.u}}{j\{X_{0\ eq\ p.u} + X_{1\ eq\ p.u} + X_{2\ eq\ p.u}\}}$

$$= \dfrac{3 \times 1.0}{j\{0.2 + 0.2 + 0.05 + 3 \times 0.08\}} = -j4.34\text{p.u.}$$

$$\text{Base current, } I_b = \frac{(VA)b}{\sqrt{3}V_b} = \frac{100 \times 10^6}{\sqrt{3} \times 20 \times 10^3} = 2886.1 \text{ A}$$

Fault current in amp, $I_{f\,amp} = I_{f\,pu} \times I_{base\,amp} = 4.34 \times 2886.1 = 12.52 \text{ KA}$

Q8. The SLG fault current in the faulted phase is 300A. Determine the zero-sequence current.

Solution:

Given data:

Fault current in the fault phase due to LG fault, I_f=300 A

Zero sequence current, $\mathbf{I_{R1\,amp.}} = \dfrac{I_{f\,amp.}}{3} = 100A$

Q9. For a 50 MVA, 11 kV, three phase synchronous generator, it is given that three phase fault current is 2000A and line to line fault current is 2600 A. The generator neutral is solidly grounded. Determine the per unit value of negative sequence reactance of the generator?

Solution:

Given data:

Line to line fault current = 2600 A

Three phase fault current = 2000 A

Base MVA = 50 MVA

Base current = 11 KV

For three phase fault, $I_{f,LLL\,amp.} = \dfrac{E_{R1\,\,volts}}{j\{X_{1\,eq\,\,ohms}\}}$

$$\Rightarrow 2000 = \frac{\dfrac{11 \times 1000}{\sqrt{3}}}{j\{X_{1\,eq\,\,ohms}\}}$$

$$\Rightarrow X_{1\,eq\,\,ohms} = 3.175\Omega$$

For line to line fault, $I_{f,LL\,amp.} = \dfrac{\sqrt{3}E_{R1\,\,volts}}{j\{X_{1\,eq\,\,ohms}+X_{2\,eq\,\,ohms}\}}$

$$\Rightarrow 2600 = \frac{\sqrt{3} \times \dfrac{11 \times 1000}{\sqrt{3}}}{j\{X_{1\,eq\,\,ohms} + X_{2\,eq\,\,ohms}\}}$$

$$X_{1\,eq\,\,ohms} + X_{2\,eq\,\,ohms} = 4.23 \ \Omega$$

$$3.175 + X_{2\,eq\,\,ohms} = 4.23\Omega$$

$$X_{2\,eq\,\,ohms} = 1.055\Omega$$

$$\text{Base reactance, } X_{b,ohms} = \frac{KV_b^2}{MVA_b} = \frac{11^2}{50} = 2.42 \text{ ohms}$$

$$X_{2\ eq\ p.u.} = \frac{X_{2\ eq\ ohms}}{X_{b\ ohms}} = \frac{1.056}{2.42} = 0.436 \text{ p. u.}$$

Q10. A 100 MVA, 20 kV, three phase synchronous generator was subjected to different types of faults. If the fault current for single line to ground (LG) fault is 4200 A and that is of 2600 A for line to line (LL) fault. What is the per unit value of zero sequence reactance of the generator?

Solution:

Given data:

Line to ground fault current = 4200 A

Line to line fault current = 2600 A

Base MVA = 100 MVA

Base current = 20 KV

For line to ground fault, $I_{f,LG\ amp.} = \dfrac{3\ E_{R1\ volts}}{j\{X_{1\ eq\ ohms} + X_{2\ eq\ ohms} + X_{0\ eq\ ohms}\}}$

$$4200 = \frac{3 \times \dfrac{11 \times 1000}{\sqrt{3}}}{j\{X_{1\ eq\ ohms} + X_{2\ eq\ ohms} + X_{0\ eq\ ohms}\}}$$

$$X_{1\ eq\ ohms} + X_{2\ eq\ ohms} + X_{0\ eq\ ohms} = 8.247\ \Omega$$

For line to line fault, $I_{f,LL\ amp.} = \dfrac{\sqrt{3}E_{R1\ volts}}{j\{X_{1\ eq\ ohms} + X_{2\ eq\ ohms}\}}$

$$2600 = \frac{\sqrt{3} \times \dfrac{20 \times 1000}{\sqrt{3}}}{j\{X_{1\ eq\ ohms} + X_{2\ eq\ ohms}\}}$$

$$X_{1\ eq\ ohms} + X_{2\ eq\ ohms} = 7.69\ \Omega$$

As $X_{1\ eq\ ohms} + X_{2\ eq\ ohms} + X_{0\ eq\ ohms} = 8.247\ \Omega$

$$7.69 + X_{0\ eq\ ohms} = 8.247\Omega$$
$$X_{2\ eq\ ohms} = 0.5547\Omega$$

Base reactance, $X_{b,ohms} = \dfrac{KV_b^2}{MVA_b} = \dfrac{20^2}{50} = 4 \text{ ohms}$

$$X_{2\ eq\ p.u.} = \frac{X_{2\ eq\ ohms}}{X_{b\ ohms}} = \frac{0.5547}{4} = 0.1386 \text{ p. u.}$$

Q11. The neutral of a three-phase, 15 MVA, 11 kV alternator is solidly grounded. The positive, negative and zero sequence impedance are j1.5 ohms, j0.8 ohms and j0.3 ohms respectively. A line to ground fault occurs on the phase a. Calculate the magnitude of fault current in that phase.

Solution:

Given data:

Positive sequence reactance, $X_1 = j1.5$ ohms

Negative sequence reactance, $X_2 = j0.8$ ohms

Zero sequence reactance, $X_0 = j\,0.3$ ohms

Base MVA = 15 MVA

Base KV = 11 KV

For line to ground fault, $I_{f,LG}$ amp.

$$= \frac{3\,E_{R1}\ \text{volts}}{j\{X_{1\ eq\ ohms} + X_{2\ eq\ ohms} + X_{0\ eq\ ohms}\}}$$

$$= \frac{3 \times \dfrac{11 \times 1000}{\sqrt{3}}}{j\{1.5 + 0.8 + 0.3\}} = 7.327\ \text{KA}$$

Q12. A 3-phase, 11kV, 12 MVA alternator has sequence reactance $X_1 = X_2 = 0.3$ p.u   and $X_0 = 0.3$ p.u..   If the generator operates at no load, then determine the ratio of fault currents for line-to-ground fault to that when all the 3-phases are short-circuited.

Solution:

Given data:

Positive sequence reactance, $X_1 = 0.3$ p.u.

Negative sequence reactance, $X_2 = 0.3$ p.u.

Zero sequence reactance, $X_0 = 0.3$ p.u.

Base MVA = 12 MVA

Base KV = 11 KV

For line to ground fault, $I_{f,LG}$ amp. $= \dfrac{3\,E_{R1}\ \text{volts}}{j\{X_{1\ eq\ ohms} + X_{2\ eq\ ohms} + X_{0\ eq\ ohms}\}}$ .....(9.16)

For three phase fault, $I_{f,LLL}$ amp. $= \dfrac{E_{R1}\ \text{volts}}{j\{X_{1\ eq\ ohms}\}} = \dfrac{3\,E_{R1}\ \text{volts}}{j\{3X_{1\ eq\ ohms}\}}$ .....(9.17)

From 9.16 and 9.17,

$$\frac{I_{f,LG}}{I_{f,LLL}} = \frac{3X_{1\ eq\ ohms}}{\{X_{1\ eq\ ohms} + X_{2\ eq\ ohms} + X_{0\ eq\ ohms}\}}$$

$$= \frac{3 \times 0.3}{\{0.3 + 0.3 + 0.1\}} = 1.285$$

Q13. The neutral of three phase Y-connected alternator is solidly grounded. A single line-to-ground fault occurs on the phase 'a' and the current in this phase is found to be 100 A. Determine the positive sequence component of current in phase b?

Solution:

Given data:

Line to ground fault current $= 100$ A

For L-G fault, $I_{f\ amp.} = I_{a\ amp.} = 3\ I_{a1\ amp.} = 100A$

$$\Rightarrow I_{a1\ amp.} = 33.33\ A$$

$$\Rightarrow I_{b1\ amp.} = 33.33\angle 240°A$$

$$\Rightarrow I_{c1\ amp.} = 33.33\angle 120°\ A$$

Q14. A single line to ground fault occurs on an unloaded generator in phase a positive, negative and zero sequence reactances of the generator are 0.25 p.u., 0.25 p.u. and 0.15 p.u. The generator neutral is grounded through a reactance of 0.05 p.u. The prefault generator terminal voltage is 1.0 p.u. Determine the positive sequence sub-transient current in p.u.

Solution:

Given data:

Positive sequence reactance, $X_1 = 0.25$ p.u

Negative sequence reactance, $X_2 = 0.25$ p.u

Zero sequence reactance, $X_0 = 0.15$ p.u

Neutral reactance $= 0.05$p.u

Pre fault terminal voltage of generator $= 1$ p.u

For L-G fault, $I_{R1\ p.u.} = \dfrac{E_{R1\ p.u}}{j\{X_{0\ eq\ p.u} + X_{1\ eq\ p.u} + X_{2\ eq\ p.u}\}}$

$$= \dfrac{1.0}{j\{0.25 + 0.25 + 0.15 + 3 \times 0.05\}} = -j1.25 \text{p. u.}$$

Q15. A 20 MVA, 6.6 kV, 3 phase alternator is connected to a 3-phase transmission line. The per unit positive sequence, negative sequence and zero-sequence impedance of the alternator are $j0.1, j0.1$ and $j0.04$ respectively. The neutral of the alternator is connected to ground through an inductive reactor of $j0.05$p.u. The per unit positive, negative and zero sequence impedances of the transmission line are $j0.1, j0.1$ and $j0.3$ respectively. All per unit values are based on the machine ratings. A solid ground fault occurs at one phase of the far end of the transmission line. Calculate the voltage of the alternator neutral with respect to the ground during the fault?

Solution:

Given data:

Alternator:

Positive sequence reactance, $X_1 = j\ 0.1$ p.u

Negative sequence reactance, $X_2 = j\ 0.1$ p.u

Zero sequence reactance, $X_0 = j\ 0.04$ p.u

Base MVA $= 20$ MVA

Base KV = 6.6 KV

Transmission line:

Positive sequence reactance, $X_1 = j\,0.1$ p.u

Negative sequence reactance, $X_2 = j\,0.1$ p.u

Zero sequence reactance, $X_0 = j\,0.3$ p.u

$$\text{For L-G fault, } I_{f\ p.u.} = \frac{3E_{R1\ p.u}}{j\{X_{0\ eq\ p.u} + X_{1\ eq\ p.u} + X_{2\ eq\ p.u}\}}$$

$$= \frac{3 \times 1.0}{j\{0.2 + 0.2 + 0.49\}} = -j3.37\,p.\,u.$$

Voltage to neutral in per unit, $V_{np.u.} = I_{np.u}X_{np.u} = 3.37 \times 0.05 = 0.168$p.u.

Voltage to neutral in volts $= V_{np.u}\,V_{base} = 0.168 \times \dfrac{6 \times 1000}{\sqrt{3}} = 642.2V$

Q16. At a 220 kV substation of a power system, it is given that the three phase fault level is 4000 MVA and single-line to ground fault level is 5000 MVA. Neglecting the resistance and the shunt susceptances of the system, determine the positive sequence driving point reactance at the bus and the zero sequence driving point reactance at the bus.

Solution:

Given data:

Base KV = 220 KV

Three phase fault level = 4000 MVA

LG fault level = 5000 MVA

$$\text{For LLL fault, Short circuit current, } I_{sc} = \frac{V_{base}}{j\{X_{1\ eq\ p.u}\}}$$

$$\text{Short circuit MVA, MVA}_{SC} = 3V_b\,I_{sc} = \frac{3V_b\,V_b}{j\{X_{1\ eq\ p.u}\}} = 4000$$

$$\text{Positive sequence driving point reactance, } X_{1eq\ p.u} = \frac{3\,V_b^2}{4000} = \frac{3 \times \left(\frac{200}{\sqrt{3}}\right)^2}{4000} = 12.1\Omega$$

Let

Positive sequence driving point reactance, $X_{1eq\ p.u} =$

Negative sequence driving point reactance, $X_{2eq\ p.u} = 12.1\Omega$

$$\text{For LG fault, Short circuit current, } I_{sc} = \frac{3V_{base}}{j\{X_{1\ eq} + X_{2\ eq} + X_{0\ eq}\}}$$

$$\text{Short circuit MVA, MVA}_{SC} = 3V_b\,I_{sc} = \frac{3V_b\,V_b}{j\{X_{1\ eq} + X_{2\ eq} + X_{0\ eq}\}} = 5000$$

$$\Rightarrow X_{1\ eq} + X_{2\ eq} + X_{0\ eq} = \frac{9\,V_b^2}{5000} = \frac{9 \times \left(\frac{200}{\sqrt{3}}\right)^2}{5000} = 29\Omega$$

$$12.1 + 12.1 + X_{0\ eq} = 29\,\Omega$$

$$X_{0\ eq} = 4.8\ \Omega$$

Q17. When a 50MVA, 11kV, 3-phase generator is subjected to a 3-phase fault, the fault current is-j5 p.u. (per unit). When it is subjected to a line-to-line fault, the positive sequence current is $-j4$ p.u. Determine the positive negative sequence reactance.

Solution:

Given data:

Three phase fault current $= -j\,5$ p.u.

Line to line fault current $= -j4$ p.u.

Base MVA = 50 MVA

Base KV = 11KV

For three phase fault, $\mathbf{I_{f\ p.u.}} = \dfrac{E_{R1\ p.u}}{j\{X_{1\ eq\ p.u}\}}$

$$-j5 = \dfrac{1.0}{j\{X_{1eq\ p.u}\}}$$

$$X_{1eq\ p.u} = j0.2\ p.\,u$$

For line to line fault, $I_{R1\ p.u} = \dfrac{E_{R1\ p.u}}{j\{X_{1eq\ p.u}+X_{2eq\ p.u}\}}$

$$-j4 = \dfrac{1.0}{j\{X_{1eq\ p.u} + X_{2eq\ p.u}\}}$$

$$X_{1eq\ p.u} + X_{2eq\ p.u} = j0.25\ p.\,u.$$

$$X_{2eq\ p.u} = j0.05\ p.\,u.$$

Q18. Four alternators, each rated at 5 MVA, 11kV with 20% reactance are working in parallel. Determine the short-circuit level at bus bars.

Solution:

Given data:

Reactance of alternator, $X_{d\ p.u}^{"} = 20\%$

Equivalent reactance of alternator, $X_{d\ eq\ p.u}^{"} = 5\%$

Base MVA = 5 MVA

Base KV = 11 KV

$$MVA_{SC} = \dfrac{MVA_b}{X_{d\,eq,p.u}^{"}} = \dfrac{5}{0.05} = 100MVA$$

Q19. The following currents are recorded in a power system under a fault condition $IR_1 = j1.653$ p.u., $IR_2 = -j0.5$ p.u., $IR_0 = -j1.153$ p.u. Identify the fault that has occurred.

Solution:

Given data:

For LLG fault, $I_{R0\ p.u} + I_{R1\ p.u} + I_{R2\ p.u} = 0$

$$\Rightarrow j1.153 - j1.653 - j0.5 = 0$$

Therefore the data refers to LLG fault.

Q20. A 20 MVA, 33kV, 3-phase alternator is subjected to different types of short circuits and the following are the values of the fault current.

3-phase short circuit – 319 A

Single Line to Ground – 659 A

Line to Line – 435 A

Determine positive, negative and zero sequence reactances in p.u.

Solution:

Given data:

Base MVA= 20 MVA

Base KV= 33 KV

$$\text{Base current, } I_b = \frac{(VA)b}{\sqrt{3}V_b} = \frac{20 \times 10^6}{\sqrt{3} \times 33 \times 10^3} = 350 \text{ A}$$

For three phase fault,

$$\text{Per unit fault current}, I_{f\,p.u.} = \frac{I_{f,actual}}{I_{b,amp}} = \frac{319}{350} = 0.911 \text{ p. u.}$$

$$I_{f,LLL\ p.u} = \frac{E_{R1\ p.u.}}{j\{X_{1\ eq\ p.u.}\}}$$

$$0.911 = \frac{1}{j\{X_{1\ eq\ p.u.}\}}$$

$$X_{1\ eq\ p.u.} = 1.09 \text{ p. u.}$$

For Line to Line fault,

$$\text{Per unit fault current}, I_{f\,p.u.} = \frac{I_{f,\ actual}}{I_{b,\ amp}} = \frac{435}{350} = 1.242 \text{ p. u.}$$

$$I_{f,LL\ p.u.} = \frac{\sqrt{3}E_{R1\ p.u.}}{j\{X_{1\ eq\ p.u.} + X_{2\ eq\ p.u}\}}$$

$$1.242 = \frac{\sqrt{3} \times 1.0}{j\{X_{1\ eq\ p.u.} + X_{2\ eq\ p.u}\}}$$

$$X_{1\ eq\ p.u.} + X_{2\ eq\ p.u} = 1.393 \text{ p. u.}$$

$$X_{2\ eq\ p.u} = 0.3 \text{p. u.}$$

For line to ground fault,

$$\text{Per unit fault current}, I_{f\,p.u.} = \frac{I_{f,actual}}{I_{b,amp}} = \frac{659}{350} = 1.882 \text{ p. u.}$$

$$I_{f,LG\ p.u.} = \frac{3\ E_{R1\ p.u.}}{j\{X_{1\ eq\ p.u.} + X_{2\ eq\ p.u} + X_{0\ eq\ p.u.}\}}$$

$$1.882 = \frac{3\ \times 1.0}{j\{1.09 + 0.3 + X_{0\ eq\ p.u.}\}}$$

$$X_{0\ eq\ p.u.} = 0.203 \text{ p. u.}$$

Q21. A 50 MVA, 11 KV synchronous generator was subjected to different types of faults. The fault currents are as follows:

L − G Fault = 4200 A

L − L Fault = 2600 A

L − L − L Fault = 2000 A

The generator neutral is solidly grounded. Determine the p.u values of 3-sequence reactances of generator.

Solution:

$$I_{f,LG\ p.u.} = \frac{3\ E_{R1\ p.u.}}{j\{X_{1\ eq\ p.u.} + X_{2\ eq\ p.u} + X_{3\ eq\ p.u.}\}}$$

$$\Rightarrow 4200 = \frac{3 \times \dfrac{11 \times 10^3}{\sqrt{3}}}{j\{3.175 + 1.056 + X_{0\ eq\ ohms}\}}$$

$$X_{0,eq\ ohms} = 0.305 \text{ ohms}$$

$$I_{F,LLL\ p.u.} = \frac{E_{R1\ p.u.}}{j\{X_{1\ eq\ p.u.}\}}$$

$$2000 = \frac{\dfrac{11 * 10^3}{\sqrt{3}}}{j\{X_{1\ eq\ ohms}\}}$$

$$X_{1\ eq\ ohms} = 3.175 \text{ ohms}$$

$$I_{F,LL\ p.u.} = \frac{\sqrt{3}E_{R1\ p.u.}}{j\{X_{1\ eq\ p.u.} + X_{2\ eq\ p.u}\}}$$

$$j\{3.175 + X_{2\ eq\ ohms}\} = 4.2307$$

$$X_{2\ eq\ ohms} = 1.058 \text{ ohms}$$

$$X_{b,ohms} = \frac{KVb^2}{MVAb} = \frac{11^2}{50} = 2.42 \text{ ohms}$$

Selecting alternator ratings as a base value.

The equivalent p.u. positive sequence reactance,

$$X_{1\,eq\ p.u.} = \frac{X_{1\,eq\ ohms}}{X_{b\ ohms}}$$

$$X_{1\,eq\ p.u.} = \frac{3.175}{2.42} = 1.312\ p.u.$$

The equivalent p.u. negative sequence reactance,

$$X_{2\,eq\ p.u.} = \frac{X_{2\,eq\ ohms}}{X_{b\ ohms}} = \frac{1.056}{2.42} = 0.436\ p.u.$$

The equivalent p.u. zero sequence reactance,

$$X_{0\,eq\ p.u.} = \frac{X_{0\,eq\ ohms}}{X_{b\ ohms}} = \frac{0.305}{2.42} = 0.126\ p.u.$$

Q22. A 50 MVA, $3-\emptyset$ alternator having its neutral solidly grounded is operating at no load, its voltage being 16 KV between lines. It has a reactance of positive sequence currents 5 ohms, the reactances to negative and zero sequence currents are 70% and 55% of positive sequence value respectively.

For a double line to ground fault, determine

1. Currents in fault lines.

2. Current flowing through ground.

Solution:

Given data:

Positive sequence reactance in ohms $\Longrightarrow X_{1\,eq\ ohms} = 5$ ohms

Negative sequence reactance in ohms $\Longrightarrow X_{2\,eq\ ohms} = 70\%$ of $X_{1\,eq\ ohms} = 3.5$ ohms

The zero sequence reactance,

$$X_{0\,eq\ p.u.} = 55\%\ of\ X_{1\,eq\ ohms} = 0.55 * 5 = 2.75\ ohms$$

Positive sequence current in Amperes:

$$I_{R1\,amp} = \frac{E_{R1\ volts}}{j\{X_{1\,eq\ ohms} + (X_{2\,eq}\ \|\ X_{0\,eq}\ )\}} = -j\,1.412\ KA = |1.41|KA$$

Zero sequence current,

$$I_{R0\,amp} = \frac{-I_{R1\,amp} \cdot X_{2\,eq\ ohms}}{X_{0\,eq\ ohms} + X_{2\,eq\ ohms}}$$

$$= \frac{-(1.41*10^3*3.5)}{2.75+3.5} = -789.6\ A = |789.6|A$$

Negative sequence current,

$$I_{R2\ amp} = \frac{-I_{R1\ amp} \cdot X_{0\ eq\ ohms}}{2.75 + 3.5}$$

$$= \frac{-(1.41 \times 10^3 \times 2.75)}{2.75 + 3.5} = -621.28 = |621.28|\ A$$

Q23. The current in 3ϕ supply are $I_a = 12 + j\,24$ A; $I_b = 16 - j\,2$ A; $I_c = -4 - j\,6$ A. The phase sequence is ABC. Calculate the sequence components of currents.

Solution:

$$I_{a0} = \frac{1}{3}(I_a + I_b + I_c)$$

$$= \frac{1}{3}\left(12 + j24 + (16.124\angle - 7.125^0)\right) + (7.211\angle - 123.69^0)$$

$$= 8 + j5.33\ A = 9.612\angle 33.67^0\ A$$

$$I_{a1} = \frac{1}{3}(I_a + KI_b + K^2 I_c)$$

$$= \frac{1}{3}\left(12 + j24 + (1\angle 120^0)(16.124\angle - 7.125^0)\right) + (1\angle 240^0)(7.211\angle - 123.69^0)$$

$$= \frac{1}{3}\{12 + j24 - 6.267 + 14.855j - 3.196 + 6.464j\} = 0.845 + 15.106j$$

$$= 15.129\angle 86.79^0\ A$$

$$I_{a2} = I_{a1} = \frac{1}{3}(I_a + K^2 I_b + KI_c)$$

$$= \frac{1}{3}\left(12 + j24 + (1\angle 240^0)(16.124\angle - 7.125^0)\right) + (1\angle 120^0)(7.211\angle - 123.69^0)$$

$$= \frac{1}{3}(12 + j24 - 9.731 - 12.855j + 7.196 - 0.464j)$$

$$= 3.155 + 3.56j = 4.756\angle 48.45^0\ A$$

OBJECTIVE QUESTIONS

1. _______________ is characterized by increase in voltage and frequency.

2. _______________ is characterized by increase in current.

3. The reactance offered by fault immediately after occurrence of fault is known as_______________.

4. _______________ do not flow for faults not involved with ground.

5. _______________ flows for unsymmetrical faults.

6. _______________ flows for every fault.

7. Neutral current and fault current are same for _______________

8. Neutral current is _______________ for faults not involved with ground.

9. Per unit sequence current components are equal for ______________

10. Per unit sequence voltage components are equal for ______________

11. Per unit positive and negative sequence current components are equal for ______________

12. Sequence networks are connected in ______________ for L-G fault.

13. Positive and negative sequence networks are connected in ______________ for L-L fault.

14. Per unit neutral current for faults involved with ground is______________.

15. The reference to the flow of zero sequence currents is______________.

16. The reference to the flow of positive and negative sequence currents is______________.

17. Positive sequence network is connected in series with parallel combination of zero and negative sequence networks for______________.

18. If reactance neutral and reactance fault are considered for faults______________ then X_{0eq} = $X_{R0}+3X_n+3X_f$.

19. If reactance neutral and reactance fault are considered for faults ______________ then sub transient reactance is $X_{1eq}+X_{0eq}+X_f$.

20. The Thevenin's equivalent networks representing different faults are interconnected based on______________.

21. Unsymmetry in the power system network is indicated by the presence of______________.

22. Symmetrical components are used to analyze an ______________ power system network.

23. Numbers of boundary conditions are equal to number of ______________irrespective of fault.

24. ______________ is the most commonly occurring fault.

25. ______________ is the most severe fault with fault on alternator terminals.

26. The ______________ sequence currents are recorded in a power system under an unbalanced condition.

27. A power system network is characterized as follows.

$$I_{a1} = -j\,1.653\ \text{p. u}$$
$$I_{a2} = j\,0.5\ \text{p. u}$$
$$I_{a0} = j\,1.153\ \text{p. u}$$

The fault that has occurred is <u>LLG fault.</u>

UNSOLVED PROBLEMS

Q1: The new generator having $E_g = 1.4\angle30^0$ p. u. [equivalent to $1.212+j\,0.70$ p.u.] and synchronous reactance X_s of 1.0 p.u. on the system base, is to be connected to a bus having voltage V, in the existing power system. This existing power system can be represented by Thevenin's voltage $E_{th} = 0.9\angle0^0$ p.u. in series with Thevenin's impedance $Z_{th} = 0.25\angle90^0$ p. u. Determine the magnitude of the voltage V_t of the system in p.u.

Ans: $0.973\angle8.27^0$ p. u.

Q2: A 25 MVA, 13.2 KV alternator with solidly grounded neutral has a sub-transient reactance of 0.25 p.u. The negative and zero sequence reactances are 0.35 p.u. and 0.1 p.u. respectively. A single line to ground fault occurs at the terminals of an unloaded alternator. Determine the fault current and line to line voltages.

Q3: Three 6.6 KV, 10 MVA, three phase generators are connected to common set of bus bars. Each machine has a reactance to positive sequence currents of 20%.Thereactances to negative and zero sequence currents are 75% and 30% of the positive sequence value. If an earth fault occurs on one bus bar, determine the fault current:

(a) If all alternators are solidly grounded.

(b) If only one of the alternator neutrals are solidly earthed and others are isolated.

(c) If only one of the alternator neutrals are earthed through a reactance of 0.3Ω and others are isolated.

Ans: (a) 19202 A   (b) 14854 A (c) 9657.8 A

Q4: A three phase AC generator developing a constant internal EMF is subjected to different types of faults. The values of fault currents for the different types are as follows:

1. Three phase, 1000A

2. Line to line, 1400A

3. Line to ground, 2200A

If the positive sequence voltage to neutral is 2KV (pre fault), determine the sequence impedances.

Ans: 2Ω, 0.475Ω, 0.25Ω

Q5: Assume a system with sustained supply voltage of 2.2 KV from line to neutral and with the total sequence impedances $Z_1= j10\ \Omega$, $Z_2=j8\ \Omega$, $Z_0=10+j12\ \Omega$. Find the currents in the faulted phases for the following faults.

1. Single line to ground fault on phase 'a'.

2. Line to Line fault on phases 'b' and 'c'

3. Line to Line fault involving fault on phases 'b' and 'c'

4. Symmetrical three phase fault.

Assume E on phase 'a' is $2200\angle0^0$KV

Ans: $214\angle36^0A, 211\angle180^0A, 161.6\angle75.1A, 220A$

Q6: A 50Hz, 13.2KV, 15 MVA alternator has, $X_d^{''} = X_2 =20\%$ and $X_0 =8\%$ and its neutral is solidly grounded through a reactor of 0.5 ohms. Determine the initial symmetrical RMS current in the line 'c' when a line to line fault involved with ground occurs on phase 'b' and 'c' and the generator voltage is 12 KV before fault occurs.

Ans: 2909A and 2959A

Q7: Four 50 MVA generators of 15% reactance each are connected through 35 MVA reactors each of 10% reactance to a common bus bar. The feeders are each connected to the junction of each alternator and its reactor. Determine the rating of each feeder circuit breaker.

Ans: 541.3 MVA

Q8: Two generating stations having short circuit capacities 1500 MVA and 1000 MVA operating at 11 KV are linked by an interconnected cable having a reactance of 0.6Ω per phase. Determine the short circuit capacity of each station.

Ans: 1667.9 MVA and 1177.8 MVA

Q9: Four bus bar sections have each a generator of 40 MVA, 10% reactance and a bus bar reactor of 8% reactance. Determine the maximum MVA fed into a fault on any bus bar section and also the maximum MVA if the number of similar bus bars in each section is very large.

Ans: 685.7 MVA and 900 MVA

Q10: Three 6.6 KV, 12 MVA, 3 phase alternators are connected to a common set of bus bars. The positive, negative and zero sequence impedances of each alternator are 15%, 12% and 4.5% respectively. If an earth fault occurs on one bus bar, determine the fault current with all the alternator neutrals solidly grounded.

Ans: 30KA

Q11: Two power stations P and Q having short circuit capacities of 800 and 600 MVA respectively operates at 11 kV each. Choose base MVA as 100 MVA and base kV is 11kV. If they are interconnected by a cable of 0.4 ohms reactance per phase, then determine the maximum short-circuit MVA.

Ans: 819.67 MVA

10 Symmetrical Fault Analysis

10.1 INTRODUCTION

In electrical power systems consisting of alternators, transformers, transmission lines and distribution circuits, most of the faults are liable to occur in the transmission circuits. When a system is subjected to a fault, viz. short circuit on one or more lines, the current flowing into the fault depends on the path met by the current and on the severity and nearness of the fault to the source of power. A fault with equal fault current flowing in all the phases is known as symmetrical fault.

Symmetrical faults are classified in to two types.

1. Line–line–line Fault (LLL)
2. Line–Line–Line–Ground Fault (LLLG)

10.2 LINE–LINE–LINE FAULT (LLL)

Circuit Representation

Consider an alternator operating under no load condition with solid neutral or solid grounding i.e. $X_n = 0$ as shown in figure 10.1.Assume solid fault i.e. $X_f = 0$. Assume LLL fault involves R, Y and B phases. Assume LLL fault involves all the three phases, LLL fault is a balanced fault and hence fault current is calculated on per phase basis.

Boundary Conditions

As number of boundary conditions for any fault is equal to number of sequence networks = 3, there exists three boundary conditions. The three boundary conditions are

(a) $I_{Rp.u.} = I_{Yp.u.} = I_{Bp.u.}$

(b) $I_R \lfloor 0^0 + I_Y \lfloor -120^0 + I_B \lfloor -240^0 = 0$

(c) $V_{Rp.u.} = V_{Yp.u.} = V_{Bp.u.}$

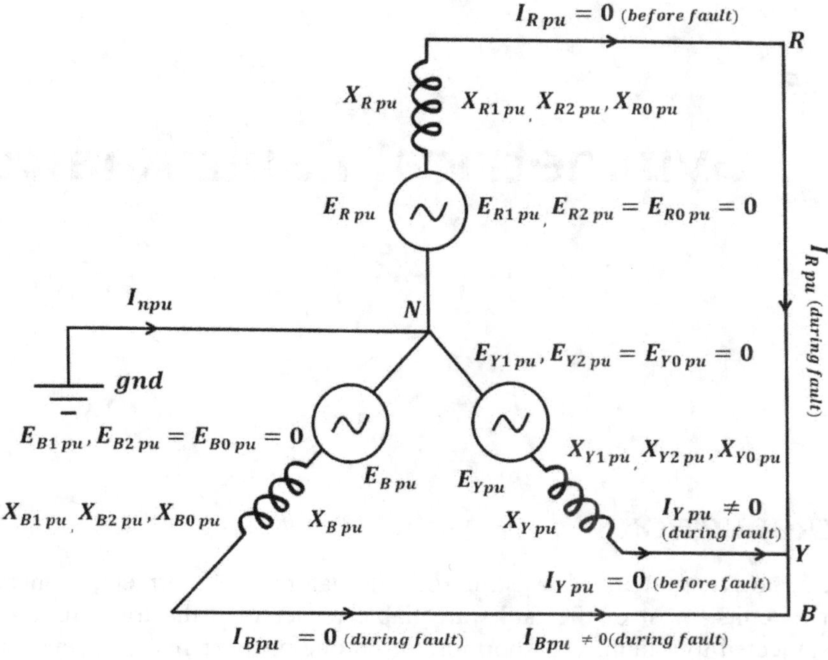

Figure 10.1 LLL fault or three phase fault on an unloaded synchronous generator

Per unit sequence voltage components of R–phase

Per unit zero sequence voltage of R–phase, $V_{R0p.u.} = \dfrac{1}{3}[V_{Rp.u.} + V_{Yp.u.} + V_{Bp.u.}]$

$$V_{R0p.u.} = \frac{1}{3}[V_{Rp.u.} + K^2 V_{Rp.u.} + K\,V_{Rp.u.}]$$

$$= \frac{V_{Rp.u.}}{2}[1 + K + K^2] = 0$$

Per unit positive sequence voltage of R–phase, $V_{R1p.u.} = \dfrac{1}{3}[V_{Rp.u.} + K\,V_{Yp.u.} + K^2 V_{Bp.u.}]$

$$= \frac{1}{3}[V_{Rp.u.} + K V_{Rp.u.} + K^2 V_{Rp.u.}] = 0$$

Per unit negative sequence voltage of R–phase, $V_{R2p.u.} = \dfrac{1}{3}[I_{Rp.u.} + K^2 V_{Yp.u.} + K\,V_{Bp.u.}]$

$$= \frac{1}{3}[V_{Rp.u.} + K^2 V_{Rp.u.} + K\,V_{Rp.u.}] = 0$$

Therefore, $V_{R0p.u.} = V_{R1p.u.} = V_{R2p.u.} = 0$ .....(10.1)

From equation (10.1) it can be concluded that the per unit sequence voltage components of R–phase are equal to zero for LLL fault

Per unit sequence current components of R–phase

As per unit zero sequence voltage of R–phase, $V_{R0p.u.} = 0$

i.e. $-j\,I_{R0p.u.}X_{R0eqp.u.} = 0$ and hence $I_{R0\,p.u.} = 0$

Therefore per unit zero sequence current is equal to zero and can be justified.

As LLL fault do not involve ground, per unit zero sequence current is equal to zero.

As per unit negative sequence voltage of R–phase, $V_{R2p.u.} = 0$

i.e. $-jI_{R2p.u.}\, X_{2eqp.u.} = 0$ and hence $I_{R2\,p.u.} = 0$

Therefore per unit negative sequence current is equal to zero and can be justified.

As LLL fault is a symmetrical fault, per unit negative sequence current is equal to zero.

As per unit positive sequence voltage of R–phase, $V_{R1p.u.} = 0$

i.e. $\qquad\qquad E_{R1p.u.} - j\,I_{R1p.u.}X_{1eqp.u.} = 0$

i.e. $\qquad\qquad I_{R1p.u.} = \dfrac{E_{R1\,p.u.}}{j(X_{1eq\,p.u.})}$ .....(10.2)

It can be observed that only positive sequence current flows due to LLL fault.

Per Unit Fault Current

As LLL fault is a balanced fault,

$$I_{fp.u.} = I_{Rp.u.} = \dfrac{E_{R1\,p.u.}}{j(X_{1eq\,p.u.})} + 0 + 0$$

$$I_{fp.u.} = \dfrac{E_{R1\,p.u.}}{j(X_{1eq\,p.u.})} \qquad .....(10.3)$$

Fault Current in Amps

Fault current in amperes, $I_f = I_{fp.u.}.I_b$

Per Unit Sub–transient Reactance

From equation (10.3) $X_{d\,p.u.}'' = j\,X_{1eq\,p.u.}$ .....(10.4)

Sequence Networks

The sequence network as shown in figure 10.2 consists of only positive sequence network with it's terminals short circuited.

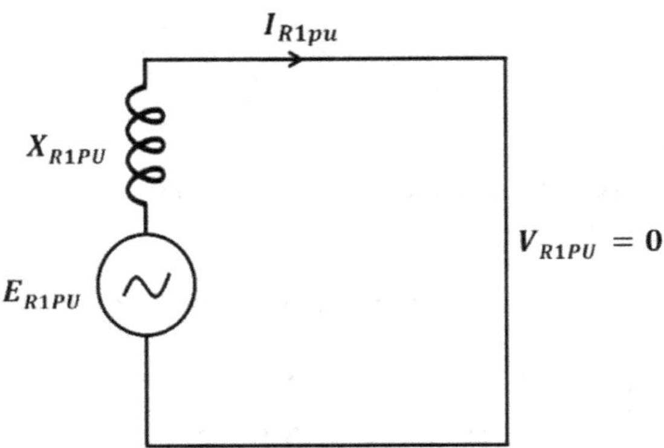

Figure 10.2 Sequence network for LLL fault

Per Unit Neutral Current

Case 1: For Balanced Condition

$$I_{np.u.} = I_{Rp.u.} + I_{Yp.u.} + I_{Bp.u.} = 0$$

Case 2: For Unbalanced condition

$$I_{np.u.} = 3\ I_{R0p.u.} = 3 \times 0 = 0$$

Per Unit Neutral Voltage

Case 1: For balanced condition

$$V_{np.u.} = I_{np.u.} \times X_{np.u.} = 0$$

Case 2: For unbalanced condition

$$V_{np.u.} = I_{np.u.} \times X_{np.u.} = 0$$

Note: For a LLLG fault, $I_{f\,p.u.} = I_{Rp.u.} + I_{Yp.u.} + I_{Bp.u.} = 0$.

10.3 REPRESENTATION OF POSITIVE X_{1EQ}, NEGATIVE X_{2EQ} AND ZERO SEQUENCE X_{0EQ} REACTANCES IN TERMS OF SELF-REACTANCE, X_S AND MUTUAL REACTANCE, X_M OF TRANSMISSION LINE

As most of the faults in power system network occurs on a transmission line and three phases of transmission line are in close proximity of each other, mutual reactance X_m is considered along with self-reactance, X_s and positive, negative and zero sequence reactances X_{1eq}, X_{2eq}, X_{0eq} are expressed in terms of self-reactance, X_s and mutual reactance $X_{m.}$ of transmission line.

Consider the transmission line as in figure 10.3

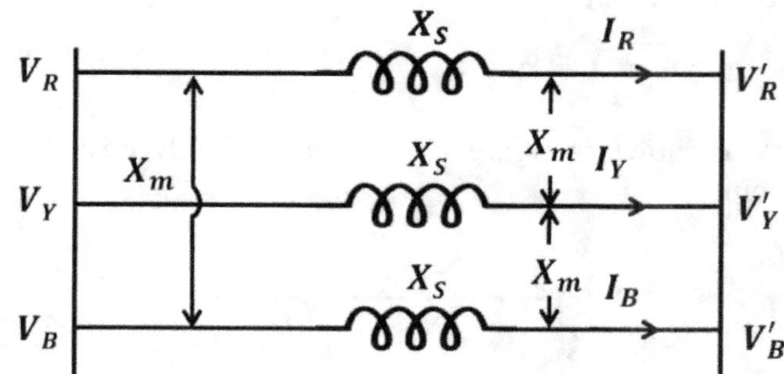

Figure 10.3 Transmission line with self and mutual reactances

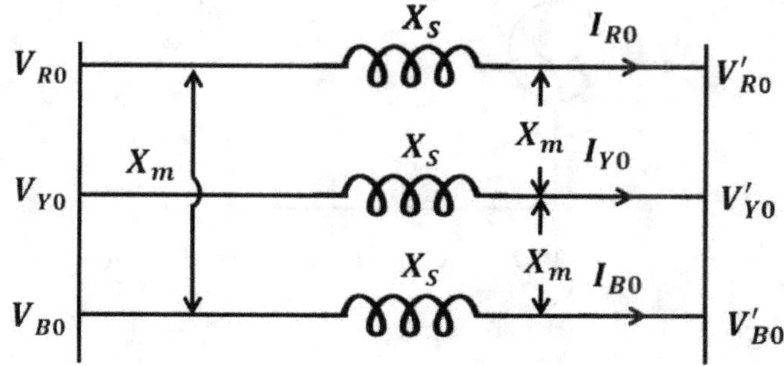

Figure 10.4 Equivalent circuit to determine zero sequence reactance, X_{0eq}

1. Zero sequence voltage, $V_{R0} = V_{R0}' + j\, I_{R0}X_s + j\, I_{Y0}X_m + jI_{B0}X_m$

$$= V_{R0}' + j\, I_{R0}\,[X_s + 2X_m]$$

As opposition to the flow of I_{R0} is X_{0eq}, zero sequence reactance, $X_{0eq} = X_s + 2X_m$

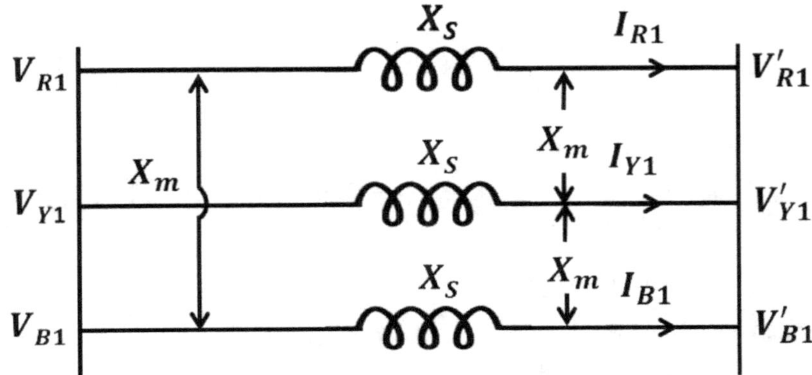

Figure 10.5 Equivalent circuit to determine positive sequence reactance, X_{1eq}

2. Positive sequence voltage, $V_{R1} = V_{R1}' + j\, I_{R1}X_s + j\, I_{y1}X_{mt} + jI_{B1}X_m$

$$= V_{R1}' + j\, I_{R1}\,[X_s + (K^2 + K)X_m]$$

$$= V_{R1}' + j\, I_{R1}\,[X_s - X_m]$$

As opposition to the flow of I_{R1} is X_{1eq}, positive sequence reactance, $X_{1eq} = X_s - X_m$

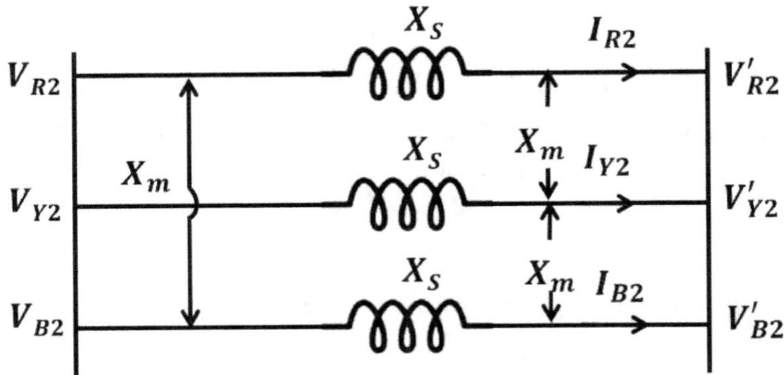

Figure 10.6 Equivalent circuit to determine negative sequence reactance, X_{2eq}

3. Negative sequence voltage, $V_{R2} = V_{R2}' + j\, I_{R2}X_s + jI_{Y2}X_m + jI_{B2}X_m$

$$= V_{R2}' + j\, I_{R2}\,[K + K^2]\,X_m + X_s$$

$$= V_{R2}' + j\, I_{R2}\,[X_s - X_m]$$

As opposition to the flow of I_{R2} is X_{2eq}, negative sequence reactance, $X_{2eq} = X_s - X_m.$

SOLVED QUESTIONS

1. The voltages across three–phase unbalanced load are $E_a = 300\underline{|-20^0}$ V, $E_b = 360\underline{|-90^0}$ V, $E_c = 500\underline{|-14^0}$ V. Determine the symmetrical components of voltages. Phase sequence is a, b, c.

Solution:

Zero sequence voltage,

$$E_{a0} = 1/3\ [E_a + E_b + E_c]$$
$$E_{a0} = 1/3\ [281.9 + j\ 102.6 + 0 + j360 - 383.022 - j321.39]$$
$$E_{a0} = -33.707 + j47.07$$
$$E_{a0} = 57.89\underline{|-125.60^0}\ \text{volts}$$

Positive sequence voltage, $E_{a1} = 1/3\ [E_a + kE_b + k^2E_c]$

$$E_{a1} = 1/3\ [281.9 + j102.6 + j360\ (-0.5 + j0.866)$$
$$(-383.022 - j321.39)\ (-0.5 - j0.866)]$$
$$E_{a1} = -38.248 + j137.56$$
$$= 142.77\underline{|-105.5^0}$$

Negative sequence voltage, $E_{a2} = 1/3\ [E_a + k^2E_b + kE_c]$

$$E_{a2} = 1/3\ [281.9 + j102.6 + j360\ (-0.5 - j0.866)\ (-383.022 - j321.39)$$
$$(-0.5 + j0.866)]$$
$$E_{a2} = 354.49 - j82.8$$
$$= 364.031\underline{|-13.14^0}\ \text{Volts}$$

2. The line currents in the three–phase supply are $I_a = 12 + j24$A, $I_b = 16 - j2$A, $I_c = -4 - j6$A. The phase sequence is abc. Calculate the symmetrical components.

Solution:

$$I_{a0} = 1/3\ [I_a + I_b + I_c]$$
$$= 1/3\ [(12 + j24) + (16 - j2) + (-4 - j6)]$$
$$= 1/3\ [24 - j16]$$
$$= 8 - j\ 5.33$$
$$= 9.61\underline{|-33.6^0}\ \text{A}$$

$$I_{a1} = 1/3\ [I_a + kI_b + k^2I_c]$$
$$= 1/3\ [(12 + j24) + (-0.5 + j0.866)\ (19 - j2) + (-0.5 - j0.866)\ (-4 - j6)]$$
$$= 0.8453 + j\ 15.106$$
$$= 15.129\underline{|86.7^0}\ \text{A}$$

$$I_{a2} = 1/3\ [I_a + k^2 I_b + kI_c]$$
$$= 1/3\ [(12 + j24) + (-0.5 - j0.866)\,(16 - j2) + (-0.5 + j0.866)\,(-4 - j6)]$$
$$= 3.15466 + j3.56$$
$$= 4.7566\ \underline{|48.45^0}\ A$$

3. A line–to–line fault and line–to–ground has occurred at the load side of transmission line. Determine the neutral–to–ground connection for which the intensity of line–to–line fault and double line–to–ground fault are equal.

Solution:

For a line to line fault, the per unit positive sequence current is

$$I_{R1p.u.} = \frac{E_{R1}p.u.}{j[X_{1eq}p.u. + X_{2eq}p.u.]}$$

For a line to line to ground fault, the per unit positive sequence current is

$$I_{R1p.u.} = \frac{E_{R1}p.u.}{X_{1eq}p.u. + X_{2eq}p.u. + X_{0eq}\,p.u.}$$

For same fault intensity,

$$X_{1eqp.u.} + X_{2eqp.u.} = X_{1eqp.u.} + (X_{2eqp.u.}\ // \ X_{0eqp.u.})$$
$$X_{2eqp.u.} = X_{2eqp.u.}\ // \ X_{0eqp.u.}$$

The above relation is satisfied with $X_{0eqp.u.} = 0$.

$$X_{0eqp.u.} = \infty$$

Therefore, $X_{Rop.u.} + 3X_{np.u.} = \infty$

The above relation is satisfied with $X_{np.u.} = \infty$

Therefore, line to line fault and line to line to ground fault are equally severe for isolated neutral.

4. A 50MVA, 11kV synchronous generator was subjected to different types of faults. The fault current are as follows:

   L–G Fault – 4000A

   L–L Fault – 2600A

   L–L–L Fault – 2000A

The generate neutral is solidly grounded. Determine the per unit values of three sequence reactance of generator.

Solution:

For a three-phase fault,

$$\text{Fault current in amperes, } I_{fLLL} = \frac{E_{R1}\ (V)}{[X_{1eq}\ (\Omega)\,]}$$

$$j[X1eq(\Omega)] = \frac{E_{R1}\ \text{in Volts}}{j[I_{f\,LLL}\ \text{in Amps}]}$$

$$= (11 \times 10^3 / \sqrt{3}) / 2000$$

Equivalent positive sequence reactance, $X_1 eq\ = 3.175\Omega$

For a line to line fault,

Fault current in amperes, $I_{fLL} = \dfrac{\sqrt{3}E_{R1}(V)}{[X_{1eq}+X_{2eq}]}$

$$X_{1eq} + X_{2eq} = (\sqrt{3} \times 11 \times 10^3 / \sqrt{3}) / 2600$$
$$X_{1eq} + X_{2eq} = 4.23$$
$$X_{2eq} = 4.23 - 3.175$$

Equivalent negative sequence reactance, $X_{2eq} = 1.056\,\Omega$

For a line to ground fault,

Fault current in amperes, $I_{fLG} = \dfrac{3E_{R1}(V)}{[X_{1eq}+X_{2eq}+X_{0eq}]}$

$$X_{1eq} + X_{2eq} + X_{0eq} = (3*11*10^3 / \sqrt{3}) / 4000$$

Equivalent negative sequence reactance, $X_{0eq} = 0.305\ \Omega$

Base reactance in ohms, $X_{b\Omega} = KV_b^2 / MVA_b = 112 / 50 = 2.42\,\Omega$

The equivalent per unit positive sequence reactance, $X_{1eqp.u.} = 3.175 / 2.42 = 1.312$ p.u.

Per unit negative sequence reactance, $X_{2eqp.u.} = 0.436$ p.u.

Per unit zero sequence reactance, $X_{0eqp.u.} = 0.1260$ p.u.

5. A synchronous generator rated three–phase, 11–KV, 100MVA has $X_1 = X_2 = j0.1$ p.u. and $X_0 = j0.04$ p.u. Determine the fault current and line–to–line voltages during the LG fault condition on the generator terminals. Assume that generator is solidly grounded and operating at no load and at rated voltage prior to the fault.

Solution:

Let the alternator ratings be the base values, $MVA_b = 100$ and $kV_b = 11$

Base current in amperes, $I_b = (100 \times 10^3 / (\sqrt{3} \times 11)$

$$= 5.248\ kA.$$

For a single line–to–ground current, perunit sequence currents are equal.

$$I_{R1}p.u. = I_{R2}p.u. = I_{R0}p.u. = \dfrac{E_{R1eq}}{X_{1eq}\ pu + X_{2eq}\ pu + X_{0eq}\ pu}$$

Per unit positive sequence reactance, $X_{1eq}p.u. = X_{R1}p.u. = X_1p.u. = j0.1$ p.u.

Per unit negative sequence reactance, $X_{2eq}p.u. = X_{R2}p.u. = X_2p.u. = j0.1$ p.u.

Per unit zero sequence reactance, $X_{0eq}p.u. = X_{R0}p.u. + 3X_np.u. = 0.04 + 3 \times 0 = 0.04$ p.u.

Sequence currents, $I_{R1}p.u. = I_{R2}p.u. = I_{R0}p.u. = \dfrac{1}{j[0.1+0.1+0.04]}$

$$= \dfrac{1}{j[0.24]}$$

$$= -j4.166\ p.u.$$

$$= 4.166\ \underline{|-90^0}\ p.u.$$

Per unit fault current $= 3I_{R1}$ p.u.

$$= 3 \times 4.166 \, \underline{|-90^0} \text{ p.u.}$$

$$= 12.5 \, \underline{|-90^0} \text{ p.u.}$$

Fault current in amperes $= I_{fp.u.} \times I_b \text{A}$

$$= 1.25 \times 5.248 \times 10^3$$

$$= 65610 \text{A}$$

The p.u. sequence voltages at the fault point,

Per unit positive sequence voltage, $V_{R1p.u.} = E_{R1} - jI_{R1p.u.} \times X_{R1eq}$

$$= 1 - j4.166 \, (0.1j) = 0.583 \text{ p.u.}$$

Per unit negative sequence voltage, $V_{R2p.u.} = -jI_{R2p.u.} \times X_{R2eq}$

$$= -(j4.166 \, \underline{|-90^0}) \, (0.1) = -0.4166 \text{ p.u.}$$

Per unit zero sequence voltage, $V_{R0p.u.} = -jI_{R0p.u.} \times X_{R0eq}$

$$= -j \, (4.166 \, \underline{|-90^0}) \, (0.4) = -0.1664 \text{ p.u.}$$

Phase voltages at the fault point in p.u.,

Voltage in R–phase, $V_{Rp.u.} = V_{R0} + V_{R1} + V_{R2}$

$$= 0.5833 - 0.4166 - 0.1664 = 0$$

Voltage in Y–phase, $V_{Yp.u.} = V_{Y0} + V_{Y1} + V_{Y2}$

$$= V_{R0} + K^2 V_{R1} + K V_{R2}$$

$$= -0.166 + (-0.5 - j0.866) \, 0.5833 + (-0.5 - j0.866) \times (-0.4166)$$

$$= (-0.24935 - j0.865) \text{p.u.}$$

Voltage in B–phase, $V_{Bp.u.} = V_{B0} + V_{B1} + V_{B2}$

$$= V_{R0} + K V_{R1} + K^2 V_{R2}$$

$$= (-0.249 + j0.144)$$

Line–to–line voltages at fault point in per unit,

$$V_{RY} = V_R - V_Y = 0 - V_Y$$

$$= 0.24935 + j0.865$$

$$V_{YB} = V_Y - V_B$$

$$= (-0.24935 - j0.865 + 0.249 - j0.144) = -j1.009$$

$$V_{BR} = V_B - V_R = V_B$$

$$= j(-0.249 + 0.144) - 0$$

Line–to–line voltages in volt,

$$V_{RY} \text{ (Volt)} = V_{RY} \text{p.u.} \times V_b$$

$$= (11 \times 10^3 / \sqrt{3})(0.24935 + j0.865)$$

$$= 5.715 \, \underline{|73.90^0}$$

$$V_{YB} \text{ (Volt)} = V_{YB \text{ p.u.}} \times V_b$$
$$= (11 \times 10^3 / \sqrt{3})(-0.249 + j0.144)$$
$$= -1581 - 3.6 + j\,914.51 - 11\underline{|-90^0}$$
$$V_{BR} \text{(Volt)} = V_{BR \text{ p.u.}} \times V_b$$
$$= 5.715\ \underline{|106.09^0}$$

6. Consider the synchronous generator of Question 5, determine the fault current and the line–to–line voltages if the double line fault occurs.

Solution:

Base current in amperes, $I_b = 100 \times 10^3 / (\sqrt{3} \times 11) = 5.248$ kA

For L–L fault, per unit positive sequence current is

$$I_{R1\text{p.u.}} = \frac{E_{R1eq\ p.u.}}{j[X_{1eq\ p.u.} + X_{2eq\ p.u.}]}$$

$$= \frac{1}{j[0.1 + 0.1]}$$

$$= \frac{1}{j[0.2]}$$

$$= \frac{j10}{2}$$

$$= -j5$$

Fault current in per unit

$$I_{f\text{p.u.}} = \sqrt{3} \times I_{R1\text{p.u.}} = \sqrt{3}\,(-j5) = -j8.66\text{p.u.}$$

Fault current in amperes,

$$I_f\,(A) = I_{f\text{p.u.}} \times I_{base} = -8.66 \times 5.248 \times 10^3 = 45454.6\underline{|-90^0}\ A$$

Per unit negative sequence current,

$$I_{R2\text{p.u.}} = -I_{R1\text{p.u.}} = j5 = 5\ \underline{|90^0}\text{ p.u.}$$

As LL fault do not involve with ground, $I_{R0} = 0$

Per unit positive sequence voltage, $V_{R1\text{p.u.}} = E_{R1\text{p.u.}} - j\,I_{R1\text{p.u.}} \times X_{R1\text{p.u.}}$
$$= 1 - j\,(-j5) \times 0.1 = 0.5\text{ p.u.}$$

Per unit negative sequence voltage, $V_{R2\text{p.u.}} = -j\,(j5)\,0.1 = 0.5$ p.u.

Therefore, $V_{R1\text{p.u.}} = V_{R2\text{p.u.}}$

As LL fault do not involve with ground, $I_{R0} = 0$ and $V_{R0} = 0$

Phase voltages in per unit,

Voltage in R–phase, $V_{R\text{p.u.}} = V_{R0} + V_{R1} + V_{R2} = 1$ p.u.

Voltage in Y–phase, $V_{Y\text{p.u.}} = V_{R0\text{p.u.}} + K^2 V_{R1\text{p.u.}} + K V_{R2\text{p.u.}} = -0.5$ p.u.

Voltage in B–phase, $V_{Bp.u.} = V_{R0p.u.} + KV_{R1p.u.} + K^2V_{R2p.u.} = -0.5$ p.u.

Line voltages in per unit,

$$V_{RY} = V_R - V_Y = 0.5 \text{ p.u.}$$

$$V_{YB} = V_Y - V_B = -0.5 + 0.5 = 0$$

$$V_{BR} = V_B - V_R = -0.5 - 1 = -1.5 \text{ p.u.}$$

Line–to–line voltages in Volts,

$$V_{RY} = 1.5 \times (11 \times 10^3 / \sqrt{3}) = 1.5 \times 6350.8 = 9526V$$

$$V_{YB} = 0$$

$$V_{BR} = -9526V$$

7. Consider synchronous generator of question 5, determine the fault current and L–L voltages if a double line–to–ground occurs.

Solution:

Base current, $I_b = 5.248$ kA

Positive sequence current in R–phase, $I_{R1p.u.} = \dfrac{E_{R1\ p.u.}}{j[X_{R1\ p.u.} + (X_{R2\ p.u.} // X_{R0\ p.u.})]}$

$$= \dfrac{1}{j\left[0.1 + \dfrac{0.1 \times 0.04}{0.1 + 0.04}\right]}$$

$$= -j7.77 \text{ p.u.}$$

Negative sequence current in R–phase, $I_{R2p.u.} = \left[\dfrac{j0.04 \times (-j7.77)}{j0.04 + j0.1}\right]$

$$= 2.222j$$

Zero sequence current in R–phase, $I_{R0p.u.} = \left[\dfrac{j0.1 \times (-j7.77)}{j0.14 + j0.1}\right]$

$$= j5.55 \text{ p.u.}$$

Fault current in p.u., $I_{fp.u.} = 3I_{R0} = 3 \times (j5.55\text{p.u.}) = j16.65$ p.u.

Fault current in Amperes, $I_f = 16.65 \times 5.248 \times 10^3 = 87379.2A$

Verification of fault current

$$I_{fp.u.} = I_{Yp.u.} + I_{Bp.u.}$$

$$I_{fp.u.} = 2I_{R0} + (K^2 + K)\, I_{R1} + (K^2 + K)\, I_{R2}$$

$$= 2I_{R0} - I_{R1} - I_{R2} = 3I_{R0} = 16.65 \text{ p.u.}$$

Voltages at fault point in p.u.,

Per unit positive sequence voltage, $V_{R1p.u.} = E_{R1p.u.} - j\, I_{R1p.u.} \times X_{R1p.u.}$

$$= 1 - j\, (-7.77)\, (0.1)$$

$$= 1 - 0.777 = 0.223 \text{ p.u.}$$

Per unit negative sequence voltage, $V_{R2p.u.} = -j\, I_{R2p.u.} \times X_{R2p.u.}$
$$= -j\,(-2.22j)\,(0.1)$$
$$= 0.222 \text{p.u.}$$

Per unit zero sequence voltage, $V_{R0p.u.} = -j\, I_{R0p.u.} \times X_{R0p.u.}$
$$= -j\,(j5.55)\,(j0.01)$$
$$= 0.222_{p.u.}$$

Phase voltages in p.u.,

Voltage in R–phase, $V_{Rp.u.} = V_{R0} + V_{R1} + V_{R2}$
$$= (0.223 + 0.222 + 0.222)$$
$$= 0.666 \text{p.u.}$$

Voltage in Y–phase, $V_{Yp.u.} = V_{Y0} + V_{Y1} + V_{Y2}$
$$= V_{R0} + K^2 V_{R1} + K V_{R2}$$
$$= V_{R0}\,(1 + K^2 + K) = 0$$

Voltage in B–phase, $V_{Bp.u.} = V_{B0} + V_{B1} + V_{B2} = V_{R0} + K V_{R1} + K^2 V_{R2} = 0$

Line–to–line voltages in p.u.,
$$V_{RY} = V_R - V_Y = 0.666 - 0 = 0.666 \text{p.u.}$$
$$V_{YB} = V_Y - V_B = 0$$
$$V_{BR} = -0.666 \text{p.u.}$$

Line–to–line voltages in volts,
$$V_{RY} = V_{RYp.u.}\,(11 \times 10^3 / \sqrt{3}) = 0.666(6350.5) = 4191.5\text{V}$$
$$V_{BR} = V_{BR\,p.u.} = 4191.5\text{V}$$
$$V_{YB} = 0$$

8. Consider the power system shown in the figure. A single line to ground fault occurs at bus 2 at each end of the transmission line. Consider the data as follows:

   Generator: Three–phase, 50MVA, 33 kV, $X_1 = X_2 = 0.2$ p.u., $X_0 = 0.1$p.u.,

   33kV line: $X_1 = X_2 = 10\Omega$, $X_0 = 20\Omega$

   Diagram:

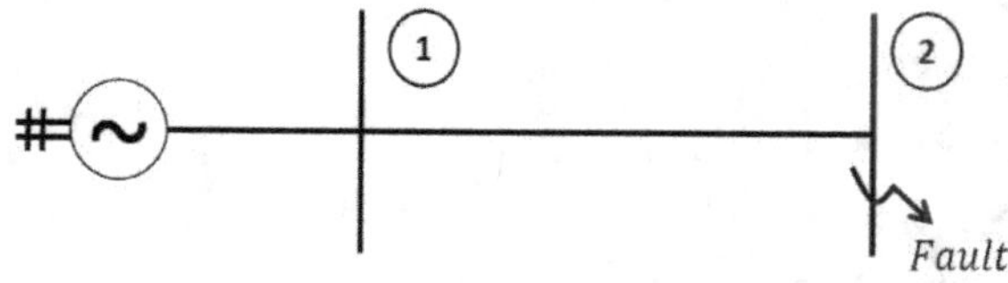

Determine the fault current magnitude,

1. If generator neutral is solidly grounded and fault impendence is zero

2. If generator neutral is solidly grounded and fault impendence is 0.1p.u. on 50MVA, 33kV

3. If generator neutral is reactance grounded with $X_n = 0.1$p.u. on 50MVA, 33kV and fault impendence is 0.1p.u. on 50MVA, 33kV.

Solution:

Base values, Let $MVA_b = 50MVA$ and $kV_b = 33kV$

Base current, $I_b = 50 \times 10^3 / (33 \times \sqrt{3}) = 0.874KV = 874V$

Base reactance, $X_b = (kV_b)^2 / MVA_b = 21.78$p.u.

Positive and negative sequence reactance of transmission line, $X_{2p.u.} = X_{1p.u.}$
$$= 10 / 21.78$$
$$= 0.4591\text{p.u.}$$

Zero sequence reactance of transmission line, $X_{0p.u.} = 20 / 21.78 = 0.918$ p.u.

1. Equivalent per unit zero sequence reactance of alternator,
$$X_{0eqp.u.\,alternator} = 3X_{np.u.} + X_{R0p.u.} = j0.1 + 3 \times 0 = j0.1\text{p.u.}$$

   Total per unit zero sequence reactance,
$$X_{0eqp.u.Total} = X_{0eqp.u.Alternator} + X_{0eqp.u.Line} = j0.1 + j0.918 = j0.2\text{p.u.}$$

   Total per unit positive sequence reactance,
$$X_{1eqp.u.Total} = X_{1eq\,p.u.Alternator} + X_{1eq\,p.u.\,Line} = j0.2 + j0.4591 = j0.66\text{p.u.}$$

   Total per unit negative sequence reactance,
$$X_{2eqp.u.Total} = X_{2\,eq\,p.u.Alternator} + X_{2\,eq\,p.u.\,Line} = j0.2 + j0.4591 = j0.66 \text{ p.u.}$$

   The per unit fault current,
$$\text{If p.u.} = \frac{E_{R1eq\,p.u.}}{X_{1eq\,p.u.} + X_{2eq\,p.u.} + X_{0eq\,p.u.}}$$
$$= \frac{3}{j[0.66 + 0.66 + 1.02]}$$
$$= -j1.282\text{p.u.}$$

   Fault current in Amperes, $I_f = (-j1.282)(874.7) = -j1.121kA$

2. Per unit fault reactance, $X_{fp.u.} = 0.1$p.u.

   Equivalent per unit zero sequence reactance of alternator,
$$X_{0eqp.u.}\text{alternator} = 3X_{np.u.} + X_{0p.u.} = j3 \times 0.1 + j0.1 = j0.4 \text{ p.u.}$$

   Total per unit positive sequence reactance,
$$X_{1eqp.u.Total} = X_{1eqp.u.Alternator} + X_{1eqp.u.Line} = j0.66 \text{ p.u.}$$

   Total per unit negative sequence reactance,
$$X_{2eqp.u.Total} = X_{2\,eqp.u.Alternator} + X_{2\,eqp.u.Line} = j0.66 \text{ p.u.}$$

   Total per unit zero sequence reactance,
$$X_{0eqp.u.Total} = X_{0eqp.u.Alternator} + X_{0eqp.u.Line} = j0.4 + j0.918 = j1.318 \text{ p.u.}$$

   Fault current in p.u.,
$$I_{f\,p.u.} = \frac{3}{j[2 \times 0.666 + 1.318]} = -j1.132 \text{ p.u.}$$

   Fault current in Amperes, $I_f = (-j1.132) \times (874.7) = -j990.22A$

3. Per unit fault reactance, $X_{fp.u.} = 0.1$ p.u.

   Per unit neutral reactance, $X_{np.u.} = 0.1$ p.u.

   Equivalent per unit zero sequence reactance of alternator,

   $$X_{0eqp.u.aleternator} = 3X_{np.u.} + X_{0p.u.} = 3 \times 0.1 + 0.1 = 0.7 \text{ p.u.}$$

   Total per unit positive sequence reactance,

   $$X_{1eqp.u.Total} = X_{1eqp.u.Alternator} + X_{1eqp.u.Line} = j0.66 \text{ p.u.}$$

   Total per unit negative sequence reactance,

   $$X_{2eqp.u.Total} = X_{2\,eqp.u.Alternator} + X_{2\,eqp.u.Line} = j0.66 \text{ p.u.}$$

   Total per unit zero sequence reactance,

   $$X_{0eqp.u.Total} = X_{0eqp.u.Alternator} + X_{0eqp.u.Line} = j\,0.7 + j0.918 = j1.618 \text{ p.u.}$$

   Fault current in p.u., $I_{fp.u.} = \dfrac{3}{j[2 \times 0.666 + 1.618]} = -j1.021$ p.u.

   Fault current in Amperes, $I_f\,(A) = -1.021 \times 874.4 = -j892.85$ Amperes

9. Consider the question no 10, if a three–phase fault occurs at the end of the line then compute the fault current.

Solution:

Three phase fault current in p.u.,

$$= \frac{E_{R1eqpu}}{j[X_{1eq}\,pu]} = \frac{1}{j[0.66]} = -j1.515 \text{ p.u.}$$

   Fault current in Amperes,

   $$I_f\,(A) = (-j1.51 \text{p.u.}) \times (874.79) = -j1325.4 A$$

10. Compare the three–phase fault current with single line to ground fault current for conditions 1 to 3 of questions 10.

Solution:

(a)
$$\frac{I_{fLLL3-phase}}{I_{fLG}} = \frac{1325.4}{1122.98} = 1.180:1$$

(b)
$$\frac{I_{fLLL3-phase}}{I_{fLG}} = \frac{1325.4}{994.83} = 1.3322:1$$

(c)
$$\frac{I_{fLLL3-phase}}{I_{fLG}} = \frac{1325.4}{893.25} = 1.483:1$$

11. The severity of LG and LLL fault are same. Determine the neutral reactance to obtain equal fault severity.

Solution:

For line to ground fault, $I_f = \dfrac{3E_{R1p.u.}}{j(X_{1eq\,p.u.} + X_{2eq\,p.u.} + X_{0eq\,p.u.})}$

For three phase fault, $I_f = \dfrac{E_{R1 p.u.}}{j[X_{1eq\ p.u.}]} = \dfrac{E_{R1}\,X_3}{j[X_{1eq}X_{3\ p.u.}]}$

As both faults must be equally severe, $3X_{1eq} = X_{1eq} + X_{2eq} + X_{3eq}$

$$X_{2eq} + X_{0eq} = 2X_{1eq}$$

Assume that $X_{1eq} = X_{2eq}$

Therefore, $X_{0eq} = X_{1eq}$

$$X_{R0\ eq} = X_{1eq} = X_{R0} + 3X_n$$

Therefore, neutral reactance is $X_n = \dfrac{X1eq + XR0}{3}$

12. The shunt capacitance of the line can be neglected if the line has positive sequence impedance 15Ω and zero sequence impedance is 48Ω. Determine the mutual impedance and self-impedance in ohms?

Solution:

Positive sequence impedance, $Z_{1eq} = Z_s - Z_m$

i.e. $\qquad\qquad\qquad 15 = Z_s - Z_m$

Zero sequence impedance, $Z_{0eq} = Z_s + 2Z_m$

i.e. $\qquad\qquad\qquad 48 = Z_s + 2Z_m$

On solving the above equations,

Mutual impedance $Z_m = 11\Omega$ and self-impedance $Z_s = 26\ \Omega$

OBJECTIVE QUESTIONS

1. _______________________ is a balanced or symmetrical fault.

2. All the _______________________ are equal to zero for LLL fault.

3. The sequence network for LLL fault consists of only _______________________ sequence network with its _______________________.

4. _______________________ fault is treated as no–fault.

5. The sub–transient reactance for LLL fault is $Xd" = $ _______________________.

6. A short circuit occurs in a transmission line when the voltage wave is going through zero, the maximum possible momentary short circuit current corresponds to _______________________.

7. Fault level means Fault _______________________.

8. Fault calculations are done by _______________________ method.

9. The rated breaking capacity of a circuit breaker is equal to the product _______________________.

10. When a line to ground fault occurs, the current in a fault phase is 100A. The zero-sequence current is _______________________.

11. Symmetrical components are used in power systems because _______________________.

12. For a power transformer, _______________________ impedances are equal.

13. Zero sequence currents can flow from a line into a transformer bank if the windings are in ________________________.

14. For a fully transposed transmission line, ________________________ impedance is larger than ________________________.

15. ________________________ network gets affected by the method of neutral grounding.

16. The ________________________ voltages are maximum at fault point towards neutral.

17. A three phase balanced system is always analyzed on ________________________ basis.

18. The most severe fault is ________________________ fault.

19. The following sequence currents were recorded in a power system under a fault condition $I_{positive} = j1.753 p.u.$, $I_{negative} = -j0.6 p.u.$, $I_{zero} = -j1.153 p.u.$ The fault is ________________________.

Power System Stability Analysis

11.1 INTRODUCTION

Apart from generating power and meeting the load demand, power system network should operate under stable conditions. Here various power system networks are analyzed with respect to stability.

Case 1: Consider a synchronous generator connected to an asynchronous load as shown in figure 11.1.

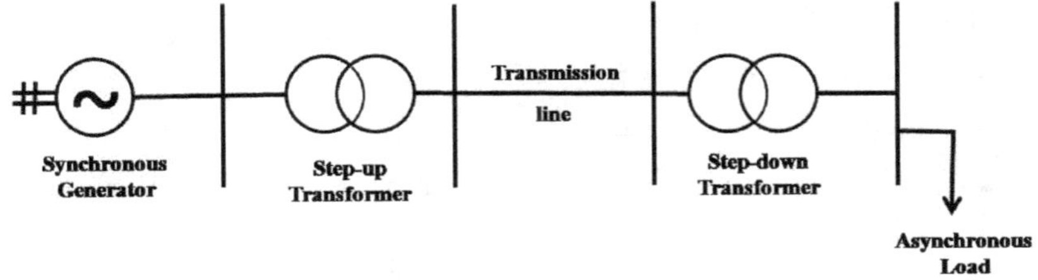

Figure 11.1 A synchronous generator is connected to an asynchronous load

As source is a synchronous machine, synchronous power is transmitted through the transmission line.

Case 2: Consider an asynchronous machine connected to an asynchronous load as shown in figure 11.2.

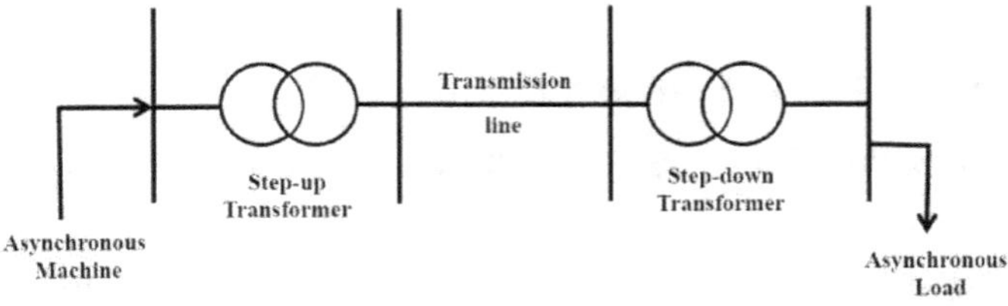

Figure 11.2 An asynchronous machine is connected to an asynchronous load

As source is an asynchronous machine, asynchronous power is transmitted through the transmission line.

Case 3: Consider a synchronous generator connected to a synchronous load as shown in figure 11.3.

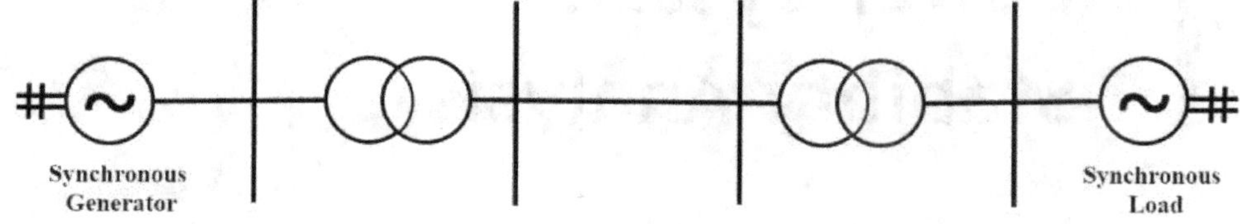

Figure 11.3 A synchronous Generator is connected to a synchronous load

As source is a synchronous machine, synchronous power is transmitted through the transmission line.

11.1.1   Power System Stability

Definition

The property of a synchronous machine to transmit maximum active power through the transmission line by maintaining synchronism with externally connected transmission line is defined as "power system stability".

Conditions for a stable power system network operation

1.   Angular velocity of rotor $\omega = \omega_s$ (or) speed of the rotor , $N = N_s$
2.   Rotor neither accelerates nor decelerates.
3.   Accelerating power $P_a = P_s - P_e = 0$ (or)$P_s = P_e$
4.   The position of the Rotor axis with respect to stator axis is constant as shown in figure 11.4.

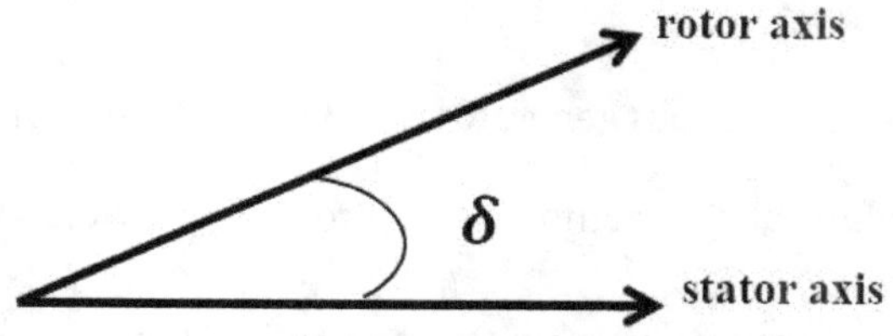

Figure 11.4 Position of rotor with respect to stator

11.1.2   Classification of Power System Stability

Power system stability is classified into three types.

They are

1. Steady state stability
2. Transient stability
3. Dynamic stability

Steady State Stability

Definition: The property of a synchronous machine to transmit maximum active power through transmission line by maintaining synchronism with externally connected transmission line for small and gradual changes in load demand.

1. Steady state stability occurs due to changes in consumer's load demand.
2. The frequency of rotor is approximately equal to 90% of supply frequency. i.e., natural frequency of rotor oscillations $\left(f_{supply} \cong 90\% \, f_n\right)$

Transient stability

Definition: The property of a synchronous machine to transmit maximum active power through transmission line by maintaining synchronism with externally connected transmission line for sudden and large changes in load demand.

1. Transient stability is due to occurrence of series faults or shunt faults.
2. The frequency of rotor is greater than supply frequency.

11.2 METHODS TO DETERMINE STEADY STATE STABILITY

Steady state stability of a system can be determined by four methods.

They are as follows.

1. Static model.
2. Synchronizing power coefficient.
3. ABCD constants.
4. Swing equation.

11.2.1   Static Model

As the changes in the load demand is slow and gradual, steady state stability can be determined by static model.

Consider an alternator connected to load through a transmission line as shown in figure 11.5.

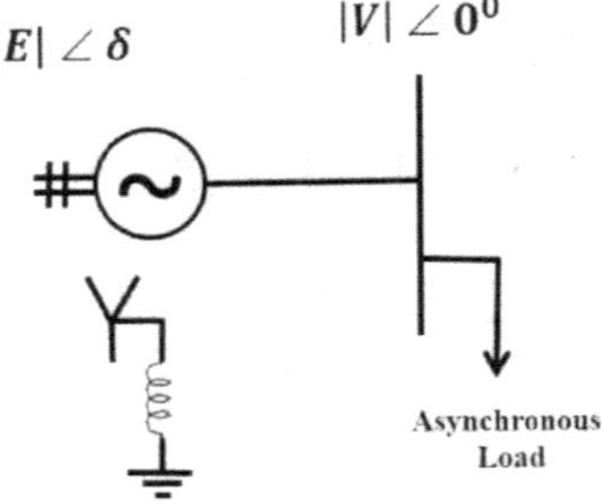

Figure 11.5 An alternator connected to load through a transmission line

The above power system network can be analyzed from per unit reactance diagram of the system.

The per unit reactance diagram is as shown in figure 11.6.

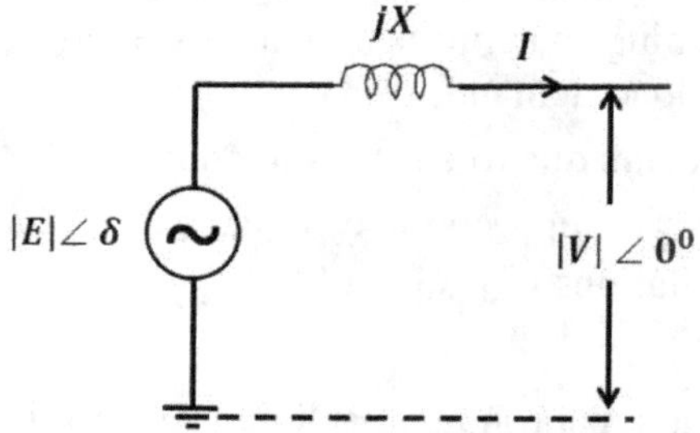

Figure 11.6 Per unit reactance diagram

From the per unit reactance diagram,
the terminal voltage is

$$V_t = E_{gp.u} - j\, I_{p.u}\big(X_{lp.u} + X_{gp.u}\big)$$
$$V_t = E_{gp.u} - j\, I_{p.u} \times X_{p.u}$$

Complex power delivered by synchronous generator to the load

$$S = V.I^*$$

$$= |V| \angle 0 \left\{ \frac{|E|\angle\delta - |V|\angle 0}{jX} \right\}^*$$

$$= |V| \angle 0 \left\{ \frac{|E|\angle\delta - |V|\angle 0}{|X|\angle 90} \right\}^*$$

$$= |V| \angle 0 \left\{ \frac{|E|}{|X|} \angle\delta - 90 - \frac{|V|}{|X|} \angle - 90 \right\}^*$$

$$= |V| \angle 0 \left\{ \frac{|E|}{|X|} \angle 90 - \delta - \frac{|V|}{|X|} \angle 90 \right\}$$

$$S = \frac{|E||V|}{|X|} \angle 90 - \delta - \frac{|V|^2}{|X|} \angle 90 \qquad\qquad .....(11.1)$$

Active power delivered by synchronous generator to the load (or) electrical power transmitted through the transmission line is

$$P_e = \frac{|E||V|}{|X|} \cos(90 - \delta) - \frac{|V|^2}{|X|} \cos 90$$

$$P_e = \frac{|E||V|}{|X|} \sin\delta \qquad\qquad .....(11.2)$$

Steady state stability limit

Definition: The maximum active power transmitted through the transmission line is known as "steady state stability limit".

$$\text{SSSL} = P_{e,max} = \frac{|E||V|}{|X|} \sin 90^0$$

$$\text{Steady State Stability Limit, SSSL} = \frac{|E||V|}{|X|} \qquad \qquad .....(11.3)$$

Power angle curve

Power angle curve gives the relation between electrical power transmitted through the line and load angle, $'\delta'$ as shown in the figure 11.7.

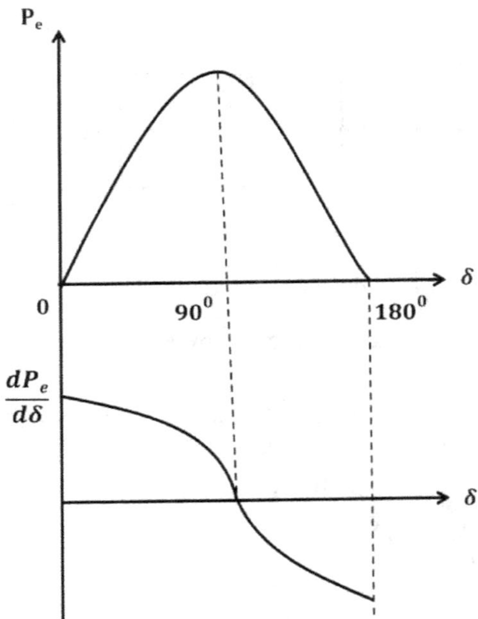

Figure 11.7 Power angle curve and curve representing synchronizing power coefficient

Synchronizing Power Coefficient

The term Synchronizing power coefficient is a measure of degree of magnetic coupling between stator axis and rotor axis.

Synchronizing Power Coefficient,

$$P_{sync} = \frac{dP_e}{d\delta} =$$

$$\frac{|E||V|}{|X|} \cos \delta \qquad \qquad .....(11.4)$$

Case 1: $0 < \delta < 90^0$

From equation 11.2.4., synchronizing power coefficient, P_{sync} is positive and rotor air gap flux is magnetized. Therefore, system operates under stable conditions.

Case 2: $\delta > 90^0$

From equation 11.2.4., synchronizing power coefficient, P_{sync} is negative and rotor air gap flux is demagnetized. Therefore, system operates under unstable conditions.

11.2.2   ABCD Constants

Consider a synchronous generator connected to an asynchronous load through a transmission line as shown in figure 11.8. The transmission line is represented by ABCD constants.

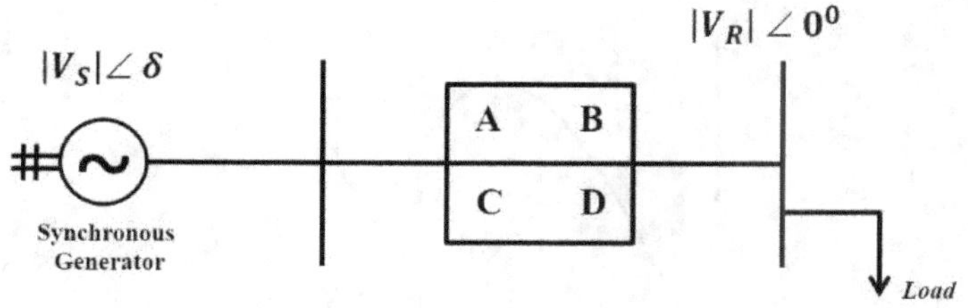

Figure 11.8 A synchronous generator connected to an asynchronous load

Expression for active power and reactive power of the sending and receiving end of transmission line

Consider a transmission line connected between synchronous generator, $|V_S| \angle \delta$ and the load, $|V_R| \angle 0$.

In the matrix form, sending end quantities are

$$\begin{bmatrix} V_S \\ I_S \end{bmatrix} = \begin{bmatrix} A & B \\ C & D \end{bmatrix} \begin{bmatrix} V_R \\ I_R \end{bmatrix}$$

$$\Rightarrow V_S = AV_R + BI_R \qquad\qquad\qquad .....(11.5)$$

$$\Rightarrow I_S = CV_R + DI_R \qquad\qquad\qquad .....(11.6)$$

In the matrix form, receiving end quantities are

$$\begin{bmatrix} V_R \\ I_R \end{bmatrix} = \begin{bmatrix} A & B \\ C & D \end{bmatrix}^{-1} \begin{bmatrix} V_S \\ I_S \end{bmatrix}$$

$$\begin{bmatrix} V_R \\ I_R \end{bmatrix} = \frac{1}{AD - BC} \begin{bmatrix} D & -B \\ -C & A \end{bmatrix} \begin{bmatrix} V_S \\ I_S \end{bmatrix}$$

As transmission lines are by default reciprocal, $AD - BC = 1$

$$\begin{bmatrix} V_R \\ I_R \end{bmatrix} = \begin{bmatrix} D & -B \\ -C & A \end{bmatrix} \begin{bmatrix} V_S \\ I_S \end{bmatrix}$$

$$\Rightarrow V_R = DV_S - BI_S \qquad\qquad\qquad .....(11.7)$$

$$\Rightarrow I_R = CV_S + AI_S \qquad\qquad\qquad .....(11.8)$$

Expressing I_S in terms of V_S and V_R

Substitute 11.8 in 11.6

$$\Rightarrow I_S = CV_R + D[CV_S + AI_S]$$

$$\Rightarrow I_S = CV_R - CDV_S + ADI_S$$

$$\Rightarrow I_S(1 - AD) = -CDV_S + CV_R$$

$$\Rightarrow I_S = \frac{-CDV_S + CV_R}{-BC}$$

$$\Rightarrow I_S = \frac{CD}{BC}V_S - \frac{1}{B}V_R$$

$$\Rightarrow I_S = \frac{D}{B}V_S - \frac{1}{B}V_R \qquad\qquad .....(11.9)$$

Expressing I_R in terms of V_S and V_R

Substitute 11.6 in 11.8

$$\Rightarrow I_R = -CV_S + A(CV_R + DI_R)$$

$$\Rightarrow I_R(1 - AD) = -CV_S + ACV_R$$

$$\Rightarrow I_R = \frac{1}{B}V_S - \frac{A}{B}V_R \qquad\qquad .....(11.10)$$

As power transmission is AC, expressing the electrical terms in equations 11.9 and 11.10 in the polar form

$$V_R = V_R\angle 0^0$$

$$V_S = V_S\angle\delta$$

$$A = |A|\angle\propto$$

$$B = |B|\angle\beta$$

$$D = |D|\angle\Delta$$

Constant 'A' is a measure of decrease in voltage wave magnitude from starting end to the receiving end of the transmission line.

Constant 'D' is a measure of decrease in current wave magnitude from starting end to the receiving end of the transmission line.

Constant 'B' is measure of change in phase angle between voltage wave and current wave from sending end to receiving end of line.

Therefore, equation 11.2.9 can be represented as

$$I_S = \frac{|D|\angle\Delta}{|B|\angle\beta}|V_S|\angle\delta - \frac{1}{|B|\angle\beta}|V_R|\angle 0$$

$$= \frac{|D||V_S|}{|B|}\angle\Delta + \delta - \beta - \frac{|V_R|}{|V_B|}\angle -\beta$$

$$I_S{}^* = \frac{|D||V_S|}{|B|}\angle\beta - \delta - \Delta - \frac{|V_R|}{|B|}\angle +\beta \qquad\qquad .....(11.11)$$

Therefore, equation 11.10 can be represented as

$$I_R = \frac{1}{|B|\angle\beta}|V_S|\angle\delta - \frac{|A|\angle\propto}{|B|\angle\beta}V_R\angle 0^0$$

$$= \frac{|V_S|}{|B|}\angle\delta - \beta - \frac{|A||V_R|}{|B|}\angle\alpha - \beta \qquad\qquad .....(11.12)$$

$$I_R^* = \frac{|V_S|}{|B|} \angle \beta - \delta - \frac{|A||V_R|}{|B|} \angle B - \alpha$$

Complex power per phase at sending end of the transmission line is

$$S_S = P_S + JQ_S = V_S I_S^*$$

$$S_S = |V_S|\angle\delta \left\{ \frac{|D||V_S|}{|B|} \angle \beta - \delta - \Delta - \frac{|V_R|}{|B|} \angle + \beta \right\}$$

Active power per phase at sending end of the transmission line is

$$P_S = \frac{|D||V_S|^2}{|B|} \cos(\beta - \Delta) - \frac{|V_S||V_R|}{|B|} \cos(\beta + \delta)$$

$$\Rightarrow \left\{ P_S - \frac{|D||V_S|^2}{|B|} \cos(\beta - \Delta) \right\} = -\frac{|V_S||V_R|}{|B|} \cos(\beta + \delta) \qquad \ldots\ldots(11.13)$$

Reactive power per phase at sending end of the transmission line is

$$Q_S = \frac{|D||V_S|^2}{|B|} \sin(\beta - \Delta) - \frac{|V_S||V_R|}{|B|} \sin(\beta + \delta)$$

$$\Rightarrow \left\{ Q_S - \frac{|D||V_S|^2}{|B|} \sin(\beta - \Delta) \right\} = -\frac{|V_S||V_R|}{|B|} \sin(\beta + \delta) \qquad \ldots\ldots(11.14)$$

Squaring and adding above equations 11.13 and 11.14

$$\Rightarrow \left\{ P_S - \frac{|D||V_S|^2}{|B|} \cos(\beta - \Delta) \right\}^2 + \left\{ Q_S - \frac{|D||V_S|^2}{|B|} \sin(\beta - \Delta) \right\}^2$$

$$= \left\{ \frac{|V_S||V_R|}{|B|} \right\}^2 \qquad \ldots\ldots(11.15)$$

Equation 11.15 represents equation of circle with centre in the 1st quadrant and radius equal to $\frac{|V_S||V_R|}{|B|}$ i.e. Characteristic impedance loading.

Complex power per phase at receiving end of the transmission line is

$$S_R = P_R + jQ_R = V_R I_R^* = V_R\angle 0^0 \left\{ \frac{|V_S|}{|B|} \angle \beta - \delta - \frac{|A||V_R|}{|B|} \angle \beta - \alpha \right\}$$

Active power per phase at receiving end of the transmission line is

$$\Rightarrow P_R = \frac{|V_S||V_R|}{|B|} \cos(\beta - \delta) - \frac{|A||V_R|^2}{|B|} \cos(\beta - \alpha)$$

For maximum active power to be transmitted through the transmission line, impedance phase angle, $\beta = \tan^{-1}\left(\frac{X_L}{R}\right)$=load angle, δ

$$\Rightarrow \left\{ P_R + \frac{|A||V_R|^2}{|B|} \cos(\beta - \alpha) \right\} = -\frac{|V_S||V_R|}{|B|} \cos(\beta - \delta) \qquad \ldots\ldots(11.16)$$

Reactive power per phase at receiving end of the transmission line is

$$\Rightarrow Q_R = \frac{|V_S||V_R|}{|B|}\sin(\beta - \delta) - \frac{|A||V_R|^2}{|B|}\sin(\beta - \alpha)$$

$$\Rightarrow \left\{Q_R + \frac{|A||V_R|^2}{|B|}\sin(\beta - \alpha)\right\} = \frac{|V_S||V_R|}{|B|}\sin(\beta - \delta) \qquad .....(11.17)$$

Squaring and adding above equations 11.16 and 11.17

$$\Rightarrow \left\{P_R + \frac{|A||V_R|^2}{|B|}\cos(\beta - \alpha)\right\}^2 + \left\{Q_R + \frac{|A||V_R|^2}{|B|}\sin(\beta - \alpha)\right\}^2$$

$$= \left\{\frac{|V_S||V_R|}{|B|}\right\}^2 \qquad .....(11.18)$$

Equation 11.18 represents equation of circle with centre in the 3rd quadrant and radius equal to $\frac{|V_S||V_R|}{|B|}$ i.e. Characteristic impedance loading.

Active power transmitted from sending end to receiving end of the line is

$$\Rightarrow P_R = \frac{|V_S||V_R|}{|B|}\cos(\beta - \delta) - \frac{|A||V_R|^2}{|B|}\cos(\beta - \alpha)$$

Analysis of a short transmission line

As the length of the short txn line is less, the time taken to transmit power is less and decreases in voltage wave magnitude from sending end to the receiving end of the line is negligible. Therefore, $|A| = 1$ and $\alpha = 0$

Impedance of transmission line, $Z = R + jX_L$

$$\Rightarrow |B| = |Z| = \sqrt{R^2 + X_L{}^2}$$

$$\Rightarrow \tan\beta = \frac{X_L}{R}$$

$$\Rightarrow \beta$$

$$= \tan^{-1}\left(\frac{X_L}{R}\right), \text{ where} \beta \text{ is the impedance phase angle of transmission line.}$$

For maximum active power to be transmitted through the line,

$$\Rightarrow P_{R_{max}} = \frac{|V_S||V_R|}{\sqrt{R^2 + X_L^2}} \; 1 - \frac{1}{\sqrt{R^2 + X_L^2}} \; \frac{|V_R|^2}{\sqrt{R^2 + X_L^2}}\cos\beta$$

$$\Rightarrow P_{R_{max}} = \frac{|V_S||V_R|}{\sqrt{R^2 + X_L^2}} - \frac{|V_R|^2}{\sqrt{R^2 + X_L^2}} \cdot \frac{R}{\sqrt{R^2 + X_L^2}}$$

The reactance of the transmission line at which the maximum power is transmitted is calculated as follows:

$$\Rightarrow \frac{d(P_{R\,max})}{dX_L} = 0$$

$$\Rightarrow \frac{d(P_{R\,max})}{dX_L} = \frac{d}{dX_L}\left[\left(|V_S||V_R|(R^2 + X_L^2)^{-1/2}\right)\right] - \frac{|V_R|^2 R}{R^2 + X_L^2}$$

$$\Rightarrow \quad = \frac{-1}{2}|V_S||V_R|(R^2 + X_L^2)^{-1/2-1}2V_L + |V_R|^2 R(R^2 + X_L^2)^{-1}$$

$$\Rightarrow \quad = -|V_S||V_R|(R^2 + X_L^2)^{-3/2}X_L + |V_R|^2 R(R^2 + X_L^2)^{-2}2X_L$$

$$\Rightarrow \quad = -|V_R|X_L\left[V_S(R^2 + X_L^2)^{3/2} + |V_R|R(R^2 + X_L^2)^2 2\right]$$

$$\Rightarrow V_S(R^2 + X_L^2)^{-3/2} + 2V_R R(R^2 + X_L^2)^{-2} = 0$$

let $V_S = V_R$

$$\Rightarrow \frac{1}{(R^2 + X_L^2)^{3/2}} + \frac{2R}{(R^2 + X_L^2)^2} = 0$$

$$\Rightarrow \frac{1}{(R^2 + X_L^2)^{3/2}} = \frac{-2R}{(R^2 + X_L^2)^2}$$

$$\Rightarrow R^2 + X_L^2 = 4R^2$$

$$\Rightarrow X_L^2 = 3R^2$$

$$\Rightarrow X_L = \sqrt{3}R$$

For maximum power to be transmitted,

$$\text{load angle, } \delta = \beta = \tan^{-1}\left(\frac{X_L}{R}\right)$$

$$= \tan^{-1}\left(\frac{\sqrt{3}R}{R}\right)$$

$$= \tan^{-1}(\sqrt{3})$$

$$\text{load angle, } \delta = \beta = 60^0$$

As load angle, $\delta < 90^0$, short transmission line is by default stable.

Analysis using long transmission lines

For a long transmission line,

1. Power carrying capacity is high

2. Size of Conductor is high

3. Area of Conductor is high

4. Resistance of conductor, R is low.

5. ↑ load angle, δ = impedance phase angle, β ↑= $\tan^{-1}\left(\frac{X_C}{R\downarrow}\right)$

Therefore load angle increases and may increase beyond 90 degrees.

If δ or β exceeds 90^0 then power system network becomes unstable.

Series compensation

The process of connecting a capacitor in series with the transmission line such that the net reactance of the line decreases, load angle δ do not exceed 90^0 and thereby ensuring stable operation of the power system network is known as "series compensation".

$$jX = jX_L + (-jX_{Cse})$$
$$= j(X_L - X_{Cse})$$

Therefore, series capacitor is considered as a power system stability device.

11.2.3   Methods to Improve Steady State Stability

1. By increasing excitation i.e., by increasing field current.
2. By increasing load voltage (or) terminal voltage.
3. By decreasing reactance of the line.
    (a) By connecting a capacitor in series with transmission line.
    (b) Using parallel transmission lines.
    (c) With symmetrical spacing of R phase, Y Phase and B Phase conductor.
    (d) By replacing a solid conductor with a bundled conductor consisting of more number of unsymmetrical spaced sub conductors.

Case 1: Symmetrical Spacing

Consider symmetrical spacing of conductors with 'd' as spacing between conductors.

$$GMD = (d \times d \times d)^{1/3} = d$$
$$GMR = 0.7788\ r$$

Case 2: UnSymmetrical Spacing

Consider unsymmetrical spacing of conductors with 'd' as spacing between conductors.

$$GMD = (d \times d \times 2d)^{1/3} = 1.26d$$
$$GMR = 0.7788\ r$$
$$L \propto \ln\left(\frac{d}{0.7788r}\right) \propto \ln\left(\frac{GMD}{GMR}\right)$$

As GMD is low for symmetrically spaced conductors, inductance is low and CIL is high.

11.2.4 Determination of Transfer Reactance

The Reactance measured between source and load due to fault at any point in the power system network is known as "Transfer Reactance".

➢ Transfer reactance is calculated only during fault.

➢ Transfer reactance is calculated between source and load by eliminating the fault point.

11.3 ANALYSIS OF TRANSIENT STABILITY

Definition: The property of the synchronous machine to transmit maximum active power through transmission line by maintaining synchronism with externally connected transmission line for sudden and large changes in load demand due to occurrence of three phasefault on one of the parallel transmission lines near bus bar.

Consider a synchronous generator connected to an asynchronous load through a stepup transformer and a transmission line as shown in figure 11.9.

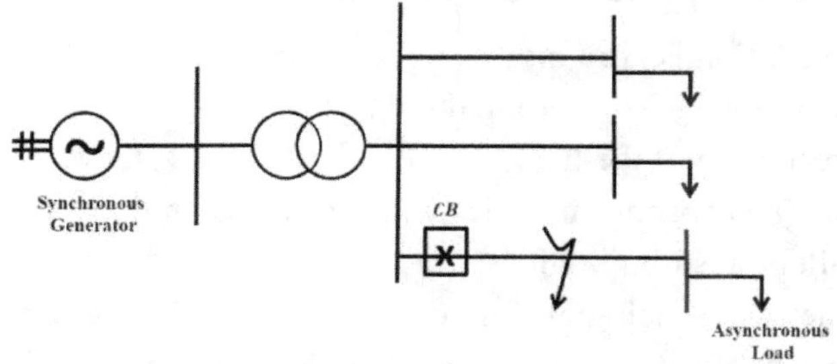

Figure 11.9 A synchronous generator connected to an asynchronous load

Electrical power transmitted before fault, $P_{e1} = P_{m1}\sin\delta_0$

where P_{m1} is the maximum electrical power transmitted before fault.

Electrical power transmitted during fault, $P_{e2} = P_{m2}\sin\delta$

where P_{m2} is the maximum electrical power transmitted during fault.

$$\text{and } P_{e2} < P_{e1}$$

11.3.1 Conditions for which Power Transmitted during Fault in Zero

Case 1: Occurrence of fault on a transmission line near bus bar

Consider a synchronous generator connected to an asynchronous load through a stepup transformer and a transmission line as shown in figure 11.10.

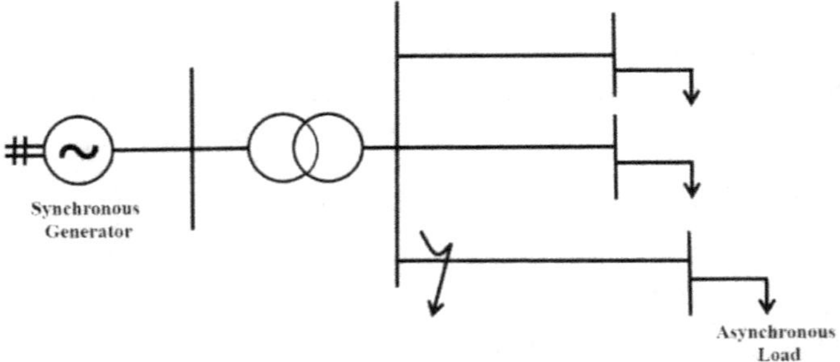

Figure 11.10 A synchronous generator connected to an asynchronous load

Occurrence fault on a transmission line near bus bar is considered as fault on bus bars power transferred during fault is zero.

Case 2: Occurrence of fault on bus bar

Consider a synchronous generator connected to an infinite bus bar through a stepup transformer and a transmission line as shown in figure 11.11.

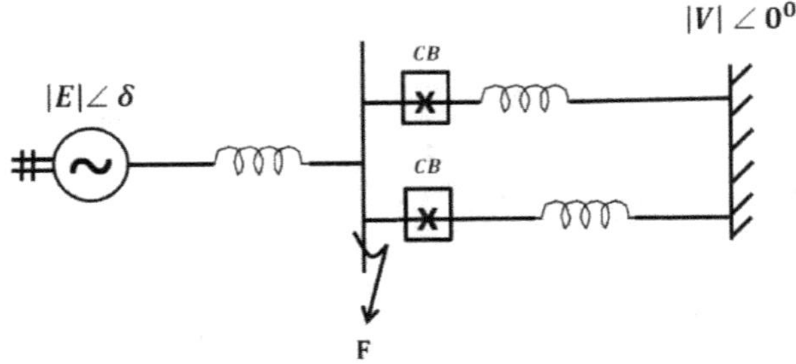

Figure 11.11 A synchronous generator connected to an infinite bus bar

Occurrence of fault on bus bar results in severe disturbances in power system network.

Rotor of alternator is subjected to severe oscillations. Therefore, dynamic mode of generator is represented to analyse the severe oscillations made by the rotor of synchronous generator.

11.3.2 Dynamic Model

Consider the dynamic model of the alternator as shown in figure 11.12.

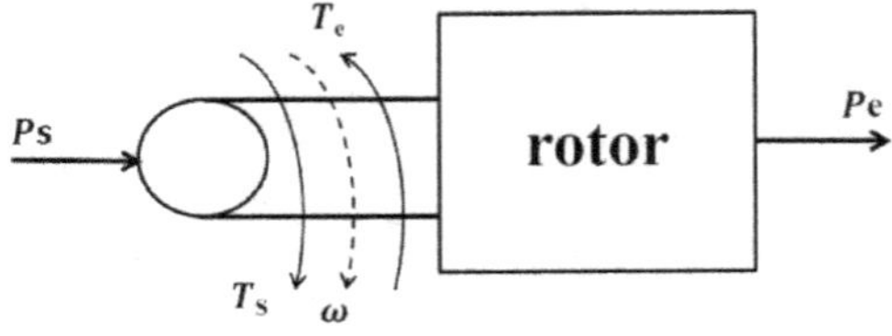

Figure 11.12 Dynamic model

Case 1: Before fault:

Accelerating power $P_a = P_s - P_e = 0$

i.e., Rotor neither accelerates nor decelerates and load angle, δ is constant i.e., $\delta = \delta_0$

Case 2: During fault:

As the electrical power transmitted during fault, P_e decreases.

$$P_a = P_s - P_e \downarrow$$
$$= \text{positive}$$

Therefore, Rotor accelerates and load angle, δ increases

Accelerating power, $P_a = T_a \cdot \omega$

where $T_a = $ Accelerating Torque

$\omega = $ Angular velocity of rotor in radians/Second

$$= \frac{2\pi NT}{60}$$

where N is actual rotor speed in rpm.

According to kinematic of rotating machine,

Accelerating torque, $T_a = I \cdot \alpha$

where I is inertia in Kg-m

α is Angular acceleration in rad/sec

Accelerating power, $P_a = (I \cdot \alpha)\,\omega = (I \cdot \omega)\alpha$

$$P_a = M \cdot \alpha$$

where $M = $ Moment of inertia with the units $\dfrac{MJ-sec}{Elec.deg}$ (or) $\dfrac{MJ-sec}{Elec-rad}$

Consider the position of rotor with respect to stator and reference axis as shown below in the figure 11.13.

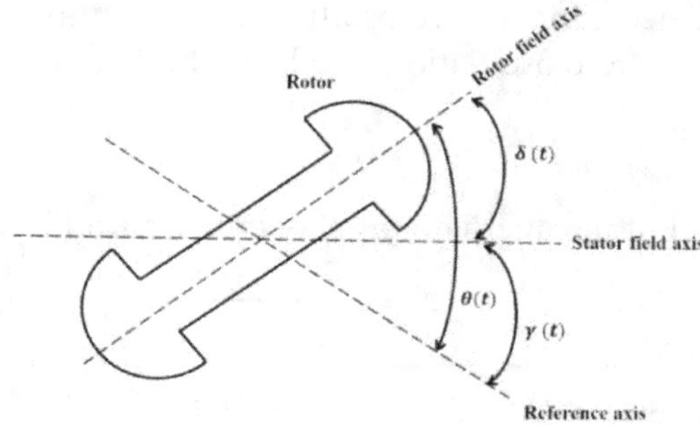

Figure 11.13 Position of rotor with respect to stator and reference axis

From the above figure

$$\theta(t) = \delta(t) + \gamma(t) = \delta(t) + \omega_r \cdot t$$

$$\frac{d\theta(t)}{dt} = \frac{d}{dt}\delta(t) + \omega_r$$

$$\text{acceleration of rotor, } \propto = \frac{d^2\theta(t)}{dt^2} = \frac{d^2\delta(t)}{dt^2} \qquad\qquad \text{.....(11.19)}$$

Substitute 11.19 in 11.18

$$P_a = M \cdot \frac{d^2\delta(t)}{dt^2}$$

$$\frac{d^2}{dt^2}\delta(t) = \frac{1}{M}P_a = \frac{1}{M}(P_s - P_e) = \frac{1}{M}\left(P_s - \frac{EV}{X}\sin\delta\right) \qquad \text{.....(11.20)}$$

Equation 11.20 is the swing equation for synchronous generator.

Swing equation is a second order differential equation

Swing equation gives the information about change in load angle, δ between rotor and stator axis for specified number of cycles during which the system is subjected to sudden large changes inload demand due to occurrence of a three-phase fault on bus-bar.

As the load demand changes, the position of rotor axis with respect to stator changes but position of stator with respect to reference axis do not change. Therefore, stator axis is considered as reference axis.

Swing equation for synchronous motor is,

$$\propto = \frac{d^2\delta(t)}{dt^2} = \frac{1}{M}(P_e - P_s) = \frac{1}{M}\left(\frac{EV}{X}\sin\delta - P_s\right)$$

11.4 ANALYSIS OF STEADY STATE STABILITY USING SWING EQUATION

Though swing equation determines the changes in the position of rotor with respect to stator for sudden and large changes in load demand, it can also determine the change in position of rotor with respect to stator for small and gradual changes in load demand (steady state stability).

Consider an alternator connected to an infinite bus through a transmission line as shown in figure 11.14.

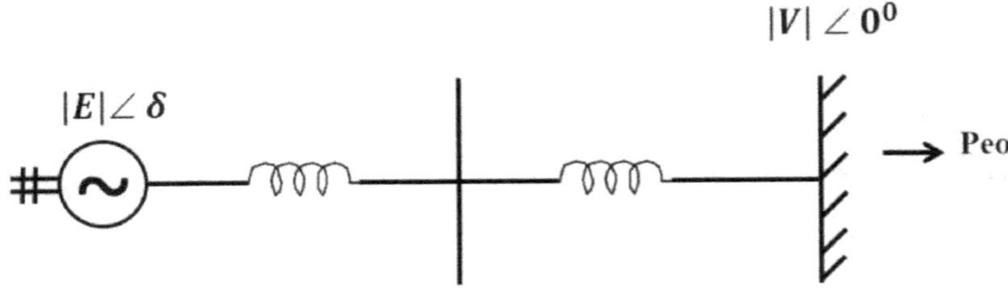

Figure 11.14 Steady state stability using swing equation

Note: Voltage and frequency are constant for infinite bus.

Let P_s and P_{eo} be the mechanical input and electrically output initially.

Case 1: Before small and gradual changes in load demand.

Accelerating power, $P_a = P_s - P_{eo} = 0$

Therefore, rotor neither accelerates nor decelerates and load angle, $\delta = \delta_0$ is constant.

Case 2: After small and gradual changes in load demand.

Let ΔP_e be the increase in electrical output due to small and gradual changes in the load demand.

Let $\Delta \delta$ be the change in the load angle.

Therefore, final electrical output $= P_{eo} + \Delta P_e$

$$P_a = P_e - (P_{eo} + \Delta P_e)$$
$$P_a = -\Delta P_e$$

According to Swing equation,

$$M \cdot \frac{d^2\delta}{dt^2} = -\Delta P_e$$

$$M \cdot \frac{d^2\delta}{dt^2} + \Delta P_e = 0 \qquad \qquad .....(11.21)$$

The solution for equation 11.21 can be obtained by using Taylor series of expansion.

After applying the Taylor series of expansion,

$$M \cdot \frac{d^2\delta}{dt^2} + \frac{\partial P_e}{\partial \delta}\bigg|_{\delta=\delta_0} \Delta\delta = 0$$

From case 2

$$\delta = \delta_0 + \Delta\delta$$

On substituting the above relation

$$M \frac{d^2}{dt^2}(\delta_0 + \Delta\delta) + \frac{\partial P_e}{\partial \delta}\bigg|_{\delta=\delta_0} \Delta\delta = 0$$

As δ_0 is constant,

$$M \frac{d^2(\Delta\delta)}{dt^2} + \frac{\partial P_e}{\partial \delta}\bigg|_{\delta=\delta_0} \Delta\delta = 0$$

$$\left(M \frac{d^2}{dt^2} + \frac{\partial P_e}{\partial \delta}\bigg|_{\delta=\delta_0} \right) \Delta\delta = 0 \qquad \qquad .....(11.22)$$

Equation 11.22 is satisfied if

$$M \frac{d^2}{dt^2} + \frac{\partial P_e}{\partial \delta}\bigg|_{\delta=\delta_0} = 0 \qquad \qquad .....(11.23)$$

Equation 11.23 represents 2nd order characteristic equation to analyze steady state stability

Let $\frac{d}{dt} = k$

$$Mk^2 + \frac{\partial P_e}{\partial \delta}\bigg|_{\delta=\delta_0} = 0$$

The roots of the characteristic equation are

$$Mk^2 = \left(\frac{\partial P_e}{\partial \delta}\bigg|_{\delta=\delta_0}\right)$$

$$k^2 = \frac{1}{M}\left(\frac{\partial P_e}{\partial \delta}\bigg|_{\delta=\delta_0}\right)$$

$$k = \pm\left\{\frac{1}{M}\left(\frac{-\partial P_e}{\partial \delta}\bigg|_{\delta=\delta_0}\right)\right\}^{1/2} \qquad \ldots\ldots(11.24)$$

Equation 11.24 represents the roots of the characteristic equation

Case 1: $\frac{\partial p_e}{\partial \delta}$ is positive

As $\frac{\partial p_e}{\partial \delta} = \frac{EV}{X}\cos\delta$ is positive, load angle δ lies between 0 and 90^0 i.e., $0 < \delta < 90^0$ and the system is stable.

As $\frac{\partial p_e}{\partial \delta}$ is positive, the roots of characteristic equation are non-repeated on $j\omega$ axis. However, stable operation is ensured by damper winding.

Case 2: $\frac{\partial p_e}{\partial \delta}$ is negative

As $\frac{\partial p_e}{\partial \delta} = \frac{EV}{X}\cos\delta$ is negative, load angle, δ is $> 90^0$ and the system is unstable.

The roots of the characteristic equation are real and symmetrical about imaginary axis.

As a pole is located to the right side of $-j\omega$ axis, the system is unstable.

Inertia constant and moment of inertia

$$\text{Inertia costant, } H = \frac{\text{Kinetic energy stored by rotor}}{\text{VA rating of synchronous machine, S}}$$

The units of H are $\frac{MJ}{MVA}$.

$$H = \frac{\frac{1}{2}I\,\omega^2}{S} = \frac{\frac{1}{2}(I\omega)(\omega)}{S}$$

$$= \frac{\frac{1}{2}(M)(2\pi f)}{S} = \frac{M\pi f}{S}$$

$$\text{Moment of inertia, } M = \frac{SH}{\pi f}$$

$$\text{Units of M are } \frac{MJ-Sec}{Elec-deg}, \frac{MJ-Sec}{Elec-rad}$$

11.5 ANALYSIS OF TRANSIENT STABILITY BASED ON NUMBER OF SYNCHRONOUS MACHINES

The transient stability of a system depends on the solution of swing equation.

The solution of the swing equation depends on the number of synchronous machines.

Case 1: One machine System

A synchronous generator is connected to an infinite bus through a transmission line as shown in figure 11.15.

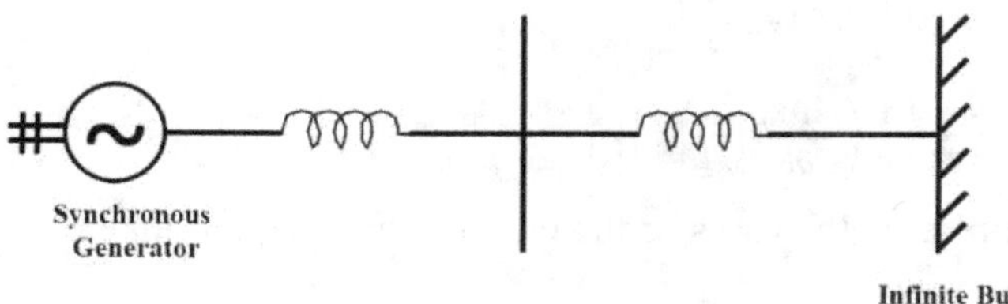

Figure 11.15 One machine system

For a single machine system, the solution of swing equation is obtained by Elliptical Integral method.

Case 2: Two Machine System

A synchronous generator is connected to a synchronous motor through a transmission line as shown in figure 11.16.

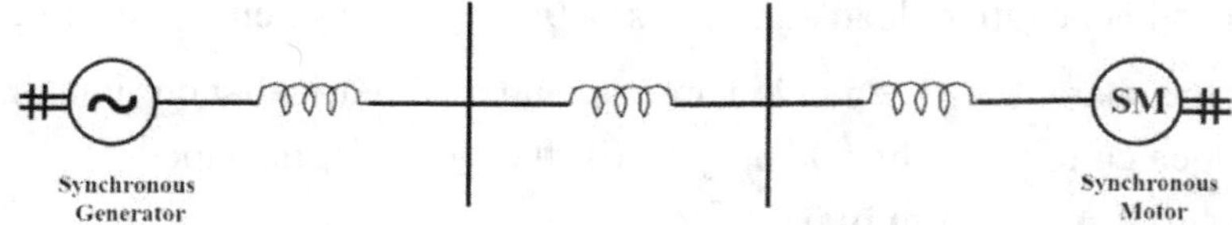

Figure 11.16 Two machine system

For a two machine system, the solution of swing equation is obtained by graphical method known as Equal Area Criteria method.

Case 3: Multi-machine system or interconnected system

Consider an interconnected system as shown in figure 11.17.

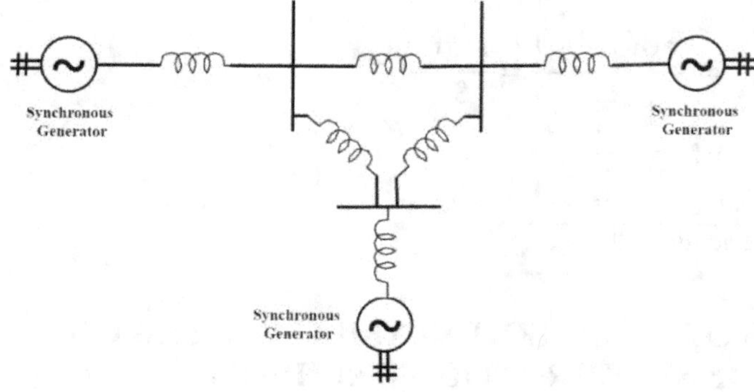

Figure 11.17 Interconnected system

The solution of swing equation is obtained by mathematical technique known as Step-by-Step method or Point-by-Point method.

1. For a two-machine system, the ratings of synchronous generator and synchronous motor are equal.
2. Therefore, the sizes of both synchronous machines are equal.
3. The moments of inertia exerted by both synchronous machines are equal.
4. Therefore, the frequency of rotor oscillations of synchronous generator and synchronous motor are equal.
5. Two machine system can be considered as single machine system and equal area criteria method is also applicable for a single machine system.

Assumptions of Transient Stability

1. The resistance and shunt elements of a transmission line are neglected.
2. The power system network is modeled as a transfer reactance i.e., the reactance measured between source and load by eliminating the fault point.
3. The mechanical input to the synchronous generator is constant.
4. The damping force provided by the rotor is negligible with respect to swinging of rotor.
5. The voltage across the reactance of synchronous machine is constant and the changes in voltages are less compared to changes in reactance of the line.
6. The angular velocity of synchronous machine is assumed constant because the changes in angular velocity during fault are negligible.

11.6 SWINGING OF SYNCHRONOUS MACHINES

11.6.1   Coherent Swinging of Synchronous Machines

In coherent swinging of synchronous machines, the increase in the load demand is equally shared by both synchronous machines.

Consider two alternators connected in parallel to a common bus bar as shown in figure 11.18.

As the load is equally shared by both synchronous machines, the change in rotor axis with respect to stator axis of both synchronous machines is equal $\delta_1 = \delta_2 = \delta$ (say) as shown in figure 11.19.

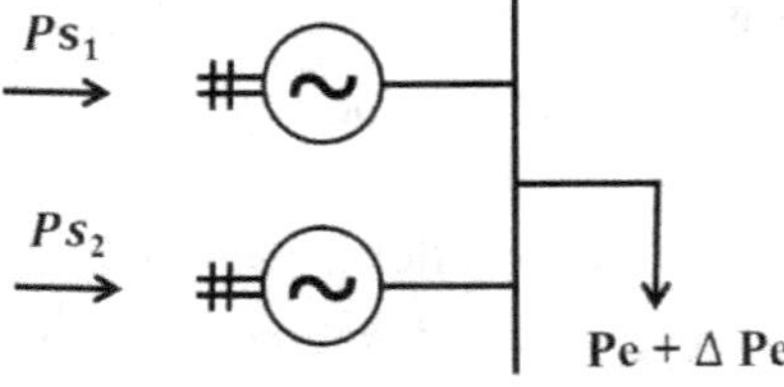

Figure 11.18 Two alternators connected in parallel to a common bus bar

Let P_{S1} and P_{S2} be the mechanical inputs to both alternators.

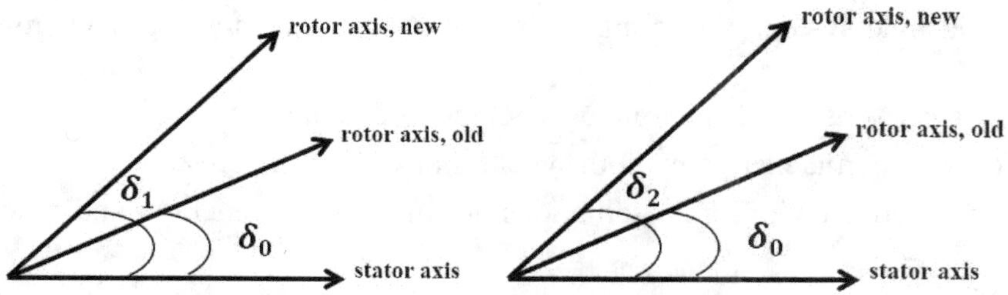

Figure 11.19 Coherent swinging of synchronous machines

Case 1: Let δ_0 be the load angle between rotor axis and stator axis before the load demand increases.

Case 2: Let δ_1 and δ_2 be the load angles of synchronous generators 1 and 2 due to increase in load demand.

As the load is equally shared, $\delta_1 = \delta_2 = \delta$ (say)

The equivalent accelerating power,

$$P_{a,eq} = P_{a1} + P_{a2}$$

According to swing equation,

$$M_{eq} = \frac{d^2\delta}{dt^2} = M_1 \frac{d^2\delta_1}{dt^2} + M_2 \frac{d^2\delta_2}{dt^2}$$

$$= M_1 \frac{d^2\delta}{dt^2} + M_2 \frac{d^2\delta}{dt^2}$$

$$= (M_1 + M_2)\frac{d^2\delta}{dt^2}$$

$$\Rightarrow M_{eq} = M_1 + M_2 \qquad\qquad .....(11.25)$$

$$\Rightarrow H_{eq} = H_1 + H_2 \qquad\qquad .....(11.26)$$

With 'n' alternators swinging coherently together,

$$\Rightarrow M_{eq} = M_1 + M_2 + - - - - - + M_n \text{ and } H_{eq} = H_1 + H_2 + - - - - - + H_n$$

11.6.2 Non-Coherent Swinging of Synchronous Machines

In non-coherent swinging of synchronous machines, the increase in the load demand is shared by only one synchronous machine.

Consider two alternators connected in parallel to a common bus bar as shown in figure 11.20.

As the load is shared by only one synchronous machine, the change in rotor axis with respect to stator axis of both synchronous machines is not equal $\delta_1 \neq \delta_2$ as shown in figure 11.21.

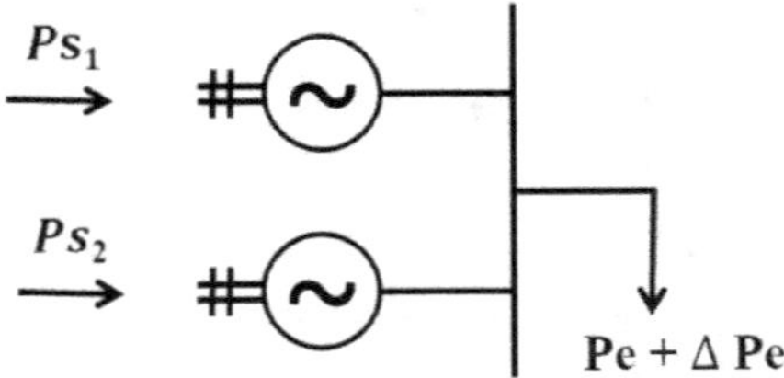

Figure 11.20 Two alternators connected in parallel to a common bus bar

Let P_{S1} and P_{S2} be the mechanical inputs to both alternators.

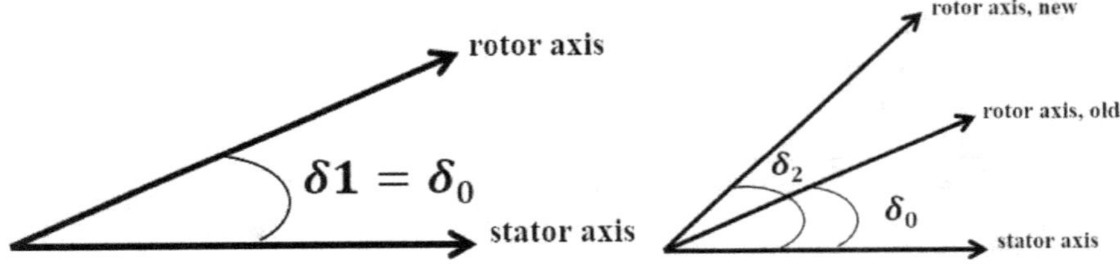

Figure 11.21 Coherent swinging of synchronous machines

Assume that the entire increase in the load demand is met by only second alternator.

Change in load angle, $\delta = \delta_2 - \delta_1$

On differentiating,

$$\Rightarrow \frac{d\delta}{dt} = \frac{d\delta_2}{dt} - \frac{d\delta_1}{dt}$$

$$\Rightarrow \frac{d^2\delta}{dt^2} = \frac{d^2\delta_2}{dt^2} - \frac{d^2\delta_1}{dt^2}$$

$$\Rightarrow \frac{d^2\delta}{dt^2} = \frac{Pa_2}{M_2} - \frac{Pa_1}{M_1}$$

$$\Rightarrow \frac{d^2\delta}{dt^2} = \frac{(Ps_2 - Pe_2)}{M_2} - \frac{(Ps_1 - Pe_1)}{M_1}$$

$$\Rightarrow \frac{d^2\delta}{dt^2} = \left(\frac{Ps_2}{M_2} - \frac{Ps_1}{M_1}\right) - \left(\frac{Pe_2}{M_2} - \frac{Pe_1}{M_1}\right)$$

Multiplying left hand side and right-hand side with $(M_1 M_2 / M_1 + M_2)$

$$\Rightarrow \left(\frac{M_1 M_2}{M_1 + M_2}\right)\frac{d^2\delta}{dt^2} = \left(\frac{M_1 M_2}{M_1 + M_2}\right)\left\{\left(\frac{Ps_2}{M_2} - \frac{Ps_1}{M_1}\right) - \left(\frac{Pe_2}{M_2} - \frac{Pe_1}{M_1}\right)\right\}$$

$$\Rightarrow \left(\frac{M_1 M_2}{M_1 + M_2}\right)\frac{d^2\delta}{dt^2} = \frac{(M_1 Ps_2 - M_2 Ps_1)}{(M_1 + M_2)} - \frac{(M_1 Pe_2 - M_2 Pe_1)}{(M_1 + M_2)}$$

In the standard form,

$$\Rightarrow M_{eq} \cdot \frac{d^2\delta}{dt^2} = P_{a,eq} = P_{s,eq} - P_{e,eq}$$

$$\therefore M_{eq} = \frac{M_1 M_2}{M_1 + M_2}$$

$$\Rightarrow \frac{1}{M_{eq}} = \frac{1}{M_1} + \frac{1}{M_2} \qquad \qquad .....(11.27)$$

$$\Rightarrow \frac{1}{H_{eq}} = \frac{1}{H_1} + \frac{1}{H_2} \qquad \qquad .....(11.28)$$

For non-coherent swinging of 'n' alternators,

$$\Rightarrow \frac{1}{M_{eq}} = \frac{1}{M_1} + \frac{1}{M_2} + \cdots .. + \frac{1}{M_n}$$

and

$$\frac{1}{H_{eq}} = \frac{1}{H_1} + \frac{1}{H_2} + \cdots .. + \frac{1}{H_n}$$

11.7 EQUAL AREA CRITERION METHOD

Equal area Criterion method gives the solution of swing equation for two machine system and one machine system.

According to swing equation,

$$\frac{d^2\delta}{dt^2} = \frac{1}{M} P_a$$

The objective is to determine the condition for stability in terms of the area enclosed by power angle curve.

Multiplying both LHS ad RHS with $2\frac{d\delta}{dt}$

$$\Rightarrow \left(2\frac{d\delta}{dt}\right)\frac{d^2\delta}{dt^2} = \frac{2}{M} P_a \frac{d\delta}{dt}$$

$$\Rightarrow \frac{d}{dt}\left(\frac{d\delta}{dt}\right)^2 = \frac{2}{M} P_a \frac{d\delta}{dt}$$

On integrating,

$$\Rightarrow \left(\frac{d\delta}{dt}\right)^2 = \frac{2}{M} \int \left(P_a \frac{d\delta}{dt}\right) dt$$

$$= \frac{2}{M} \int P_a \, d\delta$$

$$= \frac{2}{M} \int (P_s - P_e) \, d\delta \qquad \qquad .....(11.29)$$

Before any disturbance occurs load angle, $\delta = \delta_0 = $ constant

$$\therefore \frac{d\delta}{dt} = 0 \qquad \qquad .....(11.30)$$

To satisfy the above equation 11.29,

$$\Rightarrow \int (P_s - P_e)\, d\delta = 0$$

Considering part of power angle curve as shown in the figure 11.22.

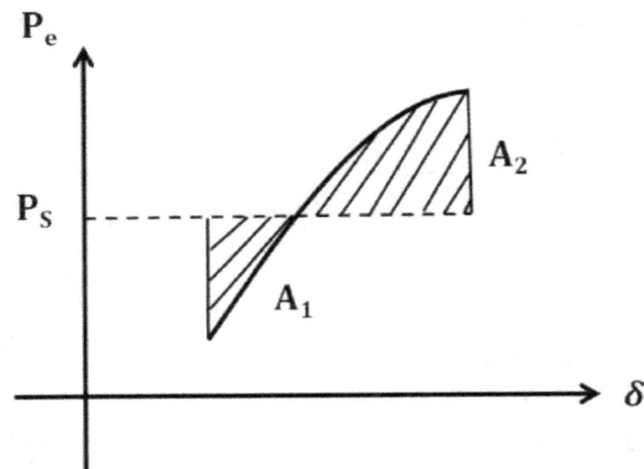

Figure 11.22 Power angle curve

Area A_1

For Area A_1, Mechanical input, P_s is greater than electrical output, P_e

Accelerating power, $P_a = P_s - P_e$ is positive, rotor accelerates and kinetic energy is stored by rotor.

Area A_2

For Area, A_2, Mechanical input, P_s is lesser than electrical output, P_e

Accelerating power, $P_a = P_s - P_e$ is negative, rotor decelerates and kinetic energy stored by rotor during acceleration is restored back to the system

As the condition 2 is satisfied with $A_1 = A_2$, hence the name Equal Area Criteria Method

1. $A_1 < A_2$ the system is stable.
2. $A_1 = A_2$ the system is critically stable.
3. $A_1 > A_2$ the system is unstable.

Assumptions of Equal Area Criterian Method

1. Stable operation: $P_a = 0$ and $\omega = \omega_s$
2. Unstable operation: $P_a \neq 0$ or $\omega \neq \omega_s$ or both.
3. The change in rotor angle is influenced by moment of inertia.

11.8 APPLICATIONS OF EQUAL AREA CRITERION METHOD

The applications of Equal area criteria method are:

1. Sudden increase in mechanical input to a Synchronous generator.

2. Sudden increase in mechanical output of a Synchronous motor.
3. Removal of one of the parallel transmission lines forcibly using fast tracking circuit breakers.
4. Occurrence of fault on one of the parallel transmission lines near bus bar.
5. Occurrence of fault at the middle of one of the parallel transmission lines.
6. Occurrence of fault on bus bar.

11.8.1 Sudden Increase in Mechanical Input to a Synchronous Generator

Consider a synchronous generator connected an infinite bus bar as shown in figure 11.23

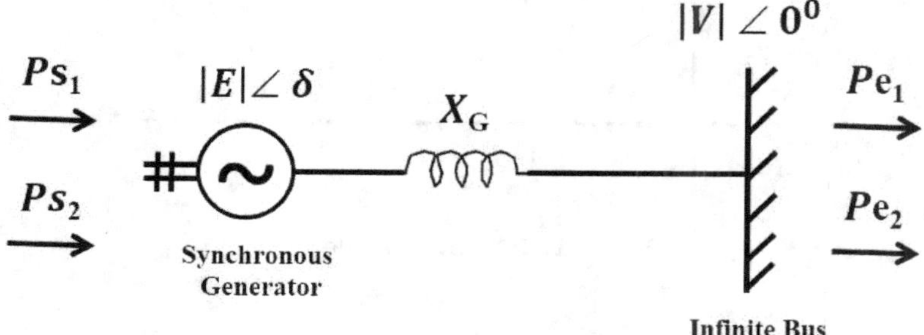

Figure 11.23 Power system network to study sudden increase in mechanical input to a synchronous generator

Let P_{s1} be the initial mechanical input

Let P_{s2} be the final mechanical input

Let P_{e1} be the initial electrical output

$$\Rightarrow P_{e1} = \frac{EV}{X_{1\,eq}} \sin \delta_0 = P_{m1} \sin \delta_0$$

where $X_{1\,eq}$ is transfer reactance before increase in mechanical input.

Let P_{e2} be the final electrical output

$$\Rightarrow P_{e2} = \frac{EV}{X_{2\,eq}} \sin \delta = P_{m2} \sin \delta$$

where $X_{2\,eq}$ is transfer reactance after increase in mechanical input.

The power angle curve for sudden increase in mechanical input to a synchronous generator is shown in figure 11.24.

Let operating point 'a' on P_e curve represents initial mechanical input P_s corresponding to load angle, δ_0.

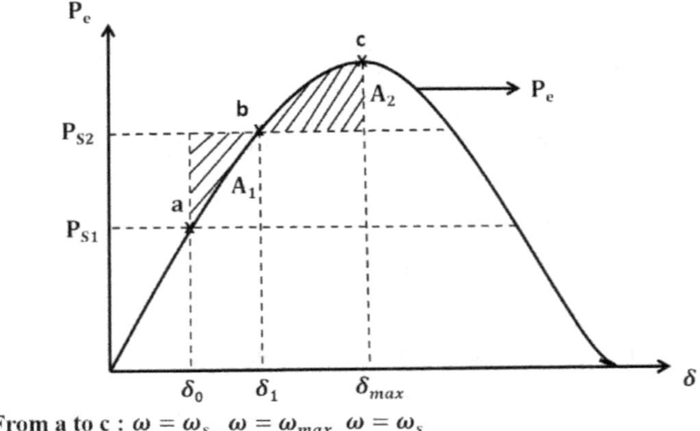

Figure 11.24 The power angle curve for sudden increase in mechanical input to a synchronous generator

At operating point 'a'

Accelerating power, $P_a = P_{s1} - P_{e1} = P_{s1} - P_{m1} \sin \delta_0 = 0$

Rotor neither accelerates nor decelerates

Rotor angular velocity, $\omega = \omega_s$ i. e. stable operating point and load angle, δ do not change

Let operating point b on P_{e2} curve represents final mechanical input P_{s2} corresponding to load angle δ_1.

From operating point 'a' to 'b'

Accelerating power, $P_a = P_{s2} - P_{e2} = P_{s2} - P_{m2} \sin \delta$; $(\delta_0 < \delta < \delta_1)$ is positive

Rotor accelerates.

Rotor angular velocity, $\omega > \omega_s$ i. e. swinging operating point and load angle, δ increases due to acceleration.

At operating point 'b'

Accelerating power, $P_a = P_{s2} - P_{e2} = P_{s2} - P_{m2} \sin \delta_1 = 0$

Rotor neither accelerates nor decelerates.

Rotor angular velocity, $\omega = \omega_{max}$ $(\omega \gg \omega_s)$

i. e. swinging operating point and load angle, δ increases due to moment of inertia.

Let load angle, δ increases till operating point 'c' is reached corresponding to maximum load angle, δ_{max}

From operating point 'b' to 'c'

Accelerating power, $P_a = P_{s2} - P_{e2} = P_{s2} - P_{m2} \sin \delta$; $(\delta_1 < \delta < \delta_{max})$ is negative

Rotor decelerates.

Rotor angular velocity, $\omega < \omega_{max}$ i. e. swinging operating point and load angle, δ increases inspite of deceleration because $\omega > \omega_s$.

At operating point 'c'

Accelerating power, $P_a = P_{s2} - P_{e2} = P_{s2} - P_{m2} \sin \delta_{max}$ is negative.

Rotor decelerates.

Rotor angular velocity, $\omega = \omega_s$ i. e. swinging operating point and load angle, δ decreases due to deceleration.

From operating point 'c' to 'b'

Accelerating power, $P_a = P_{s2} - P_{e2} = P_{s2} - P_{m2} \sin \delta$; $(\delta_{max} < \delta < \delta_1)$ is negative

Rotor decelerates.

Rotor angular velocity, $\omega < \omega_s$ i. e. swinging operating point and load angle, δ decreases due to deceleration.

At operating point 'b'

Accelerating power, $P_a = P_{s2} - P_{e2} = P_{s2} - P_{m2} \sin \delta_1 = 0$

Rotor neither accelerates nor decelerates.

Rotor angular velocity, $\omega = \omega_{min}$ $(\omega \ll \omega_s)$

i. e. swinging operating point and load angle, δ decreases due to moment of inertia.

Now load angle decreases upto a point before 'a' and increases upto 'c' and finally settles at operating point 'b'.

For the power system network to be stable, $A_1 < A_2$

11.8.2   Sudden Increase in Mechanical Output of a Synchronous Motor

Consider a synchronous motor as shown in figure 11.25.

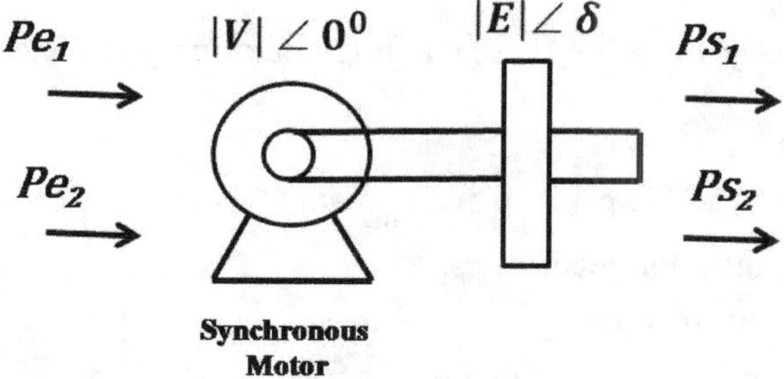

Figure 11.25 Power system network to study sudden increase in mechanical output to a synchronous motor

Let P_{s1} be the initial mechanical output

Let P_{s2} be the final mechanical output

Let P_{e1} be the initial electrical input

$$\Rightarrow P_{e1} = \frac{EV}{X_{1\,eq}} \sin \delta_0 = P_{m1} \sin \delta_0$$

where $X_{1\,eq}$ is transfer reactance before increase in mechanical output.

Let P_{e2} be the final electrical input

$$\Rightarrow P_{e2} = \frac{EV}{X_{2\,eq}} \sin \delta = P_{m2} \sin \delta$$

where $X_{2\,eq}$ is transfer reactance after increase in mechanical output.

The power angle curve for sudden increase in mechanical output to a synchronous motor is shown in figure 11.26.

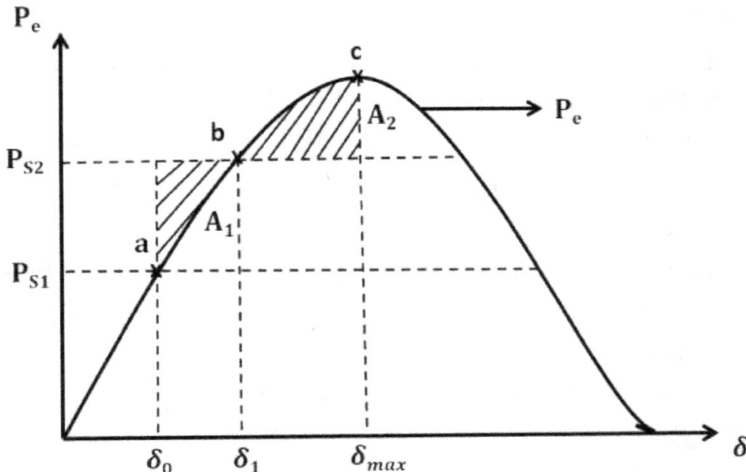

Figure 11.26 Power angle curve for sudden increase in mechanical output to a synchronous motor

Let operating point 'a' on P_e curve represents initial mechanical output P_s corresponding to load angle, δ_0

At operating point 'a'

Accelerating power, $P_a = P_a = P_{e1} - P_{s1} = P_{e1} - P_{m1} \sin \delta_0 - P_{s1} = 0$

Rotor neither accelerates nor decelerates.

Rotor angular velocity, $\omega = \omega_s$ i. e. stable operating point and load angle, δ do not change.

From operating point 'a' to 'b'

Accelerating power, $P_a = P_{s2} - P_{e2} = P_{s2} - P_{m2} \sin \delta;$ $(\delta_0 < \delta < \delta_1)$ is positive

Rotor accelerates.

Rotor angular velocity, $\omega > \omega_s$ i. e. swinging operating point and load angle, δ increases due to acceleration.

At operating point 'b'

Accelerating power, $P_a = P_{e2} - P_{s2} = P_{m2} \sin \delta - P_{s2} = 0$

Rotor neither accelerates nor decelerates.

Rotor angular velocity, $\omega = \omega_{min}$ and $\omega \ll \omega_s$

i. e. swinging operating point and load angle, δ increases due to moment of inertia.

Let load angle, δ increases till operating point 'c' is reached corresponding to maximum load angle, δ_{max}.

From operating point 'b' to 'c'

Accelerating power, $P_a = P_{e2} - P_{s2} = P_{m2} \sin \delta - P_{s2}$ is positive and $\delta_1 < \delta < \delta_{max}$

Rotor accelerates.

Rotor angular velocity, $\omega > \omega_{min}$ i. e. swinging operating point and load angle, δ increases inspite of acceleration because $\omega < \omega_s$.

At operating point 'c'

Accelerating power, $P_a = P_{e2} - P_{s2} = P_{m2} \sin \delta_{max} - P_{s2}$ is positive

Rotor accelerates.

Rotor angular velocity, $\omega = \omega_s$ i. e. swinging operating point and load angle, δ decreases due to acceleration.

From operating point 'c' to 'b'

Accelerating power, $P_a = P_{e2} - P_{s2} = P_{m2} \sin \delta - P_{s2}$ $(\delta_{max} < \delta < \delta_1)$ is positive

Rotor accelerates.

Rotor angular velocity, $\omega > \omega_s$ i. e. swinging operating point and load angle, δ decreases due to acceleration.

At operating point 'b'

Accelerating power, $P_a = P_{e2} - P_{s2} = P_{m2} \sin \delta_1 - P_{s2} = 0$

Rotor neither accelerates nor decelerates.

Rotor angular velocity, $\omega = \omega_{max}$ and $\omega \gg \omega_s$)

i. e. swinging operating point and load angle, δ decreases due to moment of inertia.

Now load angle decreases upto a point before 'a' and increases upto 'c' and finally settles at operating point 'b'.

For power system network to be stable, $A_1 < A_2$

Therefore, Equal Area Criteria method is a measure of area stability.

11.8.3 Removal of One of the Parallel Transmission Lines Forcibly using Fast Acting Circuit Breakers

Consider a synchronous generator connected to an infinite bus bar through a parallel transmission line as shown in figure 11.27.

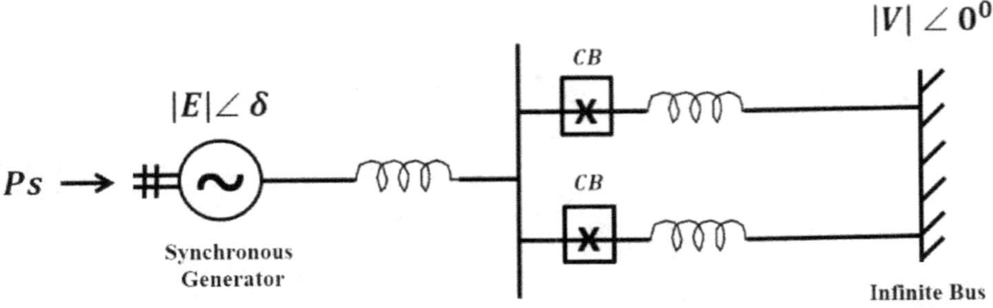

Figure 11.27 Power system network to study removal of one of the parallel transmission lines forcibly using fast acting circuit breakers

As the load demand increases, power generation increases and electrical power transmitted increases. Therefore, accelerating power, $P_a = P_s - P_e$ is negative i.e., rotor decelerates and frequency decreases.

Considering moment of inertia, the permissible variation in frequency of supply is $\pm 5\%$ i.e., 47.5 Hz to 52.5 Hz.If the frequency decreases below 47.5 Hz then power system network becomes unstable.

In order to ensure stable operation of power system network, one of the parallel transmission lines has to be forcibly disconnected by fast acting circuit breakers.

This method of ensuring stable operation of power system network by disconnecting one of the parallel transmission lines forcibly is known as Load Management or Load Curtailment.

Let the mechanical input, P_s be constant.

Let P_{e1} be the electrical output before disconnecting one of the parallel transmission lines.

$$\Rightarrow P_{e1} = \frac{EV}{X_{1\ eq}} \sin \delta_0 = P_{m1} \sin \delta_0$$

where $X_{1\ eq\ p.u.} = X_{G\ p.u.} + \frac{X_{line}}{2} p.\,u.$

Let P_{e2} be the electrical output after disconnecting one of the parallel transmission lines forcibly using fast acting circuit breakers

$$\Rightarrow P_{e2} = \frac{EV}{X_{2\ eq}} \sin \delta = P_{m2} \sin \delta$$

where $X_{2\ eq\ p.u.} = X_{G\ p.u.} + X_{line\ p.u.}$

and $\qquad\qquad P_{e2} < P_{e1}$

The power angle curve for removal of one of the parallel transmission lines forcibly using fast acting circuit breakers is shown in figure 11.28.

Let operating point 'a' on P_{e1} curve represents initial mechanical input P_s corresponding to load angle, δ_0.

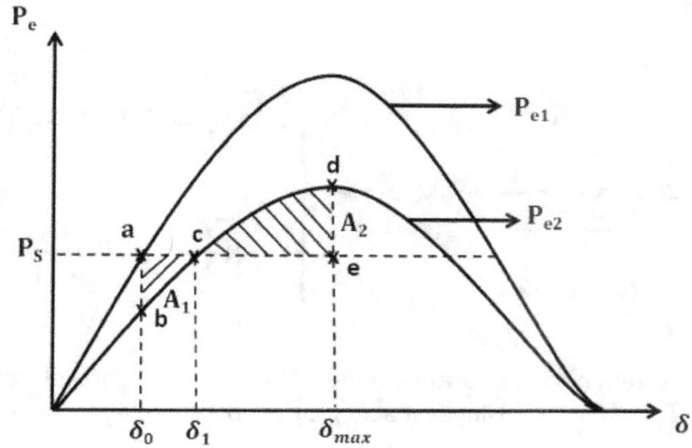

Figure 11.28 Power system network to study removal of one of the parallel transmission lines forcibly using fast acting circuit breakers

At operating point 'a'

Accelerating power, $P_a = P_s - P_{e1} = P_s - P_{m1} \sin \delta_0 = 0$

Rotor neither accelerates nor decelerates

Rotor angular velocity, $\omega = \omega_s$ i.e. stable operating point and load angle, δ do not change

Let operating point b on P_{e2} curve represents the instant at which one of the parallel transmission lines is forcibly disconnected using fast acting circuit breakers, corresponding to load angle, δ_0.

At operating point 'b'

Accelerating power, $P_a = P_s - P_{e2} = P_s - P_{m2} \sin \delta_0$ is positive

Rotor accelerates.

Rotor angular velocity, $\omega > \omega_s$

i.e. swinging operating point and load angle, δ increases due to acceleration.

Let load angle, δ increases till operating point 'c' is reached corresponding to constant mechanical input.

From operating point 'b' to 'c'

Accelerating power, $P_a = P_s - P_{e2} = P_s - P_{m2} \sin \delta;\quad (\delta_0 < \delta < \delta_1)$ is positive

Rotor accelerates.

Rotor angular velocity, $\omega \gg \omega_s$ i.e. swinging operating point and load angle, δ increases due to acceleration.

At operating point 'c'

Accelerating power, $P_a = P_s - P_{e2} = P_s - P_{m2} \sin \delta_1 = 0$.

Rotor neither accelerates nor decelerates.

Rotor angular velocity, $\omega = \omega_{max}$ i.e. swinging operating point and load angle, δ decreases due to moment of inertia.

Let load angle, δ increases till operating point 'd' P_{e2} curve is reached corresponding to maximum load angle, δ_{max}.

From operating point 'c' to 'd'

Accelerating power, $P_a = P_s - P_{e2} = P_s - P_{m2} \sin \delta$; $(\delta_1 < \delta < \delta_{max})$ is negative

Rotor decelerates.

Rotor angular velocity, $\omega < \omega_{max}$ and $\omega >>> \omega_s$ i.e. swinging operating point and load angle, δ increases inspite of deceleration because $\omega >>> \omega_s$.

At operating point 'd'

Accelerating power, $P_a = P_s - P_{e2} = P_s - P_{m2} \sin \delta_{max}$ = negative

Rotor decelerates.

Rotor angular velocity, $\omega \ll \omega_{max}$ and $\omega \gg \omega_s$ i.e. swinging operating point and load angle, δ increases inspite of deceleration because $\omega > \omega_s$.

From operating point 'd' to 'e'

Accelerating power, $P_a = P_s - P_{e2} = P_s - P_{m2} \sin \delta$; $(\delta_1 < \delta < \delta_{max})$ is negative

Rotor decelerates.

Rotor angular velocity, $\omega <<< \omega_{max}$ and $\omega > \omega_s$ i.e. swinging operating point and load angle, δ increases inspite of deceleration because $\omega > \omega_s$.

At operating point 'e'

Accelerating power, $P_a = P_s - P_{e2} = P_s - P_{m2} \sin \delta_{max} = 0$

Rotor neither accelerates nor decelerates.

Rotor angular velocity, $\omega = \omega_s$ i.e. stable operating point and load angle, δ do not change.

For power system network to be stable, $A_1 < A_2$.

11.8.4   Occurrence of Fault at the Middle of One of the Parallel Transmission Line

Consider a synchronous generator connected to an infinite bus bar through a parallel transmission line and consider a fault at the middle of one of the parallel transmission line as shown in figure 11.29.

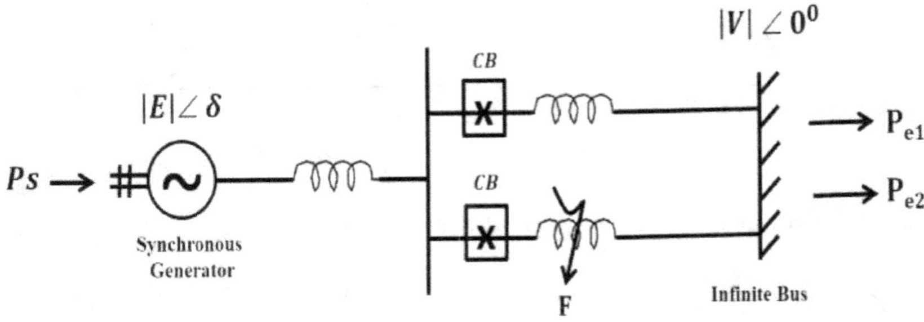

Figure 11.29 Power system network to study fault at the middle of one of the parallel transmission line

Let P_s be the constant mechanical input.

Assume a three phase fault at middle of one of the parallel transmission lines as shown in figure 11.30.

Let P_{e1} be the electrical power transmitted before fault

$$\Rightarrow P_{e1} = \frac{EV}{X_{1\,eq}} \sin \delta_0 = P_{m1} \sin \delta_0$$

and X_{1eq} is transfer reactance before fault

Let P_{e2} be the electrical power transmitted during fault

$$\Rightarrow P_{e2} = \frac{EV}{X_{2\,eq}} \sin \delta = P_{m2} \sin \delta$$

and $X_{2\,eq}$ is transfer reactance during fault

Let P_{e3} be the electrical power transmitted after fault

$$\Rightarrow P_{e3} = \frac{EV}{X_{3\,eq}} \sin \delta = P_{m3} \sin \delta$$

where $X_{3\,eq}$ is transfer reactance after fault

The power angle curve for fault at the middle of one of the parallel transmission line is shown in figure 11.30.

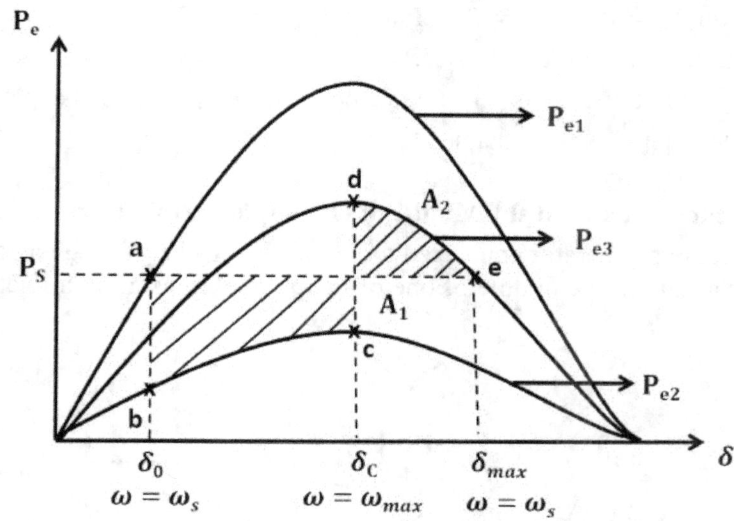

Figure 11.30 Power angle curve for fault at the middle of one of the parallel transmission line

Let operating point 'a' on P_{e1} curve represents initial mechanical input P_s corresponding to load angle, δ_0.

Assume a three phase fault as occurred on one of the parallel transmission lines corresponding to operating point 'a' on P_{e1} curve.

At operating point 'a'

Accelerating power, $P_a = P_s - P_{e1} = P_s - P_{m1} \sin \delta_0 = 0$

Rotor neither accelerates nor decelerates.

Rotor angular velocity, $\omega = \omega_s$ i. e. stable operating point and load angle, δ do not change

Let operating point b on P_{e2} curve represents the instant at which one of the parallel transmission lines is forcibly disconnected using fast acting circuit breakers, corresponding to load angle, δ_0.

At operating point 'b'

Accelerating power, $P_a = P_s - P_{e2} = P_s - P_{m2} \sin \delta_0$ is positive

Rotor accelerates.

Rotor angular velocity, $\omega > \omega_s$

i. e. swinging operating point and load angle, δ increases due to acceleration.

Let load angle, δ increases till operating point 'c' is reached corresponding to the instant fault is cleared by circuit breaker on P_{e2} curve.

Assume that fault is cleared by circuit breaker corresponding to operating point 'c' on P_{e2} curve at load angle, $\delta = \delta_c$

Critical clearing angle, δ_c

The load angle, δ at which fault is cleared by circuit breaker is known as Critical Clearing Angle, δ_c.

From operating point 'b' to 'c'

Accelerating power, $P_a = P_s - P_{e2} = P_s - P_{m2} \sin \delta;$ $(\delta_0 < \delta < \delta_c)$ is positive

Rotor accelerates.

Rotor angular velocity, $\omega \gg \omega_s$ i. e. swinging operating point and load angle, δ increases due to acceleration.

At operating point 'c'

Accelerating power, $P_a = P_s - P_{e2} = P_s - P_{m2} \sin \delta_c$ is positive.

Rotor accelerates.

Rotor angular velocity, $\omega = \omega_{max}$ i. e. swinging operating point and load angle, δ increases due to acceleration.

Let load angle, δ increases till operating point 'd' on P_{e3} curve is reached corresponding to load angle, δ_1.

As the fault is cleared by circuit breaker operating point 'c' on P_{e2} curve, operating point 'c' on P_{e2} curve shifts to 'd' on P_{e3} curve.

At operating point 'd'

Accelerating power, $P_a = P_s - P_{e3} = P_s - P_{m3} \sin \delta_c$ is negative.

Rotor decelerates.

Rotor angular velocity, $\omega \gg \omega_s$ i.e. swinging operating point and load angle, δ increases inspite of deceleration because $\omega \gg \omega_s$.

From operating point 'd' to 'e'

Accelerating power, $P_a = P_s - P_{e3} = P_s - P_{m3} \sin \delta$; $(\delta_c < \delta < \delta_{max})$ is negative

Rotor decelerates.

Rotor angular velocity, $\omega > \omega_s$ i.e. swinging operating point and load angle, δ increases inspite of deceleration because $\omega > \omega_s$

At operating point 'e'

Accelerating power, $P_a = P_s - P_{e3} = P_s - P_{m3} \sin \delta_{max} = 0$

Rotor neither accelerates nor decelerates.

Rotor angular velocity, $\omega = \omega_s$ i.e. stable operating point and load angle, δ do not change

Calculation of critical clearing angle , δ_c

$$\int_{\delta_0}^{\delta_{max}} P_a d\delta = 0 \quad \text{i.e.} \int_{\delta_0}^{\delta_c} P_{a2} d\delta + \int_{\delta_c}^{\delta_{max}} P_{a3} d\delta = 0$$

$$\Rightarrow \int_{\delta_0}^{\delta_c} (P_s - P_{e2}) d\delta + \int_{\delta_c}^{\delta_{max}} (P_s - P_{e3}) d\delta = 0$$

$$\Rightarrow \int_{\delta_0}^{\delta_c} (P_s - P_{m2} \sin \delta) d\delta + \int_{\delta_c}^{\delta_{max}} (P_s - P_{m3} \sin \delta) d\delta = 0$$

$$\Rightarrow P_s(\delta_c - \delta_0) + P_{m2}(\cos\delta_c - \cos\delta_0) + P_s(\delta_{max} - \delta_c) + P_{m3}(\cos\delta_{max} - \cos\delta_c) = 0$$

$$\Rightarrow P_s(\delta_c - \delta_0 + \delta_{max} - \delta_c) + \cos\delta_c(P_{m2} - P_{m3}) - P_{m2}\cos\delta_0 + P_{m3}\cos\delta_{max} = 0$$

$$\Rightarrow P_s(\delta_{max} - \delta_0) + \cos\delta_c(P_{m2} - P_{m3}) + P_{m3}\cos\delta_{max} - P_{m2}\cos\delta_0 = 0$$

$$\Rightarrow \cos\delta_c(P_{m3} - P_{m2}) = P_s(\delta_{max} - \delta_0) + P_{m3}\cos\delta_{max} - P_{m2}\cos\delta_0$$

$$\Rightarrow \cos\delta_c = \frac{(\delta_{max} - \delta_0) + P_{m3}\cos\delta_{max} - P_{m2}\cos\delta_0}{(P_{m3} - P_{m2})}$$

$$\Rightarrow \delta_c = \cos^{-1}\left[\frac{(\delta_{max} - \delta_0) + P_{m3}\cos\delta_{max} - P_{m2}\cos\delta_0}{(P_{m3} - P_{m2})}\right]$$

where δ_{max} and δ_0 are in radians in the calculation of $(\delta_{max} - \delta_0)$ and

δ_{max} and δ_0 are in degrees in the calculation of $P_{m3}\cos\delta_{max} - P_{m2}\cos\delta_0$.

At operating point 'a':

$$P_s = P_{e1} = P_{m1} \sin \delta_0$$

$$\delta_0 = \sin^{-1}\left(\frac{P_s}{P_{m1}}\right) \text{ Elec. deg} \qquad .....(11.31)$$

At operating point 'e':

$$P_s = P_{e3} = P_{m3} \sin \delta_{max} = P_{m3} \sin(\pi - \delta_{max})$$

$$\delta_{max} = \pi - \sin^{-1}\left(\frac{P_s}{P_{m3}}\right) \text{ Elec. deg} \qquad .....(11.32)$$

11.8.5   Occurrence of Fault on One of the Parallel Transmission Lines near Bus Bar

Consider a synchronous generator connected to an infinite bus bar through a parallel transmission line and consider a fault on one of the parallel transmission line near bus bar as shown in figure 11.31.

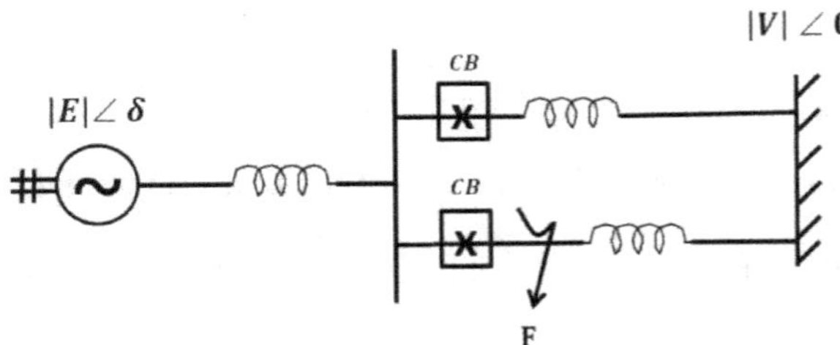

Figure 11.31 Power system network to study fault on one of the parallel transmission lines near bus bar

Let P_s be the constant mechanical input.

Assume a three-phase fault on one of the parallel transmission lines near bus bar as shown in figure 11.31.

Let P_{e1} be the electrical power transmitted before fault

$$\Rightarrow P_{e1} = \frac{EV}{X_{1\,eq}} \sin \delta_0 = P_{m1} \sin \delta_0$$

and X_{1eq} is transfer reactance before fault

Let P_{e2} be the electrical power transmitted during fault

$$\Rightarrow P_{e2} = \frac{EV}{X_{2\,eq}} \sin \delta = P_{m2} \sin \delta$$

and $X_{2\,eq}$ is transfer reactance during fault

As fault on the busbar is considered as fault on the busbar, electrical power transmitted during fault is zero.

$$\Rightarrow P_{e2} = 0$$

Let P_{e3} be the electrical power transmitted after fault

$$\Rightarrow P_{e3} = \frac{EV}{X_{3\,eq}} \sin \delta = P_{m3} \sin \delta$$

where $X_{3\,eq}$ is transfer reactance after fault

The power angle curve for fault at the middle of one of the parallel transmission line is shown in figure 11.32.

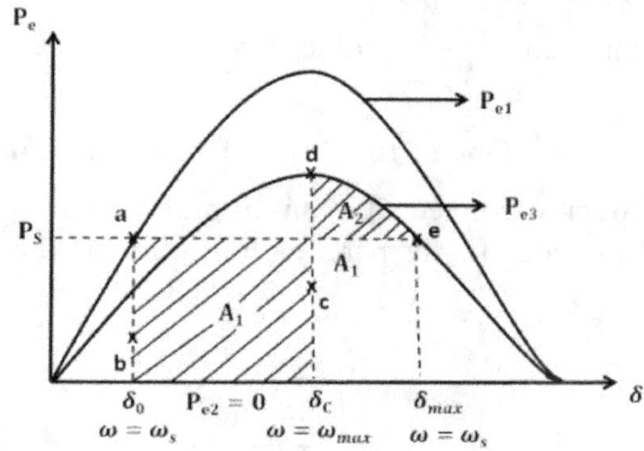

Figure 11.32 Power angle curve for fault on one of the parallel transmission lines near bus bar

Let operating point 'a' on P_{e1} curve represents initial mechanical input P_s corresponding to load angle, δ_0.

Assume a three phase fault as occurred on one of the parallel transmission lines corresponding to operating point 'a' on P_{e1} curve

At operating point 'a'

Accelerating power, $P_a = P_s - P_{e1} = P_s - P_{m1} \sin \delta_0 = 0$

Rotor neither accelerates nor decelerates

Rotor angular velocity, $\omega = \omega_s$ i. e. stable operating point and load angle, δ do not change

Let operating point b on P_{e2} curve represents the instant at which one of the parallel transmission lines is forcibly disconnected using fast acting circuit breakers, corresponding to load angle δ_0.

At operating point 'b'

Accelerating power, $P_a = P_s - P_{e2} = P_s - P_{m2} \sin \delta = P_s$ is positive

Rotor accelerates.

Rotor angular velocity, $\omega > \omega_s$

i. e. swinging operating point and load angle, δ increases due to acceleration.

Let load angle, δ increases till operating point 'c' is reached corresponding to the instant fault is cleared by circuit breaker on P_{e2} curve located on X − axis.

Assume that fault is cleared by circuit breaker corresponding to operating point 'c' on P_{e2} curve at load angle, $\delta = \delta_c$

Critical clearing angle δ_c

The load angle, δ at which fault is cleared by circuit breaker is known as Critical Clearing Angle, δ_c.

From operating point 'b' to 'c'

Accelerating power, $P_a = P_s - P_{e2} = P_s - P_{m2} \sin \delta = P_s$; $(\delta_0 < \delta < \delta_c)$ is positive.

Rotor accelerates.

Rotor angular velocity, $\omega \gg \omega_s$ i.e. swinging operating point and load angle, δ increases due to acceleration.

At operating point 'c'

Accelerating power, $P_a = P_s - P_{e2} = P_s - P_{m2} \sin \delta_c = P_s$ is positive.

Rotor accelerates.

Rotor angular velocity, $\omega = \omega_{max}$i. e. swinging operating point and load angle, δ increases due to acceleration.

Let load angle, δ increases till operating point'd' on P_{e3} curve is reached corresponding to load angle,δ_1.

As the fault is cleared by circuit breaker operating point 'c' on P_{e2} curve on $X -$ axis, operating point 'c' on P_{e2} curve shifts to 'd' on P_{e3} curve.

At operating point 'd'

Accelerating power, $P_a = P_s - P_{e3} = P_s - P_{m3} \sin \delta_c$ is negative.

Rotor decelerates.

Rotor angular velocity, $\omega \gg \omega_s$i. e. swinging operating point and load angle, δ increases inspite of deceleration because $\omega \gg \omega_s$.

From operating point 'd' to 'e'

Accelerating power, $P_a = P_s - P_{e3} = P_s - P_{m3} \sin \delta$; $(\delta_c < \delta < \delta_{max})$ is negative

Rotor decelerates.

Rotor angular velocity, $\omega > \omega_s$ i. e. swinging operating point and load angle, δ increases inspite of deceleration because $\omega > \omega_s$

At operating point 'e'

Accelerating power, $P_a = P_s - P_{e3} = P_s - P_{m3} \sin \delta_{max} = 0$

Rotor neither accelerates nor decelerates.

Rotor angular velocity, $\omega = \omega_s$i. e. stable operating point and load angle, δ do not change

Calculation of critical clearing angle , δ_c

$$\int_{\delta_0}^{\delta_{max}} P_a d\delta = 0 \quad \text{i. e.} \quad \int_{\delta_0}^{\delta_c} P_{a2} d\delta + \int_{\delta_c}^{\delta_{max}} P_{a3} d\delta = 0$$

At operating point 'e':

$$P_s = P_{e3} = P_{m3} \sin \delta_{max} = P_{m3} \sin(\pi - \delta_{max})$$

$$\delta_{max} = \pi - \sin^{-1}\left(\frac{P_s}{P_{m3}}\right) \text{ Elec. deg} \qquad \qquad .....(11.33)$$

11.8.6   Occurrence of Fault on Bus Bar

Consider a synchronous generator connected to an infinite bus bar through a parallel transmission line and consider a fault on busbar as shown in figure 11.33.

Let P_s be the constant mechanical input.

Assume a three-phase fault on one of the parallel transmission lines near bus bar as shown in figure 11.33.

Let P_{e1} be the electrical power transmitted before fault

$$\Rightarrow P_{e1} = \frac{EV}{X_{1eq}} \sin \delta_0 = P_{m1} \sin \delta_0$$

and X_{1eq} is transfer reactance before fault

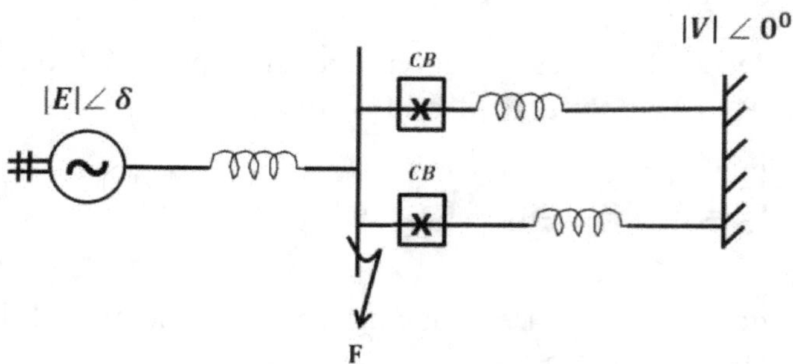

Figure 11.33 Power system network to study fault on busbar

Let P_{e2} be the electrical power transmitted during fault

$$\Rightarrow P_{e2} = \frac{EV}{X_{2eq}} \sin \delta = P_{m2} \sin \delta$$

and $X_{2\,eq}$ is transfer reactance during fault

As fault on the busbar is considered as fault on the busbar, electrical power transmitted during fault is zero.

$$\Rightarrow P_{e2} = 0$$

Let P_{e3} be the electrical power transmitted after fault

$$\Rightarrow P_{e3} = \frac{EV}{X_{3\,eq}} \sin \delta = P_{m3} \sin \delta$$

where $X_{3\,eq}$ is transfer reactance after fault

As fault is on the bus bar, after the fault on the bus bar is cleared by circuit breaker, electrical power transmitted after fault= electrical power transmitted before fault i.e. $P_{e3} = P_{e1}$

The power angle curve for fault on bus bar is shown in figure 11.34.

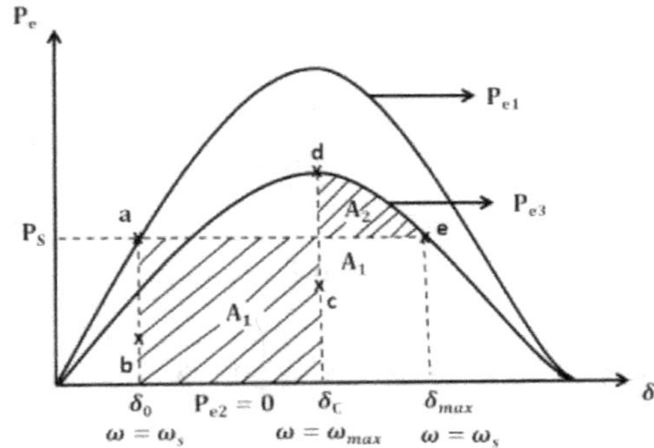

Figure 11.34 Power angle curve for fault on bus bar

Let operating point 'a' on P_{e1} curve represents initial mechanical input P_s corresponding to load angle, δ_0.

Assume a three phase fault as occurred on one of the parallel transmission lines corresponding to operating point 'a' on P_{e1} curve

At operating point 'a'

Accelerating power, $P_a = P_s - P_{e1} = P_s - P_{m1} \sin \delta_0 = 0$

Rotor neither accelerates nor decelerates

Rotor angular velocity, $\omega = \omega_s$ i.e. stable operating point and load angle, δ do not change

Let operating point b on P_{e2} curve represents the instant at which one of the parallel transmission lines is forcibly disconnected using fast acting circuit breakers, corresponding to load angle δ_0.

At operating point 'b'

Accelerating power, $P_a = P_s - P_{e2} = P_s - P_{m2} \sin \delta = P_s$ is positive

Rotor accelerates.

Rotor angular velocity, $\omega > \omega_s$

i.e. swinging operating point and load angle, δ increases due to acceleration.

Let load angle, δ increases till operating point 'c' is reached corresponding to the instant fault is cleared by circuit breaker on P_{e2} curve located on $X - $ axis.

Assume that fault is cleared by circuit breaker corresponding to operating point 'c' on P_{e2} curve at load angle, $\delta = \delta_c$

Critical clearing angle, δ_c

The load angle, δ at which fault is cleared by circuit breaker is known as Critical Clearing Angle, δ_c.

From operating point 'b' to 'c'

Accelerating power, $P_a = P_s - P_{e2} = P_s - P_{m2} \sin \delta = P_s;$ $(\delta_o < \delta < \delta_c)$ is positive

Rotor accelerates.

Rotor angular velocity, $\omega \gg \omega_s$ i.e. swinging operating point and load angle, δ increases due to acceleration.

At operating point 'c'

Accelerating power, $P_a = P_s - P_{e2} = P_s - P_{m2} \sin \delta_c = P_s$ is positive.

Rotor accelerates.

Rotor angular velocity, $\omega = \omega_{max}$ i.e. swinging operating point and load angle, δ increases due to acceleration.

Let load angle, δ increases till operating point 'd' on P_{e3} curve is reached corresponding to load angle, δ_1.

As the fault is cleared by circuit breaker operating point 'c' on P_{e2} curve on $X - axis$, operating point 'c' on P_{e2} curve shifts to 'd' on P_{e3} curve.

At operating point 'd'

Accelerating power, $P_a = P_s - P_{e3} = P_s - P_{e1}$ is negative

Rotor decelerates.

Rotor angular velocity, $\omega \gg \omega_s$ i.e. swinging operating point and load angle, δ increases inspite of deceleration because $\omega \gg \omega_s$.

From operating point 'd' to 'e'

Accelerating power, $P_a = P_s - P_{e3} = P_s - P_{e3} = P_s - P_{e1}$ is negative

Rotor decelerates.

Rotor angular velocity, $\omega > \omega_s$ i.e. swinging operating point and load angle, δ increases inspite of deceleration because $\omega > \omega_s$

At operating point 'e'

Accelerating power, $P_a = P_s - P_{e3} = P_s - P_{e1} = 0$

Rotor neither accelerates nor decelerates.

Rotor angular velocity, $\omega = \omega_s$ i.e. stable operating point and load angle, δ do not change .

11.8.7 Critical Clearing Time, t_c

The time taken by circuit breaker to clear the fault is known as Critical Clearing time.

As electrical power transmitted during fault is zero and according to swing equation,

$$\Rightarrow M \cdot \frac{d^2\delta}{dt^2} = P_a = P_s - P_{e2} = P_s - 0 = P_s$$

$$\Rightarrow \frac{d^2\delta}{dt^2} = \frac{1}{M} P_s \Rightarrow constant$$

Integrating with respect to time

$$\Rightarrow \frac{d\delta}{dt} = \frac{P_s}{M} t + K_1$$

Integrating with respect to time

$$\Rightarrow \delta = \frac{P_s}{M} \cdot \frac{t^2}{2} + K_1 t + K_2$$

Let $K_1 t + K_2 = K$

K is an unknown constant obtained from initial conditions

If $\delta = \delta_c$ then $t = t_c$

$$\Rightarrow \delta_c = \frac{P_s}{M} \cdot \frac{t_c^2}{2} + K \qquad\qquad .....(11.34)$$

At $t_c = 0$

$$\Rightarrow \delta_0 = \frac{P_s}{M} \cdot \frac{0^2}{2} + K$$

$$K = \delta_0 \qquad\qquad .....(11.35)$$

Substituting 11.35 in 11.34

$$\Rightarrow \delta_c = \frac{P_s}{M} \cdot \frac{t_c^2}{2} + \delta_0$$

$$\Rightarrow \frac{P_s}{M} \cdot \frac{t_c^2}{2} = \delta_c - \delta_0$$

$$\Rightarrow t_c^2 = \frac{2M}{P_s}(\delta_c - \delta_0)$$

$$\Rightarrow t_c = \left\{\frac{2M}{P_s}(\delta_c - \delta_0)\right\}^{1/2}$$

where δ_c and δ_0 are in radians.

Therefore, critical clearing time depends on moment of inertia.

11.9   STEP BY STEP METHOD OR POINT BY POINT METHOD

It gives the solution of swing equation for multi-machine system. It provides the information about change of the rotor angle $\Delta\delta$ for short duration of time Δt. After calculating the change of angle $\Delta\delta$ the actual angle will be calculated in order to know the position of the rotor i.e., δ

Assumptions

1. The accelerating power calculated at the beginning of the interval is assumed to be constant from the middle of the previous interval to the middle of the next interval.

2. The angular velocity calculated at the middle of the interval is assumed to be constant throughout the interval.

3. The change in rotor angle calculated is having a smooth variation till the end of the interval.

Considering the above assumptions,

(a) The calculation of change of the angle is required for the time period of 0.05 sec or 10 intervals

(b) If the actual angle at any interval is less than the previous angle then the Synchronous machine is stable.

(c) If the actual angle is not less than the previous angle, the end of the 10th interval then Synchronous machine is unstable.

(d) The calculation of rotor angle is during the fault and after the fault because there will be swinging of the machine even after the fault due to moment of inertia.

Calculation procedure

The change in angular velocity during 1st interval

$$\Delta\omega_1 = \alpha \cdot \Delta t$$

Angular velocity at the end of the 1st interval

$$\omega_1 = \omega_0 + \Delta\omega_1$$

The change in rotor angle during 1st interval

$$\Delta\delta_1 = \omega_1 \cdot \Delta t$$
$$\Delta\delta_1 = \Delta\delta_0 + \alpha(\Delta t)^2$$

The actual rotor angle at the end of the 1st interval

$$\delta_1 = \delta_0 + \Delta\delta_1$$

The change in angular velocity during 2nd interval

$$\Delta\omega_2 = \alpha \cdot \Delta t$$

Angular velocity at the end of the 2nd interval

$$\omega_2 = \omega_1 + \Delta\omega_2$$

The change in rotor angle during 2nd interval

$$\Delta\delta_2 = \omega_2 \cdot \Delta t$$
$$\Delta\delta_2 = \Delta\delta_1 + \alpha(\Delta t)^2$$

The actual rotor angle at the end of the 2nd interval

$$\delta_2 = \delta_1 + \Delta\delta_2$$

In general, the change in rotor angle during n^{th} interval

$$\delta_n = \delta_{n-1} + \Delta\delta_n$$

Assumptions in calculation procedure

The accelerating power will be substituted as average accelerating power when the system is having a sudden change i.e., fault occurring point and fault clearing point.

For the remaining intervals will be considered as accelerating power only.

During fault occurring point,

$$P_{avg} = \frac{(P_s - P_{e1}) + (P_s - P_{e2})}{2}$$

At fault clearing time,

$$P_{a\ avg} = \frac{(P_s - P_{e2}) + (P_s - P_{e3})}{2}$$

SOLVED PROBLEMS

Q1. A synchronous generator is having a synchronous reactance of 1.2 p.u. and is connected to infinite bus through a transformer and transmission line. The infinite bus voltage is 1 p.u. and excitation voltage is 1.2 p.u. The reactance of transformer and transmission line is 0.6 p.u. The inertia constant of the synchronous machine is 4 MJ/MVA. The alternator delivers 50% of max. power initially. Calculate the frequency of the rotor oscillations.

Solution:

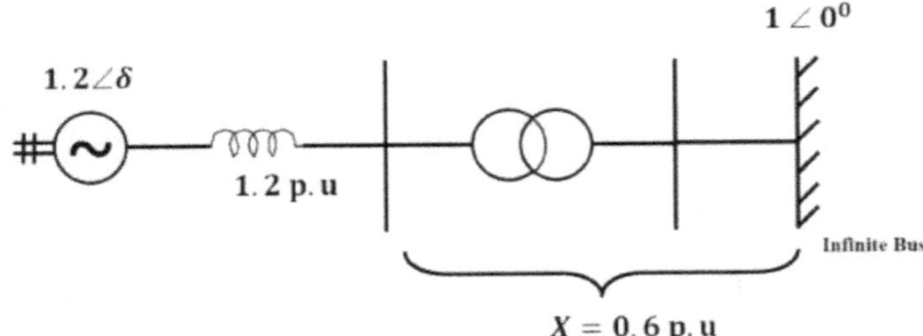

Given data:

Reactance of synchronous generator = 1.2 p.u.

Reactance of transformer and transmission line = 0.6 p.u.

$$X_{TF\ p.u.} + X_{TL\ p.u} = 0.6\ p.u.$$

Voltage at infinite bus = 1 p.u.

Excitation voltage = 1.3.p.u.

Inertia constant of machine = 4 MJ/MVA.

Power delivered by alternator initially=50%

Moment of inertia, $M = \dfrac{SH}{\pi f} = \dfrac{1.0 \times 4}{\pi \times 50} = 0.02546\ MJ - Sec/Elec - rad$

Initially, mechanical input, $P_S = P_{eo} = P_m \sin \delta_0 = 0.5\ Pm$

$$\Rightarrow \sin \delta_0 = 0.5$$

$$\Rightarrow \delta_0 = 30^0$$

Synchronizing power coefficient,

$$\left.\frac{\partial P_e}{\partial \delta}\right|_{\delta=\delta_0} = \frac{EV}{X}\cos\delta_0 = \frac{1.2 \times 1.0}{1.8}\cos(30^0) = 0.577$$

Frequency of rotor oscillations,

$$K = \pm\left\{\left(\frac{1}{M}\right)\left(\left.\frac{-\partial P_e}{\partial \delta}\right|_{\delta=\delta_0}\right)\right\}^{1/2} \frac{rad}{sec}$$

$$= \pm\left\{\left(\frac{1}{0.0254}\right)(-0.577)\right\}^{1/2}$$

$$= 4.8 \ rad/sec$$

$$= \frac{4.8}{2\pi} = 0.764 \ Hz$$

Frequency of rotor oscillations = 0.764 Hz.

Q2. Calculate the frequency of oscillations, if alternator delivers 80% of max. power initially with rest of the data as in Q1.

Solution:

Given data:

Reactance of synchronous generator = 1.2 p.u.

Reactance of transformer and transmission line = 0.6 p.u.

$$X_{TF \ p.u.} + X_{TL \ p.u} = 0.6 \ p.u.$$

Voltage at infinite bus = 1 p.u.

Excitation voltage = 1.3.p.u.

Inertia constant of machine = 4 MJ/MVA.

Power delivered by alternator initially=50%

Moment of inertia, $M = \dfrac{SH}{\pi f} = \dfrac{1.0*4}{\pi*50} = 0.02546 \ MJ - Sec/Elec - rad$

Initially, Mechanical input, $P_S = P_{eo} = P_m \sin\delta_0 = 0.8 \ Pm$

$$\Rightarrow \sin\delta_0 = 0.8$$

$$\Rightarrow \delta_0 = 53.13^0$$

Synchronizing power coefficient,

$$\left.\frac{\partial P_e}{\partial \delta}\right|_{\delta=\delta_0} = \frac{EV}{X}\cos\delta_0 = \frac{1.2 \times 1.0}{1.8}\cos(53.13^0) = 0.4$$

Frequency of rotor oscillations,

$$K = \pm \left\{ \left(\frac{1}{M}\right)\left(\frac{-\partial P_e}{\partial \delta}\bigg|_{\delta=\delta_0}\right) \right\}^{1/2} \frac{rad}{sec}$$

$$= \pm \left\{ \left(\frac{1}{0.0254}\right)(-0.4) \right\}^{1/2}$$

$$= 3.968 \ rad/sec \ = \frac{3.968}{2\pi} = 0.63 \ Hz$$

Frequency of rotor oscillations $= 0.63$ Hz.

From above, it can be concluded that as the loading on the alternator increases, the frequency of the rotor oscillations decreases.

Q3. For the given p.s. network, a three phase fault occurs at the middle of one of the parallel transmission lines. Calculate the maximum powers transmitted through the line.

(a) Before fault

(b) During fault

(c) After fault

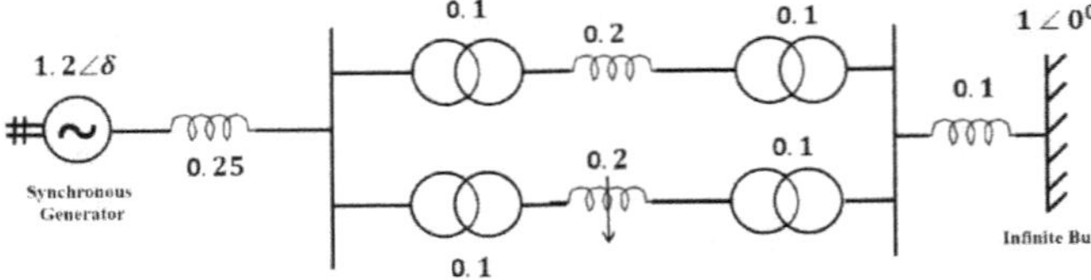

Solution:

Electrical power transmitted before fault

$$Pe_1 = \frac{EV}{X_{1\ eq}} \sin \delta_0 = P_{m1} \sin \delta_0$$

where $X_{1\ eq}$ is transfer reactance before fault.

P_{m1} is maximum electrical power transmitted before fault.

Electrical power transmitted during fault

$$Pe_2 = \frac{EV}{X_{2\ eq}} \sin \delta = P_{m2} \sin \delta$$

where $X_{2\ eq}$ is transfer reactance during fault.

P_{m2} is maximum electrical power transmitted during fault.

Electrical power transmitted after fault

$$Pe_3 = \frac{EV}{X_{1\,eq}} \sin \delta = P_{m3} \sin \delta$$

For any fault, $P_{m2} < P_{m3} < P_{m1}$

$$P_{m2} = \frac{EV}{X_{2\,eq}} \cdot \frac{X_{1\,eq}}{X_{1\,eq}} = \left(\frac{X_{1\,eq}}{X_{2\,eq}}\right) P_{m1} = r_1 \cdot P_{m1} = r_1 < 1$$

$$P_{m3} = \frac{EV}{X_{3\,eq}} \cdot \frac{X_{1\,eq}}{X_{1\,eq}} = \left(\frac{X_{1\,eq}}{X_{2\,eq}}\right) P_{m1} = r_2 \cdot P_{m1} = r_2 < 1$$

$$r_1 < r_2$$

Before fault: Transfer reactance before fault,

$$X_{1\,eq\,p.u.} = 0.25 + (0.4 \parallel 0.4) + 0.1 = 0.55 \text{ p. u.}$$

$$P_{m1} = \frac{EV}{X_{1\,eq}} = \frac{1.2 \times 1.0}{0.55} = 2.18 \text{ p. u.}$$

After fault: Even though fault is cleared, it is assumed that transmission line is not synchronized with alternator

$$X_{3\,eq\,p.u.} = j\{0.25 + (0.4) + 0.1\} = j\,0.75 \text{ p. u.}$$

$$P_{m3} = \frac{EV}{X_{3\,eq}} = \frac{1.2 \times 1.0}{0.75} = 1.6 \text{ p. u.}$$

During fault: Reducing the power system network as follows:

Step 1: Identifying reactance w.r.t fault point on either side of the transmission line.

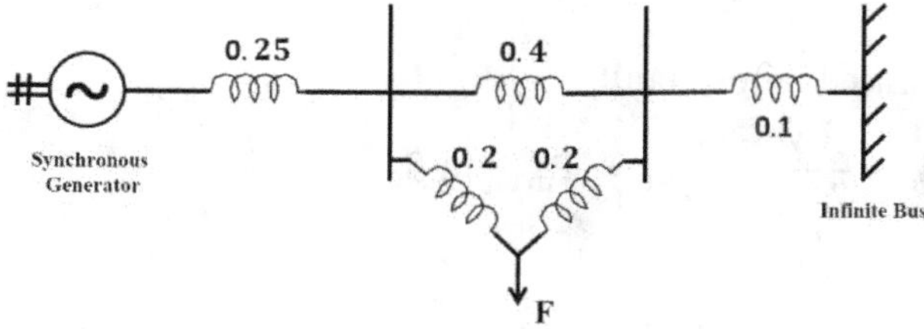

Step 2: Converting delta into star network.

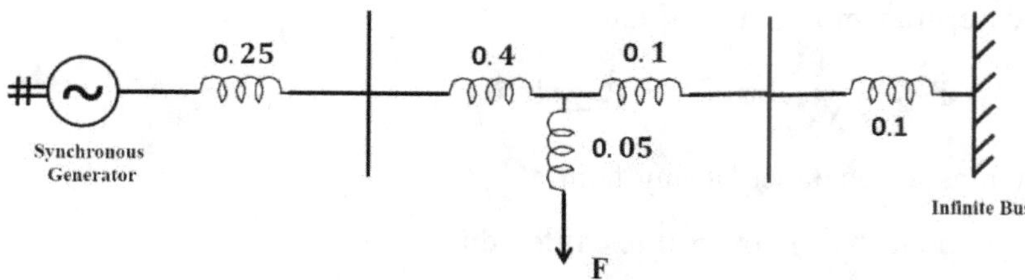

Step 3: Finding the transfer reactance

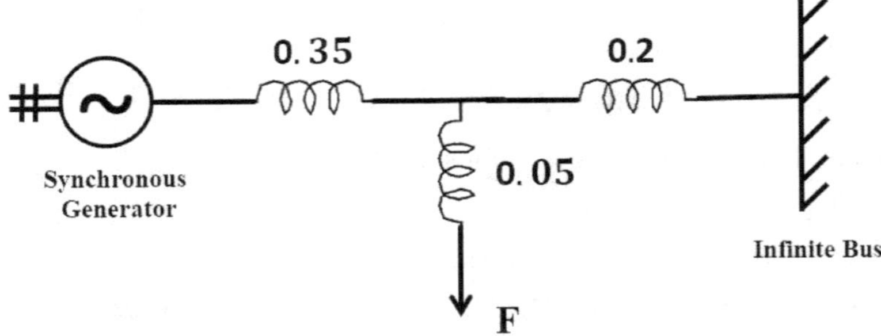

$$X_{2\ eq\ p.u.} = 0.35 + 0.2 + \frac{0.35 \times 0.2}{0.05} = 1.9\ p.u.$$

$$P_{m2} = \frac{EV}{X_{2\ eq}} = \frac{1.2 \times 1.0}{1.95} = 0.615\ p.u.$$

It can be observed that, $P_{m2} < P_{m3} < P_{m1}$.

Q4. Two 50 Hz generating units operating in parallel within the power plants and have the following ratings.

Unit 1: 500 MVA, 0.85 P.F., 20 KV, 3000 rpm, $H_1 = 5$ MJ/MVA

Unit 2: 200 MVA, 0.9 P.F., 20 KV, 1500 rpm, $H_2 = 5$ MJ/MVA

Calculate the equivalent inertia constant on 100 MVA common base.

Solution:

$$\text{Moment of inertia,} M = \frac{SH}{\pi f}$$

Unit 1:

$$S_1 \propto \frac{1}{H_1}$$

$$500 \propto \frac{1}{5}\ \text{(Based on ratings)}$$

$$100 \propto \frac{1}{H_{1,new}}\ \text{(MVA base)}$$

$$H_{1,new} * 100 = 500 \times 5$$

$$H_{1,new} = 25\ \text{MJ/MVA}$$

Unit 2:

$$S_2 \propto \frac{1}{H_2}$$

$$200 \propto \frac{1}{5}\ \text{(Based on ratings)}$$

$$100 \propto \frac{1}{H_{2,new}} \text{ (MVA base)}$$

$$H_{2,new} * 100 = 500 \times 5$$

$$H_{2,,new} = 25 \text{ MJ/MVA}$$

Assuming coherent swinging of synchronous machines,

$$H_{eq} = H_{1,new} + H_{2,new}$$

$$= 25 + 10 = 35 \text{ MJ/MVA}$$

Equivalent inertia constant on 100 MVA common base is 35 MJ/MVA .

Q5. Station A has four identical generators each rated 100 MVA and having inertia constant 9 MJ/MVA.Solution B has three identical alternators each rated 200 MVA and having inertia constant 4MJ/MVA. Calculate the equivalent inertia constants of both plants on 150 base MVA.

Solution:

Given data:

Base MVA = 150 MVA

$$\text{Moment of inertia}, M = \frac{SH}{\pi f}$$

Station A: Assuming coherent swinging of synchronous machines in station A,

$$H_{A,old} = 4 * 9 = 36 \text{ MJ/MVA}$$

$$S_A \propto \frac{1}{H_A}$$

$$100 \propto \frac{1}{36} \text{(based on ratings)}$$

$$150 \propto \frac{1}{H_{A,new}} \text{(MVA base)}$$

$$H_{A,new} * 150 = 100 * 36$$

$$H_{A,new} = 24 \text{ MJ/MVA}$$

Station B: Assuming coherent swinging of synchronous machines in station B,

$$H_{B\,old} = 3 * 4 = 12 \text{ MJ/MVA}$$

$$S_B \propto \frac{1}{H_B}$$

$$200 \propto \frac{1}{12} \text{(based on ratings)}$$

$$150 \propto \frac{1}{H_{B,new}} \text{(MVA base)}$$

$$H_{B,new} * 150 = 200 * 12$$

$$H_{B,new} = 16 \text{ MJ/MVA}$$

Assuming coherent swinging of synchronous machines,

$$H_{eq} = H_{1,new} + H_{2,new}$$
$$= 24 + 16 = 40 \text{ MJ/MVA}$$

Equivalent inertia constant on 150 MVA common base is 40 MJ/MVA .

Q6. A generator and a Synchronous motor having inertia constants6 and 4 MJ/MVA are connected through a network of reactances. The generator is delivering power of 1 p.u. when a fault occurs reduces the delivered power. The system frequency is 60 Hz. Calculate the angular acceleration of the generator with respect to motor at the instant when reduced power delivered is 0.6 p.u.

Solution:

Given data:

Inertia constants of generator and synchronous motor = 6 and 4 MJ/MVA

Mechanical input, $P_s = P_{e1} = 1.0 \text{ p.u}$

System frequency = 60 Hz.

Reduced power, $P_{e2} = 0.6$ p.u.

Accelerating power, $P_{a2} = P_\delta - P_{e2} = 1 - 0.6 = 0.4$ p.u.

$$\text{Moment of inertia, M} = \frac{S\,H_{eq}}{\pi f} = \frac{S(H_1 + H_2)}{\pi f} = \frac{1.0(6+4)}{180 * 60} = \left(\frac{1}{1080}\right) \frac{\text{MJ} - \text{Sec}}{\text{Elec. Deg}}$$

According to swing equation, angular acceleration of generator with respect to motor is,

$$\alpha = \frac{d^2\delta}{dt^2} = \frac{1}{M}P_{a2} = \frac{1}{\left(\frac{1}{1080}\right)}(0.4) = 432 \, \frac{\text{Elec. Deg}}{\text{Sec}^2}$$

Q7. A 2-pole, 50 Hz, 11 KV turbo generator has a rating of 60 MW, 0.85 lagging p.f. Its rotor has a moment of inertia 8800 Kgm². Calculate the inertia constant in MJ/MVA and its momentum in MJ − Sec/elec. deg.

Solution:

Given data:

MW rating of turbo generator = 60 MW

Operating power factor = 0.85 lagging

Moment of inertia = 8800 Kg − m²

$$\text{VA rating of generator} = \frac{P}{\cos \emptyset} = \frac{60 \text{ MW}}{0.85} = 70.58 \text{ MVA}$$

(a)   Inertia constant

$$H = \frac{\frac{1}{2} I \omega^2}{S} = \frac{\frac{1}{2}(8800)(2\pi f)^2}{70.58} = 6.152 \text{ MJ/MVA}$$

(b)   Momentum in MJ − Sec/Elec − deg

$$M = \frac{SH}{\pi f} = \frac{70.58 \times 6.152}{180 \times 50} = 0.04825 \; \frac{MJ - Sec}{Elec.\,Deg}$$

Q8. A 500 MW, 21 KV, 50Hz, $3 - \emptyset$, $2 -$ pole Synchronous generator has a rated p.f. equal to 0.9 and moment of inertia equal to 27.5×10^3 Kgm^2. Calculate the inertia constant.

Solution:

Given data:

$$\text{MW rating of turbo generator} = 500 \; MW$$

$$\text{MVA rating of turbo generator,}\; S = \frac{500}{0.9} = 555.55 \; MVA$$

$$\text{Inertia constant,}\; H = \frac{\frac{1}{2} I\, \omega^2}{S} = \frac{\frac{1}{2}(27.5 \times 10^3)(2\pi f)^2}{555.55} = 2.4 \; MJ/MVA$$

Q9. Inertia constant of 100 MVA, 50 Hz, 4-pole generator is 10 MJ/MVA. The mechanical input to the machine is suddenly increased from 50 MW to 75 MW. Calculate the rotor acceleration in Elec. degrees/sec^2

Solution:

Given data:

MVA rating of turbo generator = 100 MVA

Initial mechanical input , $P_{s1} = P_{e1} = 50 \; MW$

Initial Acceleration power, $P_{a1} = P_{s1} - P_{e1} = 0$

Final Mechanical input $P_{s2} = 75 \; MW$

$\therefore$ Final Acceleration Power $P_{a2} = P_{s2} - P_{e1} = 75 - 50 = 25 \; MW$

Moment of Inertia $M = \dfrac{SH}{\pi f}$

$$= \frac{100 * 10}{180 * 50} = \frac{1}{9} \frac{MJ - Sec}{Elec.\,Deg} = 0.11 \frac{MJ - Sec}{Elec.\,Deg}$$

According to swing equation,

$$\text{angular accleration,}\; \alpha = \frac{d^2\delta}{dt^2} = \frac{1}{M} P_{a2} = \left(\frac{1}{(1/9)}\right)(25) = 225 \; \text{electrical degrees/sec}^2$$

Q10. A generator delivers power of 1 p.u. to an infinite bus through a purely reactive network. The maximum power that should be delivered by the generator is 2 p.u. A three phase fault occurs at the terminals of generator which reduces the generator output to zero. The fault is cleared after t_c seconds. The original network is then restored. The maximum swing of the rotor angle is found to be δ_{max} equal to 110 Elec. Deg. Calculate the rotor angle in elec. degrees at $t = t_c$.

Solution:

Given data:

Maximum power delivered before fault, $P_{m1} = 2$ p. u.

As fault is on the generator terminals, power delivered during fault, $P_{e2} = P_{m2} \sin \delta = 0$

As the original network is restored, electrical Power transmitted after fault,

$$P_{m1} = 2 \text{ p. u} = P_{m3}$$

Maximum load angle, $\delta_{max} = 110^0 = 1.919$ rad

Mechanical input $P_s = P_{e1} = P_{m1} \sin \delta_0 = 1$ p. u.

$$\delta_0 = \sin^{-1}\left(\frac{1}{2}\right) = 30^0 = 0.523 \text{ rad}$$

Maximum load angle , $\delta_{max} = \pi - \sin^{-1}\left(\frac{P_s}{P_{m1}}\right)$ (if not given in the data)

Rotor angle at $t = t_c$ is the critical clearing angle, δ_c.

Critical clearing angle, $\delta_c = \cos^{-1}\left\{\dfrac{P_s(\delta_{max} - \delta_0) + P_{m3} \cos \delta_{max} - P_{m2} \cos \delta_0}{P_{m3} - P_{m2}}\right\}$

$$= \cos^{-1}\left\{\dfrac{(1)(1.919 - 0.523) + (2) \cos(110)}{2}\right\} = 1.876 = 69.14^0$$

Q11. A Synchronous generator is connected through an infinite bus through loss less double circuit. The generator is delivering 1 p.u. power at the load angle of 30^0, When a sudden falls reduces the peak power that can be transmitted through 0.5 p.u. After clearance of fault, the peak power that can be transmitted becomes 1.5 p.u. Find the crtical clearing angle.

Solution:

Given data:

Initial load angle, $\delta_0 = 30^0 = 0.523$ rad

Mechanical input , $P_s = P_{e1} = P_{m1} \sin \delta = P_{m1} \sin 30^0 = 1$ p. u. $\Rightarrow P_{m1} = 2$ p. u

Maximum power delivered before fault, $P_{m1} = 2$ p. u.

Maximum power delivered during fault, $P_{m2} = 0.5$ p. u.

Maximum electrical power transmitted after fault, $P_{m3} = 1.5$ p. u

Maximum load angle, $\delta_{max} = \pi - \delta_0 = \pi - 30 = 138.1^0 = 2.41$ rad

Critical clearing angle,

Critical clearing angle, $\delta_c = \cos^{-1}\left\{\dfrac{P_s(\delta_{max} - \delta_0) + P_{m3} \cos \delta_{max} - P_{m2} \cos \delta_0}{P_{m3} - P_{m2}}\right\}$

$$\delta_c = \cos^{-1}\left\{\frac{(2.41 - 0.523) + (1.5\cos 138.1^0) - 0.5\cos(30)}{1.5 - 0.5}\right\} = 70.27^0 \simeq 70.3^0$$

Q12. A three-phase generator delivers 1 p.u. power to an infinite bus through a transmission network. When a fault occurs the maximum power which can be transferred during prefault, during fault and post fault conditions is 1.75 p.u., 0.4 p.u. and 1.25 p.u. Find critical clearing angle.

Solution:

Given data:

Mechanical input, $P_s = P_{e1} = P_{m1}\sin\delta_o = 1$ p. u.

Maximum power delivered before fault, $P_{m1} = 1.75$ p. u.

Initial load angle,

$$\delta_0 = \sin^{-1}\left(\frac{P_s}{P_{m1}}\right) = \sin^{-1}\left(\frac{1}{1.75}\right) = 34.84^0 = 0.608 \text{ rad}$$

Maximum power delivered during fault, $P_{m2} = 0.4$ p. u.

Maximum electrical power transmitted after fault, $P_{m3} = 1.25$ p. u

Maximum load angle, $\delta_{max} = \pi - \sin^{-1}\left(\frac{P_s}{P_{m3}}\right) = \pi - \sin^{-1}\left(\frac{1}{1.25}\right) = 126.86^0 = 2.214$ rad

Critical clearing angle, $\delta_c = \cos^{-1}\left\{\dfrac{P_s(\delta_{max} - \delta_0) + P_{m3}\cos\delta_{max} - P_{m2}\cos\delta_0}{P_{m3} - P_{m2}}\right\}$

$$\delta_c = \cos^{-1}\left\{\frac{1.0(2.214 - 0.608) + 1.25\cos(126.86^0) - 0.4\cos(34.84^0)}{1.25 - 0.4}\right\} = 51.51^0$$

Q13. The power system network is characterized by $P_s = 1$ p. u. $P_{m1} = 1.736$ p. u., $X_{1eq} = 0.72$ p. u., $X_{2eq} = 3$ p. u., $X_{3eq} = 1$ p. u. Find the critical clearing angle of system.

Solution:

Given data:

Mechanical input, $P_s = P_{e1} = P_{m1}\sin\delta_o = 1$ p. u.

Maximum power delivered before fault, $P_{m1} = 1.736$ p. u.

Initial load angle,

$$\delta_0 = \sin^{-1}\left(\frac{P_s}{P_{m1}}\right) = \sin^{-1}\left(\frac{1}{1.736}\right) = 35.17^0 = 0.6138 \text{ rad}$$

Maximum power delivered during fault,

$$P_{m2} = \frac{EV}{X_{2\,eq}} = \left(\frac{X_{1\,eq}}{X_{2\,eq}}\right)\left(\frac{EV}{X_{1\,eq}}\right) = \left(\frac{0.72}{3}\right)(1.736) = 0.41664$$

Maximum electrical power transmitted after fault, $P_{m3} = \left(\dfrac{X_{1\,eq}}{X_{3\,eq}}\right)(P_{m1})$

$$= \left(\dfrac{X_{1\,eq}}{X_{3\,eq}}\right)(P_{m1})$$

$$= \left(\dfrac{0.72}{1}\right)(1.736) = 1.24 \text{ p. u}$$

Maximum load angle, $\delta_{max} = \pi - \sin^{-1}\left(\dfrac{P_s}{P_{m3}}\right) = \pi - \sin^{-1}\left(\dfrac{1}{1.24}\right) = 126.25^0 = 2.24 \text{ rad}$

Critical clearing angle, $\delta_c = \cos^{-1}\left\{\dfrac{P_s(\delta_{max} - \delta_0) + P_{m3}\cos\delta_{max} - P_{m2}\cos\delta_0}{P_{m3} - P_{m2}}\right\}$

$\delta_c = \cos^{-1}\left\{\dfrac{1.0(2.24 - 0.6138) + 1.24\cos(126.25^0) - 0.41664\cos(35.17^0)}{1.25 - 0.4}\right\} = 51.4^0$

Q14. The 50Hz alternator supplies 40% maximum power i.e., capable of delivering through a txn line to an infinite bus. A fault occurs that increases the reactance between generator and infinite bus to 600% of the value before fault. The maximum power that can be delivered i.e., 80% of original maximum value, after fault is cleared. Determine the Critical angle of the system.

Solution:

Given data:

Mechanical input , $P_s = P_{e1} = P_{m1}\sin\delta_0 = 0.4\ P_{m1}\text{p. u.}$

$$\Rightarrow \sin\delta_0 = 0.4$$

$$\Rightarrow \delta_0 = 23.57^0 = 0.41 \text{ rad}$$

Equivalent transfer reactance during fault

$$\Rightarrow X_{2\,eq} = 600\%\ \ X_{1\,eq} = 6\ X_{1\,eq}$$

Maximum power delivered after fault

$$\Rightarrow P_{m3} = 80\%\ P_{m1} = 0.8\ P_{m1}$$

Maximum power transmitted during fault

$$\Rightarrow P_{m2} = \dfrac{X_{1\,eq}}{X_{2\,eq}}P_{m1} = \dfrac{1}{6}P_{m1} = 0.166\ P_{m1}$$

Maximum load angle

$$\Rightarrow \delta_{max} = \pi - \sin^{-1}\left(\dfrac{0.4}{0.8}\right)\dfrac{P_s}{P_{m3}} = 150^0 = 2.61 \text{ rad}$$

Critical clearing angle, $\delta_c = \cos^{-1}\left\{\dfrac{P_s(\delta_{max} - \delta_0) + P_{m3}\cos\delta_{max} - P_{m2}\cos\delta_0}{P_{m3} - P_{m2}}\right\}$

$\delta_c = \cos^{-1}\left(\dfrac{0.4P_{m1}(2.61 - 0.41) + 0.8P_{m1}\cos(150^0) - 0.167\cos(23.57)}{0.8\ P_{m1} - 0.167\ P_{m1}}\right) = 86.83^0$

Q15. A Synchronous generator having an inertia constant of 6 MJ/MVA is delivering power of 1 p.u. to an infinite bus through a purely reactive network. Suddenly a fault occurs reducing the generator output power to zero. The maximum power that could be delivered is 2.5 p.u. Calculate δ_c and Critical clearing time, t_c.

Solution:

Given data:

Maximum power delivered before fault, $P_{m1} = 2.5$

Maximum power delivered during fault, $P_{m2} = 0$

As fault is on generator maximum power delivered after fault, $P_{m3} = P_{m1} = 2.5 \text{p.u.}$

Mechanical input, $P_s = P_{e1} = P_{m1}\sin\delta_0 = 1$

Initial load angle, $\delta_0 = 23.57^0 = 0.41$ rad

Maximum load angle, $\delta_{max} = \pi - \sin^{-1}\left(\dfrac{P_s}{P_{m3}}\right) = \pi - \sin^{-1}\left(\dfrac{1}{2.5}\right) = 156.42^0 = 2.73$ rad

Critical clearing angle, $\delta_c = \cos^{-1}\left\{\dfrac{P_s(\delta_{max} - \delta_0) + P_{m3}\cos\delta_{max} - P_{m2}\cos\delta_0}{P_{m3} - P_{m2}}\right\}$

$$\delta_c = \cos^{-1}\left(\dfrac{1.0(2.73 - 0.41) + 2.5\cos(156.42^0) - 0}{2.5 - 0}\right) = 89^0 = 1.558 \text{ rad.}$$

$$\text{Moment of inertia, } M = \dfrac{SH}{\pi f} = \dfrac{6 \times 1 \, p.u}{180^0 \times 50} = 0.0382 \quad \dfrac{MJ - sec}{elect - degree}$$

$$\text{Critical clearing time, } t_c = \left\{\dfrac{2M}{P_s}(\delta_c - \delta_0)\right\}^{1/2}$$

$$= \left\{\dfrac{2 \times 0.0382}{1.0}(1.558 - 0.41)\right\}^{1/2} = 0.296 \text{ sec}$$

UNSOLVED PROBLEMS

1. An alternator is initially delivering 1 p.u. to an infinite bus. The maximum power transmitted before fault, during fault and after fault are 2.18 p.u., 0.615 p.u and 1.6 p.u. The inertia constant is4 MJ/MVA. The circuit breaker clears the fault after 5 cycles. Prove that whether the system is stable or unstable using point by point method or step by step method.

    Ans: Stable

2. A round rotor generator with internal voltage $E_1=2$ p.u. and $X=1.1$ p.u is connected to a round rotor synchronous motor with internal voltage $E_2=1.3$ p.u. and $X=1.2$ p.u. The reactance of the line connecting the generator to motor is 0.5p.u.When the generator supplies 0.5 p.u. power, determine the rotor angle difference between the machines.

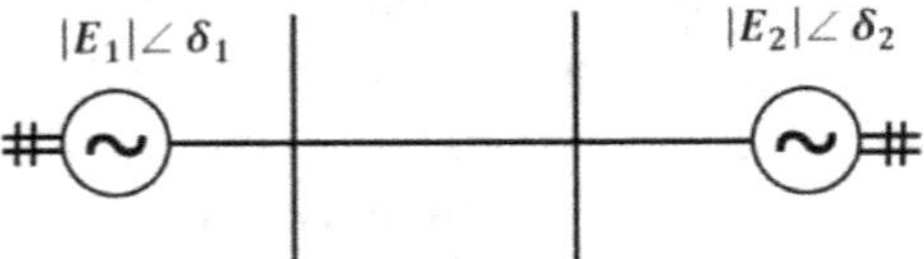

Ans: 32.58^0

3. A generator with constant 1.0 p.u. terminal voltage supplies power through a step-up transformer of 0.12 p.u. reactance and a double circuit line to an infinite bus bar as shown in the figure. The infinite bus voltage is 1.0 p.u. The steady state stability limit of the system is 6.25 p.u. With one of the double circuit tripped, determine the resulting steady state stability power limit?

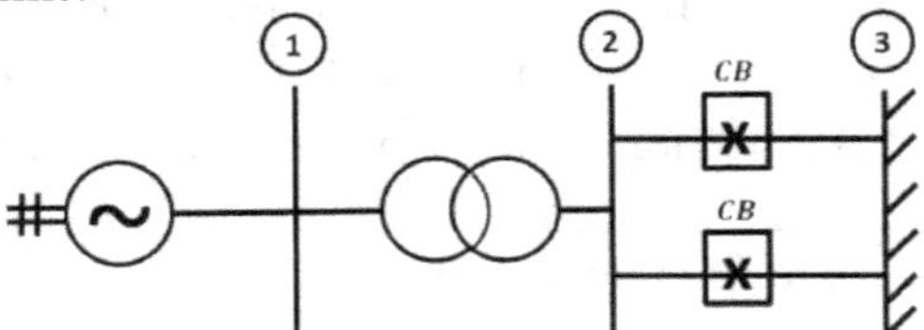

Ans: 5 p.u.

4. For the system shown below, S_{D1} and S_{D2} are complex power demands at bus1 and bus2 respectively.

    If $|V_2| = 1$ p.u. then determine the VAR rating of the capacitor connected at bus2.

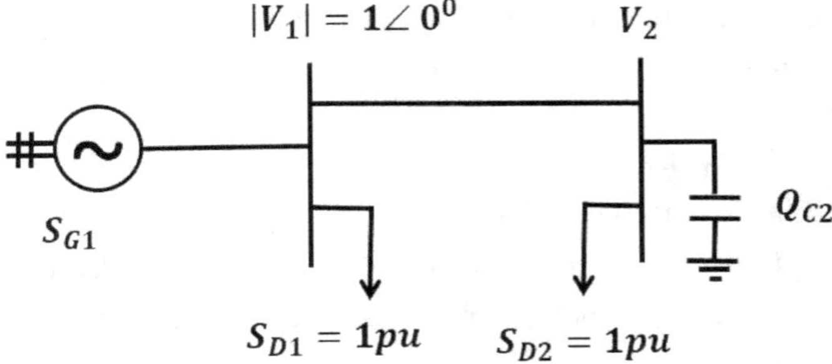

Ans: 0.268 p.u.

OBJECTIVE QUESTIONS

1. The power angle characteristic of a machine-infinite bus system is Pe=2sinδ p.u. It is operating at δ=30°. The synchronizing power coefficient at the operating point is

    ____

2. A synchronous generator is connected to a 11KV infinite bus through a transmission line.

The reactances of the generator and transmission line are 1.2Ω and 0.8 Ω. The terminal voltage of the synchronous generator is 15 KV. If the generator delivers 70MW power to an infinite bus the load angle is _______________

3. A synchronous generator with a synchronous reactance of 1.3 p.u. is connected to an infinite bus whose voltage is 1 p.u. through an equivalent reactance of 0.2 p.u. For maximum output of 1.2 p.u. the alternator emf must be _______________

4. An infinite bus of 1 p.u. is fed from a synchronous machine having E=1.1 p.u. If the transfer reactance between them is 0.5 p.u. then steady state power limit will be _______________

5. For 800 MJ stored energy in the rotor at synchronous speed, what is the inertia constant, H in MJ/MVA for a 50 Hz, 4 pole turbo generator rated 100 MVA, 11 KV is _______________.

6. If a generator of 250 MVA rating has an inertia constant of 6MJ/MVA then inertia constant on 100 MVA base is _______________.

7. If inertia constant H is 8 MJ/MVA for a 50 MVA generator then the kinetic energy stored by the rotor is _______________.

8. An alternator having an induced emf of 1.6 p.u. is connected to an infinite bus of 1p.u. If the bus bar reactance is 0.6 p.u. and alternator reactance is 0.2 p.u. then maximum power that can be transferred is _______________

9. A three-phase transmission line has an inductive reactance of 100Ω per phase. If the sending end and receiving end voltages are maintained at 110 KV then the maximum power that can be transmitted is _______________

10. If P_{max} is the maximum power transmitted through the line then with 60% of series capacitor compensation the maximum power transfer becomes _______________

11. The moment of inertia of 200 MVA generator is 10 p.u. then the moment of inertia for 400 MVA generator is _______________

12. Power system stability deals with _______________

13. Equal area stability criterion deals with _______________

14. _______________ transmission line is most stable among the transmission lines.

15. _______________ increase the power transmission capacity.

16. Power transmission _______________ with symmetrical spacing of transmission line conductors.

17. Transient stability limit is _______________ than steady state stability limit.

18. Swing equation is applicable to _______________ machine systems.

12 Voltage Control and Power Factor Improvement Methods

12.1 INTRODUCTION

The objective of this chapter is to discuss about the devices connected at the receiving end of the transmission line to control the receiving end voltage by controlling the reactive power. The voltage at the receiving end of the line can be controlled by controlling impedance voltage drop in the transmission line. Shunt controlled devices control voltage by controlling current and Series controlled devices control voltage by controlling impedance.

Consider a basic power system network consisting of an alternator, step up power transformer and a three-phase transmission line connected to three loads as shown in the figure 12.1.

Only one of the three transmission lines are replaced by resistance and inductance connected in series, capacitance and conductance connected in parallel for convenience.

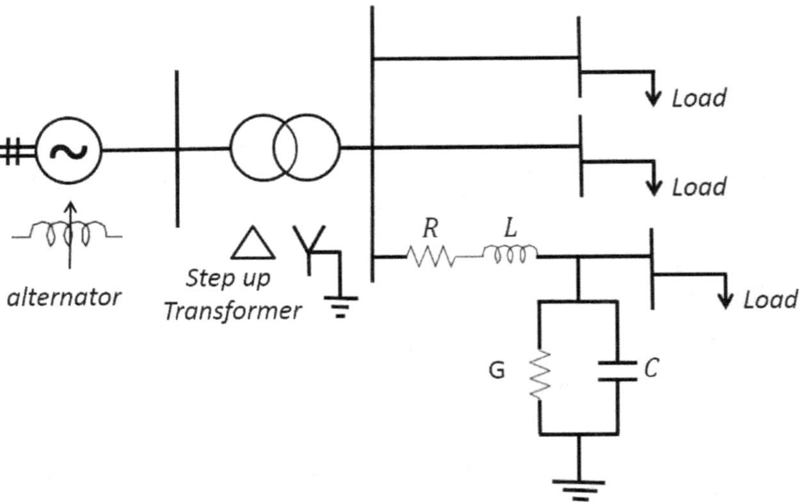

Figure 12.1 A basic power system network

The Voltage at the receiving end of a transmission line can be controlled by two methods.

1. Internal Voltage Control Method

2. External Voltage Control Method

1. **Internal voltage control method:** The method of increasing voltage at receiving end of all the three transmission lines by increasing excitation of an alternator is known as "internal voltage control method".

   Limitation: The limitation of internal voltage control method is that voltage at the specified receiving end of transmission line alone cannot be controlled.

2. **External voltage control method:** The method of controlling voltage at the receiving end of the transmission line by connecting a voltage control device at the specified receiving end of the transmission line is known as "external voltage control method".

   The voltage control devices that can be connected at the specified receiving end of the transmission line to control receiving end voltage are

1. Shunt capacitor

2. Shunt inductor

3. Series capacitor

4. Series inductor

5. Synchronous capacitor

6. Synchronous inductor

7. Synchronous phase modifier

The first six devices are loss less devices and hence active power, $P = 0$

$$\Rightarrow S = P + jQ$$
$$\Rightarrow S = \sqrt{P^2 + Q^2}$$
$$S = Q$$

The voltage at the specified receiving end of a transmission line can be controlled by controlling reactive power i.e., either by absorbing reactive power or by delivering reactive power.

12.1.1 Sign of the Net Reactive Power

As inductor and capacitor are reactive elements, the sign of the reactive powers corresponding to reactive elements has to be determined.

Inductor

Consider a lumped inductor connected to a synchronous machine(source) as shown in the figure 12.2.

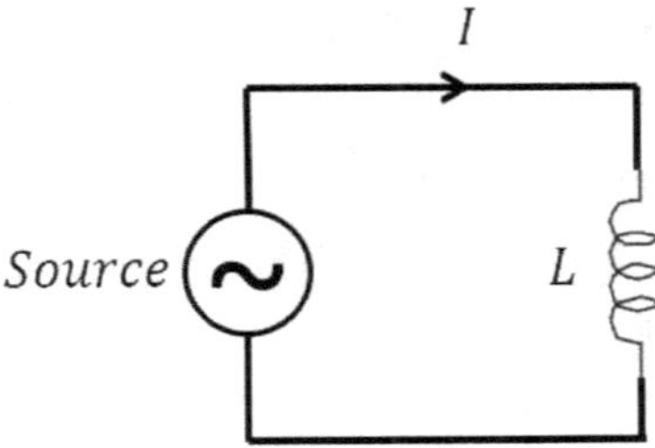

Figure 12.2 A lumped inductor connected to a synchronous machine

1. An inductor absorbs lagging reactive power.

2. The synchronous machine delivering lagging reactive power to an inductor is known as a "synchronous capacitor".

3. As power flow is from source to load, the lagging reactive power absorbed by inductor is assigned positive sign.

Capacitor

Consider a lumped capacitor connected to a synchronous machine (source) as shown in the figure 12.3.

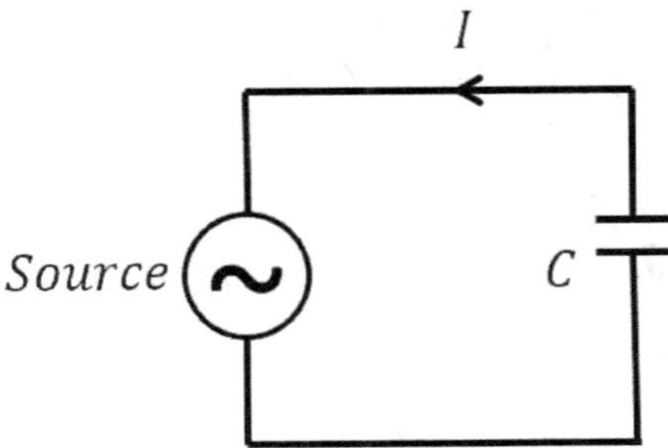

Figure 12.3 A lumped capacitor connected to a synchronous machine

1. A capacitor delivers leading reactive power.

2. The synchronous machine absorbing leading reactive power delivered by a capacitor is known as "synchronous inductor".

3. As power flow is from load to source, the leading reactive power delivered by capacitor is assigned negative sign.

Note: The sign of net reactive power is based the direction of power flow but not based on opposite nature elements inductor and capacitor.

12.1.2 Approximate Expression for Voltage at the Receiving End of Transmission Line

Consider an alternator connected to load through a transmission line as shown in the figure 12.4.

Assumption: Resistance of the transmission line is neglected.

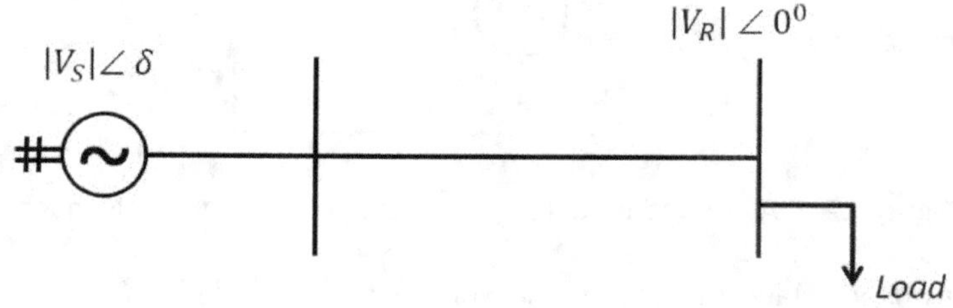

Figure 12.4 An alternator connected to load through transmission line

Voltage at the receiving end of the transmission line, V_R

$$= V_S - \Delta V \qquad \qquad \qquad .....(12.1)$$

(where ΔV is the reactive voltage drop in the transmission line, $\Delta V = I.\,jX$)

Voltage at the receiving end of the transmission line, $V_R = V_S - j\,I.\,X \qquad .....(12.2)$

Complex power delivered through transmission line, $S = P + jQ = V_S.\,I^* = V_R.\,I^* \quad .....(12.3)$

where V_S is constant and V_R is variable.

From equation 12.1.3,

$$\Rightarrow I^* = \frac{P + jQ}{|V_S|}$$

$$\Rightarrow I = \frac{P - jQ}{|V_S|^*}$$

$$\Rightarrow I = \frac{P - jQ}{|V_S|} \qquad \qquad \qquad .....(12.4)$$

Sub equation 12.1.4 in equation 12.1.2

$$\Rightarrow |V_R| = |V_S| - \frac{P - jQ}{|V_S|}(jX)$$

$$\Rightarrow |V_R| = \left[|V_S| - \frac{QX}{|V_S|}\right] - j\,\frac{PX}{|V_S|} \qquad \qquad .....(12.5)$$

The vector diagram based on the equation 12.5 is shown in figure 12.5.

The approximation is to neglect imaginary term

$$V_R \simeq |V_S| - \frac{QX}{|V_S|} \qquad \qquad \qquad .....(12.6)$$

The vector diagram based on the equation 12.6 is shown in figure 12.6

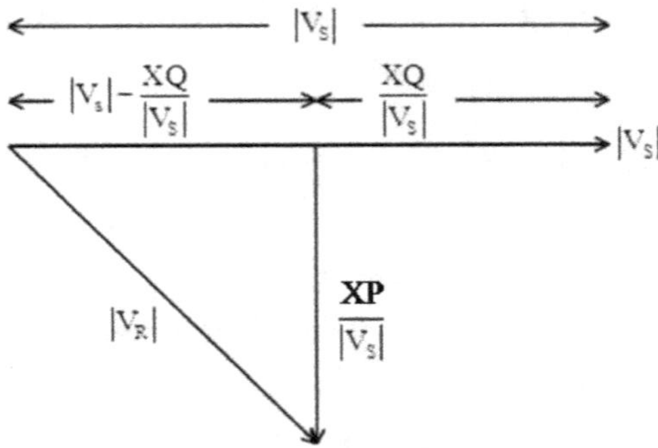

Figure 12.5 Vector diagram before neglecting imaginary term

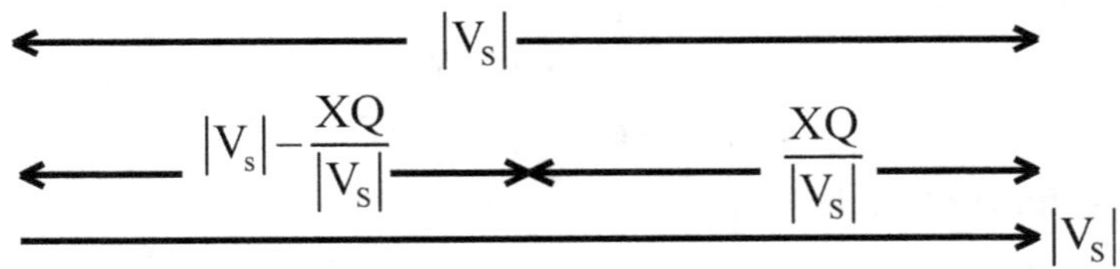

Figure 12.6 Vector diagram after neglecting imaginary term

12.1.3   Reactive Power Balance Equation

Consider an alternator connected to load through a transmission line. Connect a voltage control device connected at the receiving end of the transmission line as shown in figure 12.7.

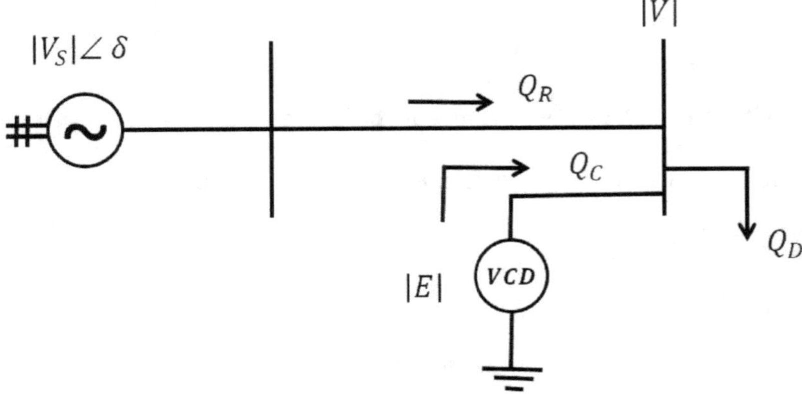

Figure 12.7 Power system network for reactive power balance equation

From figure 12.7, the reactive power balance equation is $Q_R + Q_C = Q_D$.   .....(12.7)

12.2 SHUNT CAPACITOR / POWER FACTOR IMPROVEMENT DEVICE

Analysis

If actual loading on a transmission line is greater than the surge impedance loading of the transmission line then

   (i)   Current flowing through transmission line increases.
   (ii) Electromagnetic energy stored by inductor in the magnetic field increases & electrostatic energy stored by capacitor in the electric field remains constant.
   (iii) Inductor is dominant, power factor is lagging, $|V_R| < |V_S|$ and net reactive power is lagging.
   (iv) The voltage at the receiving end of the line can be increased by connecting shunt capacitor at the receiving end of the transmission line.
   (v)   Shunt capacitor delivers capacitive reactance power & compensates the excess inductive reactive power such that the net reactine power, $Q_{net} = 0$ and $|V_R| = |V_S|$

Before connecting shunt capacitor

Consider an alternator connected to load through a transmission line before connecting a shunt capacitor at the receiving end of the transmission line as shown in figure 12.8.

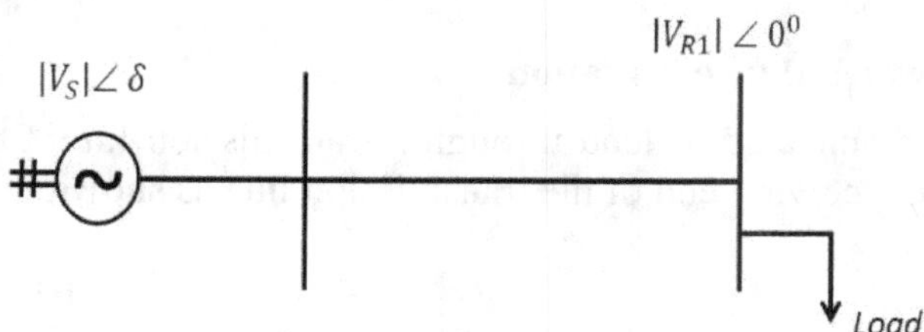

Figure 12.8 Power system network before connecting shunt capacitor

Voltage at the receiving end before connecting shunt capacitor, $\left|V_{R_1}\right|$

$$\simeq |V_S| - \frac{XQ_{net}}{|V_S|} \qquad \qquad .....(12.8)$$

$$|V_S| - \frac{X}{|V_S|}[Q_L - Q_C] \qquad \qquad .....(12.9)$$

$$\simeq |V_S| - \frac{X}{|V_S|}[Q_{absorbed} - Q_{delivered}] \qquad \qquad .....(12.10)$$

As Q_L is dominating, Q_L-Q_C is positive or $Q_{absorbed}$ –$Q_{delivered}$ is positive.

$$\therefore \ |V_{R_1}| < |V_S|$$

The equivalent circuit before connecting shunt capacitor is shown in figure 12.9.

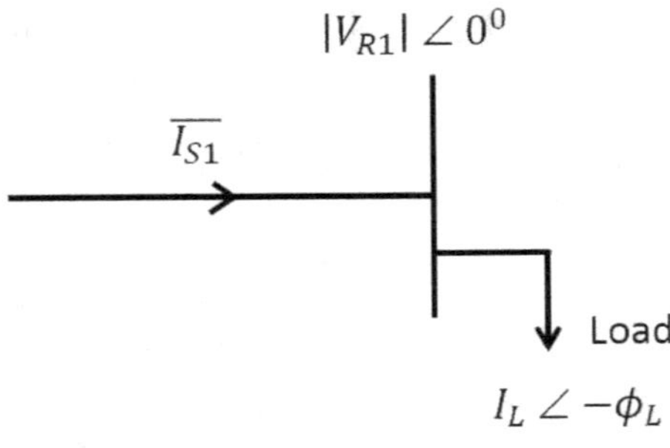

$$\overline{I_{S1}} = I_L\angle-\phi_L$$

Figure 12.9 Equivalent circuit before connecting shunt capacitor

The vector diagram before connecting shunt capacitor is shown in figure 12.10.

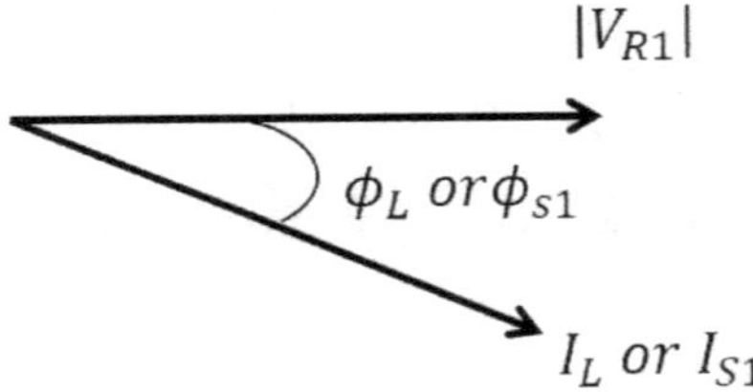

Figure 12.10 Vector diagram before connecting shunt capacitor

After connecting shunt capacitor

Consider an alternator connected to load through a transmission line before connecting a shunt capacitor at the receiving end of the transmission line as shown in figure 12.11.

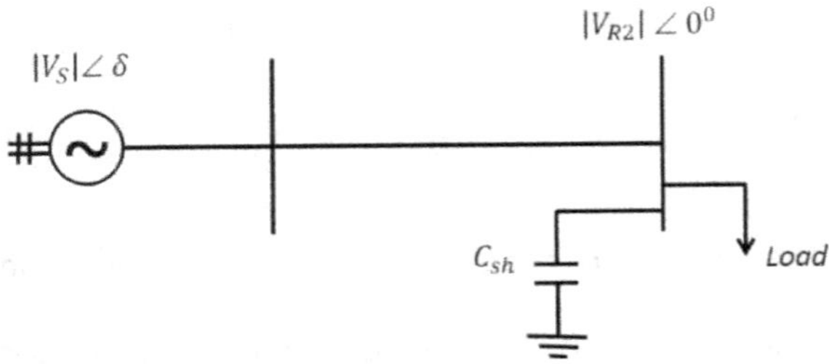

Figure 12.11 Power system network after connecting shunt capacitor

Voltage at the receiving end after connecting shunt capacitor, $\left|V_{R_2}\right|$

$$\cong |V_S| - \frac{Q_{net}}{|V_S|} \qquad\qquad .....(12.11)$$

$$\cong |V_S| - \frac{X}{|V_S|}\left[Q_L - (Q_c + Q_{\text{shunt cap.}})\right] \qquad\qquad .....(12.12)$$

$$\therefore \left|V_{R_2}\right| > \left|V_{R_1}\right| \,\&\, \left|V_{R_2}\right| < |V_S|$$

The equivalent circuit after connecting shunt capacitor is shown in figure 12.12.

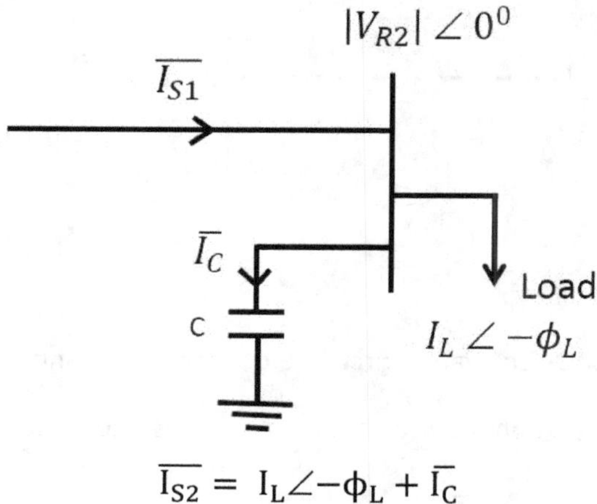

Figure 12.12 Equivalent circuit before connecting shunt capacitor

The vector diagram after connecting shunt capacitor is shown in figure 12.13.

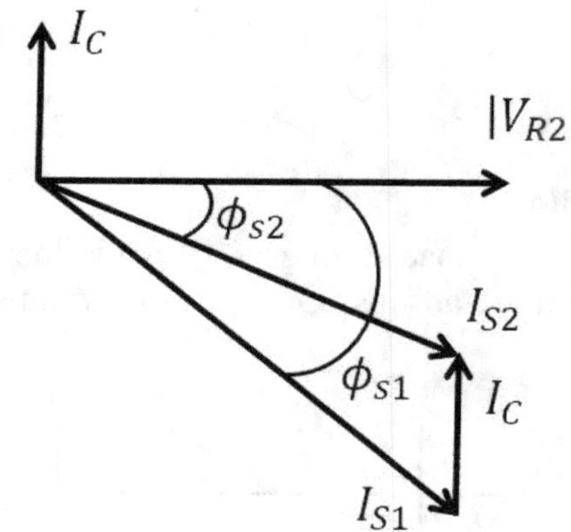

Figure 12.13 Vector diagram after connecting shunt capacitor

For Receiving end voltage after connecting shunt capacitor to sending end voltage

For $V_{R_2} = V_S$

$$Q_L - Q_c - Q_{shunt\ cap} = 0$$

$$Q_{sh\ cap} = Q_L - Q_c$$

The vector diagram for before and after connecting shunt capacitor is shown in figure 12.14.

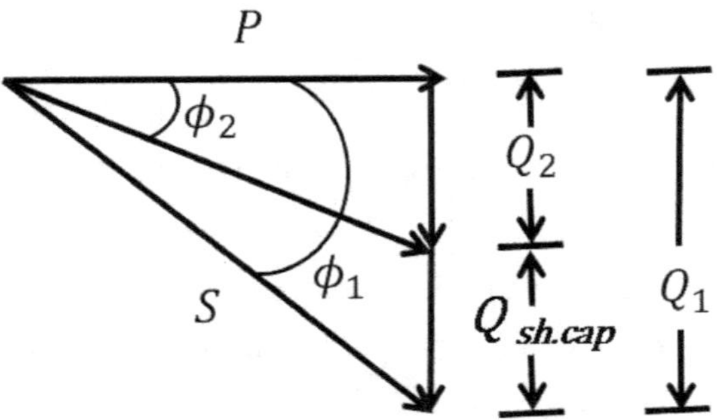

Figure 12.14 Vector diagram for before and after connecting shunt capacitor

From vector diagram it can be observed that $\Phi_2 < \Phi_1$,$\cos\Phi_2 > \cos\Phi_1$ and hence power factor of the system increases. Therefore, shunt capacitor is considered as power factor improvement device.

Reactive power delivered by shunt capacitor

The reactive power delivered by shunt capacitor is

$$Q_{\text{shunt cap.}(3-\emptyset)} = Q_1 - Q_2$$
$$= P_1 \tan\emptyset_1 - P_2 \tan\emptyset_2$$
$$= P(\tan\emptyset_1 - \tan\emptyset_2) \qquad \qquad .....(12.13)$$

Reactive power delivered by shunt capacitor per phase,

$$Q_{\text{shunt cap.}} = \frac{P(\tan\emptyset_1 - \tan\emptyset_2)}{3} \qquad \qquad .....(12.14)$$

In terms of voltage and current, reactive power delivered by shunt capacitor per phase is

$$Q_{\text{shunt cap/phase}} = V_{\text{phase}} I_{\text{shunt cap./phase}}$$
$$= V_{\text{phase}} \frac{V_{\text{ph}}}{X_{\text{shunt cap./phase}}}$$
$$= \omega C_{\text{shunt cap./phase}} V_{\text{ph}}^2 \qquad \qquad .....(12.15)$$

$$Q_{\text{sh.cap/phase}} \propto \omega V_{\text{ph}}^2$$

Shunt capacitance per phase

$$C_{\text{sh/phase}} = \frac{Q_{\text{shunt cap./phase}}}{\omega V_{\text{ph}}^2} \qquad \qquad .....(12.16)$$

12.3   SHUNT INDUCTOR

Analysis

If actual loading of a transmission line is less than the surge impedance loading of a transmission line then

(i)   Current flowing through transmission line decreases.

(ii)   Electromagnetic energy stored by inductor in the magnetic field decreases and electrostatic energy stored by capacitor in the electric field $= \frac{1}{2}CV^2$, remains constant.

(iii)   Capacitor is dominant, power factor is leading, $|V_R| > |V_S|$, the net reactive power is leading and Ferranti effect occurs.

(iv)   The voltage at the receiving end of the line can be decreased by connecting shunt inductor at the receiving end of the transmission line.

(v)   Shunt inductor absorbs the excess capacitive reactive power such that net reactive power, $Q_{net} = 0\ \&|V_R| = |V_S|$

Before connecting shunt inductor

Consider an alternator connected to load through a transmission line before connecting a shunt inductor at the receiving end of the transmission line as shown in figure 12.15.

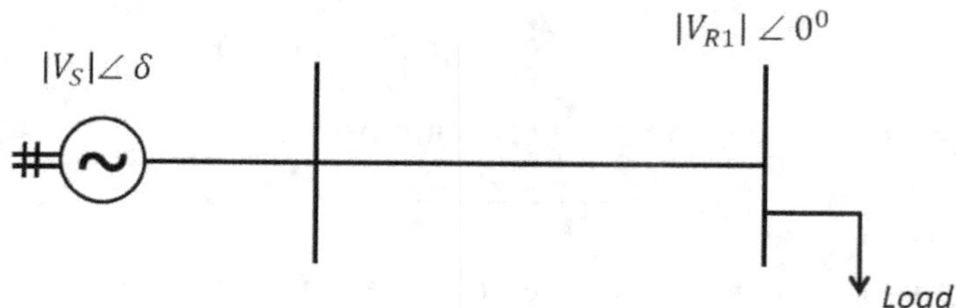

Figure 12.15 Power system network before connecting shunt inductor

Voltage at the receiving end before connecting shunt inductor, $|V_{R_1}|$

$$\simeq |V_S| - \frac{XQ_{net}}{|V_S|} \qquad .....(12.17)$$

$$\simeq |V_S| - \frac{X}{|V_S|}[Q_{absorbed} - Q_{delivered}] \qquad .....(12.18)$$

$$\simeq |V_S| - \frac{X}{|V_S|}[Q_L - Q_C] \qquad .....(12.19)$$

As Q_C is dominating, Q_L-Q_C is negative or $Q_{absorbed} - Q_{delivered}$ is negative.

$$\therefore\ |V_{R_1}| > |V_S|$$

The equivalent circuit before connecting shunt inductor is shown in figure 12.16.

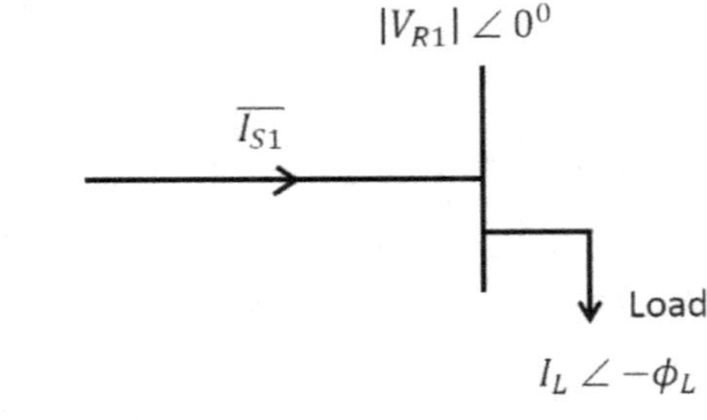

$$\overline{I_{S1}} = I_L \angle \phi_L$$

Figure 12.16 Equivalent circuit before connecting shunt inductor

The vector diagram before connecting shunt inductor is shown in figure 12.17.

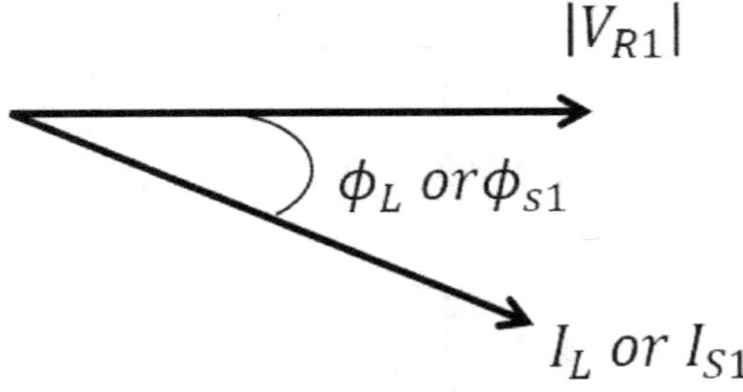

Figure 12.17 Vector diagram before connecting shunt inductor

After connecting shunt inductor

Consider an alternator connected to load through a transmission line before connecting a shunt inductor at the receiving end of the transmission line as shown in figure 12.18.

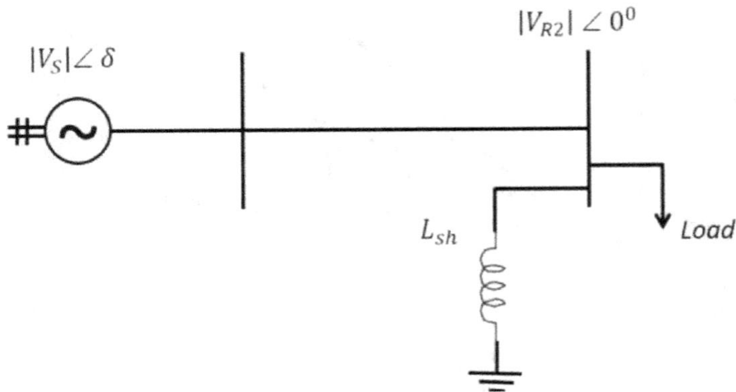

Figure 12.18 Power system network after connecting shunt inductor

Voltage at the receiving end after connecting shunt capacitor, $|V_{R_2}|$

$$|V_{R_2}| \cong |V_S| - \frac{X Q_{net}}{|V_S|} \qquad \qquad .....(12.20)$$

$$\cong |V_S| - \frac{X}{|V_S|}[Q_L + Q_{shunt\ ind} - (Q_c)] \qquad .....(12.21)$$

$$|V_{R2}| < |V_{R1}| \text{ but } |V_{R2}| > |V_S|$$

The equivalent circuit after connecting shunt inductor is shown in figure 12.19.

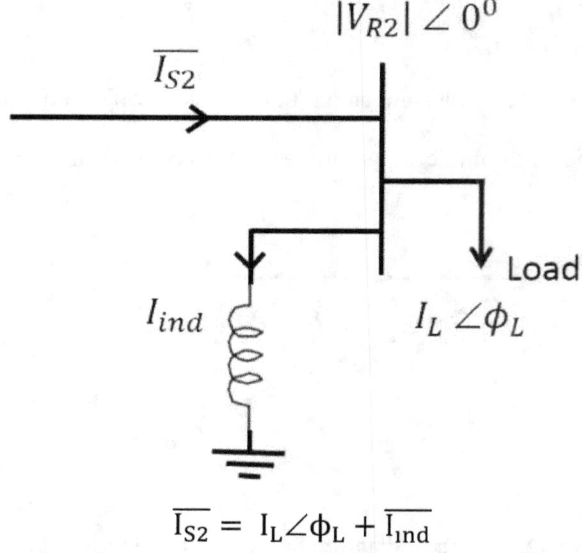

Figure 12.19 Equivalent circuit before connecting shunt inductor

The vector diagram after connecting shunt inductor is shown in figure 12.20.

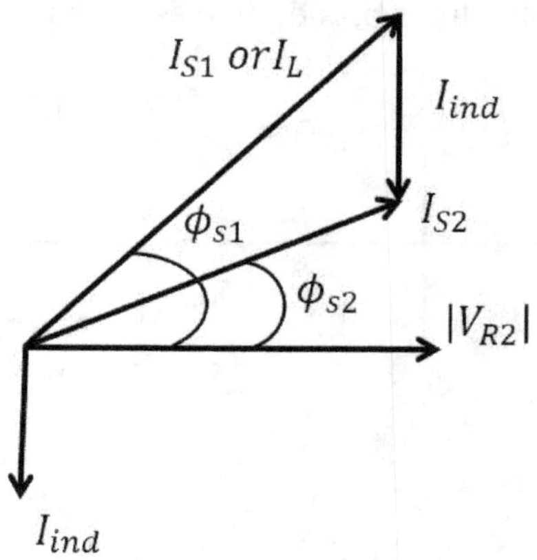

Figure 12.20 Equivalent circuit after connecting shunt inductor

For Receiving end voltage after connecting shunt capacitor to sending end voltage

For $\qquad V_{R_2} = V_S$

$$Q_L - Q_c + Q_{\text{shunt ind.}} = 0$$

$$Q_{\text{shunt ind}} = Q_C - Q_L$$

The vector diagram for before and after connecting shunt inductor is shown in figure 12.21.

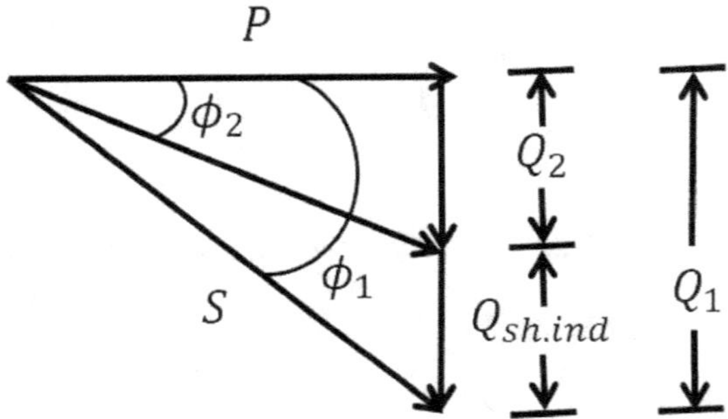

Figure 12.21 Vector diagram for before and after connecting shunt inductor

Reactive power absorbed by shunt inductor

Reactive power absorbed by shunt inductor is

$$Q_{\text{sh ind}_{(3-\emptyset)}} = Q_1 - Q_2$$

$$= P_1 \tan \emptyset_1 - P_2 \tan \emptyset_2$$

$$= P(\tan \emptyset_1 - \tan \emptyset_2) \qquad \qquad \dots..(12.22)$$

Reactive power absorbed by shunt inductor per phase is

$$Q_{\text{shunt ind.}} = \frac{P(\tan\emptyset_1 - \tan\emptyset_2)}{3} \qquad \qquad \dots..(12.23)$$

In terms of voltage & Current, reactive power absorbed by shunt inductor per phase is

$$Q_{\text{sh ind/phase}} = V_{ph} I_{\text{shunt ind/phase}} \sin 90^0 = V_{ph} \frac{V_{ph}}{X_{\text{shunt ind/phase}}}$$

$$= \frac{V_{ph}^2}{\omega\, L_{\text{shunt ind/phase}}} \qquad \qquad \dots..(12.24)$$

$$Q_{\text{sh ind/phase}} \propto \frac{V_{ph}^2}{f}$$

Shunt inductance per phase $L_{\text{sh/phase}} = \dfrac{V_{ph}^2}{\omega Q_{\text{sh ind/phase}}} \qquad \qquad \dots..(12.25)$

Shunt inductor is considered as an over voltage protection device and eliminates Ferranti effect.

12.4 SERIES CAPACITOR / POWER SYSTEM STABILITY DEVICE

Before connecting series capacitor

Consider an alternator connected to load through a transmission line before connecting a series capacitor at the receiving end of the transmission line as shown in figure 12.22.

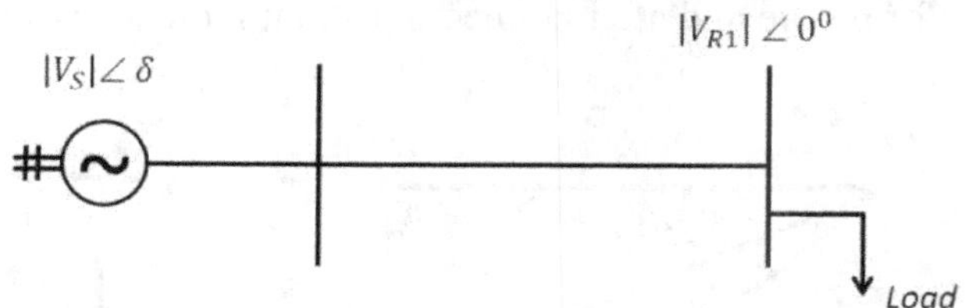

Figure 12.22 Before connecting a series capacitor

The voltage drop in the transmission line before connecting series capacitor is

$$|\Delta V_1| = |V_S| - |V_{R1}| = IR \cos \emptyset_R + IX_L \sin \emptyset_R$$
$$|V_{R1}| < |V_S| \qquad\qquad .....(12.26)$$

After connecting series capacitor

Consider an alternator connected to load through a transmission line after connecting a series capacitor at the receiving end of the transmission line as shown in figure 12.23.

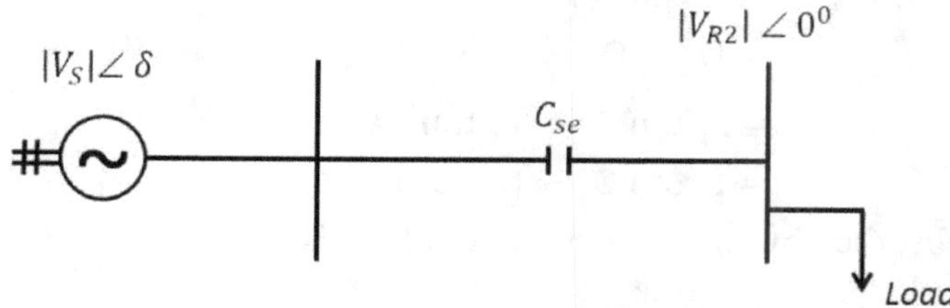

Figure 12.23 After connecting a series capacitor

The voltage drop in the transmission line before connecting series capacitor is

$$|\Delta V_2| = |V_S| - |V_{R2}| = IR \cos \emptyset_R + I(X_L - X_C)\sin \emptyset_R$$
$$|V_{R2}| < |V_S| \qquad\qquad .....(12.27)$$

From the equations 12.26 and 12.27,

$$|\Delta V_2| < |\Delta V_1| \& |V_{R2}| > |V_{R1}|$$

Therefore, a series capacitor increases voltage at receiving end of the transmission line.

Increment in voltage magnitude at the receiving end of the transmission line

$$|\Delta V_C| = |V_1| - |V_2| = I X_C \sin \emptyset_R$$

Degree of compensation (or) percentage compensation $= \dfrac{X_C}{X_L} * 100$

Reactive power delivered by series capacitor $= I^2 X_C \qquad\qquad .....(12.28)$

Reactive power delivered by series capacitor for a three-phase transmission line $= 3\ I^2\ X_C$

$$= \frac{3\ I^2}{\omega\ C_{se\ /\ ph}}$$

$$Q_{se\ cap} \propto \frac{I^2}{f}$$

Effects of connecting series capacitor in Transmission lines

1. With respect to power system stability

Case 1: Before connecting series capacitor

Consider a transmission line without series capacitor and connected to load as shown in the figure 12.24.

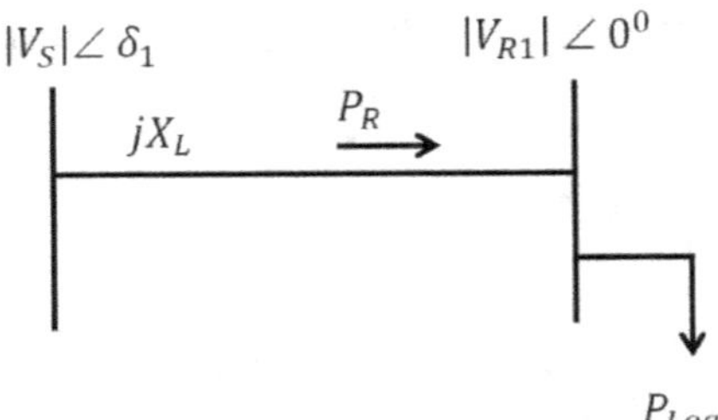

Figure 12.24 Before connecting series capacitor

$$P_{e1} = \frac{|V_S||V_{R1}|}{X_L}\ \sin(\delta_1) = P_{max1}\ \sin\delta_1 \qquad \dots\dots(12.29)$$

Case 2: After connecting series capacitor

Consider a transmission line with series capacitor and connected to load as shown in the figure 12.25.

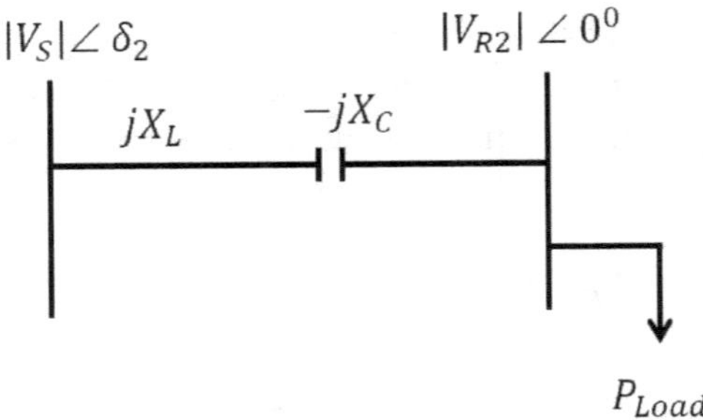

Figure 12.25 After connecting series capacitor

$$P_{e2} = \frac{|V_S||V_{R2}|}{X_L - X_C}\ \sin(\delta_2) = P_{max2}\ \sin\delta_2$$

$$P_{max2} = \frac{|V_S||V_R|}{(X_L - X_c)}$$

Assuming power transmitted through the transmission line before and after connecting the series capacitor as equal

$$P_{e1} = P_{e2}$$
$$P_{max1} \sin \delta1 = P_{max2} \sin \delta2$$
$$\Rightarrow \sin \delta_1 > \sin \delta_2$$
$$\Rightarrow \delta_1 > \delta_2 \text{ or} \Rightarrow \delta_2 < \delta_1$$

i.e., the angular stability of the power system increases.

2. With respect to the fault level

Case 1: Before connecting series capacitor

Consider a transmission line without series capacitor and connected to load as shown in the figure 12.26. Let I_{f1} be the fault current flowing in the line without series capacitor due to a fault on the transmission line.

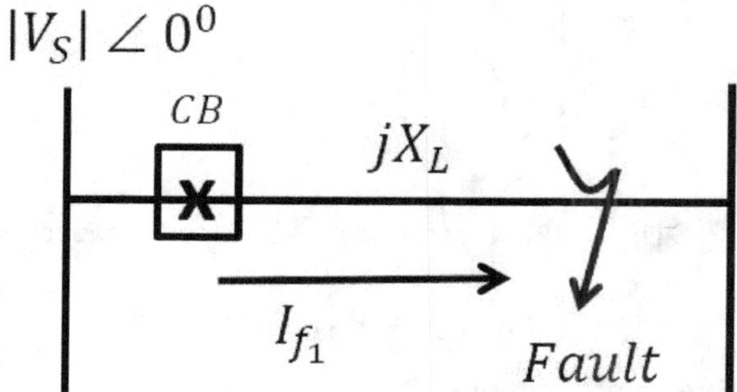

Figure 12.26 Before connecting series capacitor

Fault current before connecting series capacitor, $|I_{F1}| = \frac{|V_S|}{X_L}$

Case 2: After connecting series capacitor

Consider a transmission line with series capacitor and connected to load as shown in the figure 12.27. Let I_{f2} be the fault current flowing in the line with series capacitor due to a fault on the transmission line.

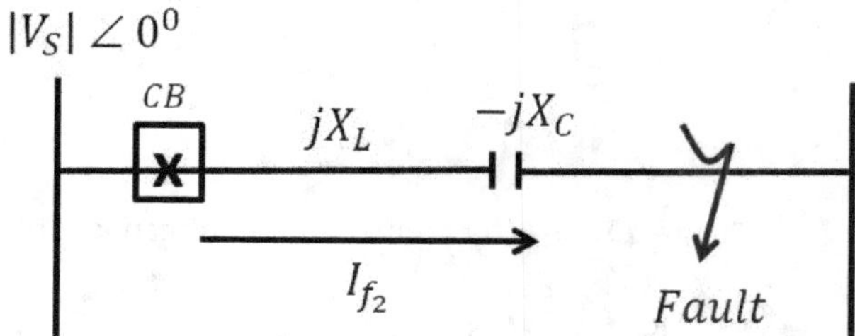

Figure 12.27 After connecting series capacitor

Fault current after connecting series capacitor, $|I_{F2}| = \dfrac{|V_S|}{(X_L - X_C)}$ .....(12.30)

From the equations 12.28 and 12.29, as the net reactance decreases, the fault current increases. Therefore I_{F2} increases

As I_{F2} increases, fault level increases, breaking capacity of circuit breakers increases, size of circuit breakers increases and overall cost increases.

Voltage across series capacitor, $V_C = |I_{F2}|X_C$

As V_C is high during fault, capacitor may be damaged due to excess current rating and deficient voltage ratings.

The capacitor is protected by considering an over voltage protection device as shown in the figure 12.28.

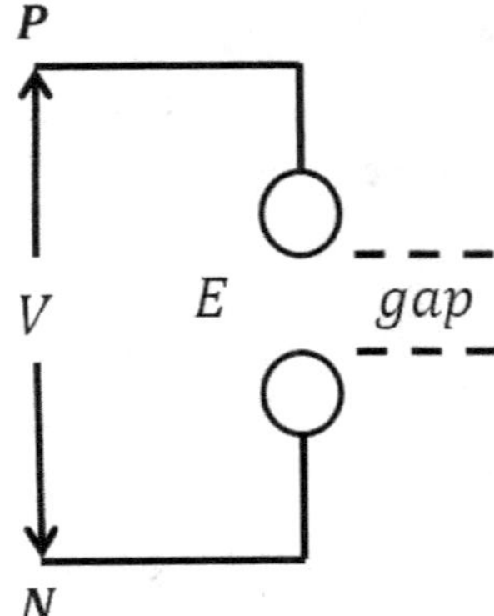

Figure 12.28 Over Voltage Protection Device

For normal voltages, E is low, gap will not ionize and resistance of the gap is infinity.

For abnormal over voltages, E is high, gap will ionize and resistance of gap is zero.

Therefore, sphere gap is connected across series capacitor from over voltages.

Figure 12.29 represents a sphere gap connected across a capacitor.

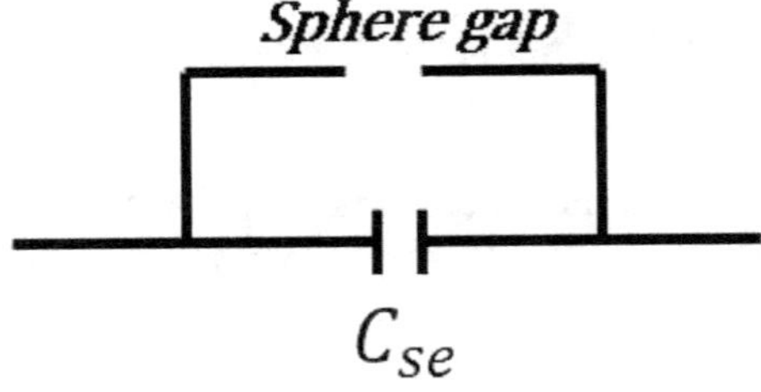

Figure 12.29 A sphere gap connected across capacitor

3. With respect to resonance

Consider an alternator connected to load through transmission line with a series capacitor as shown in the figure 12.30.

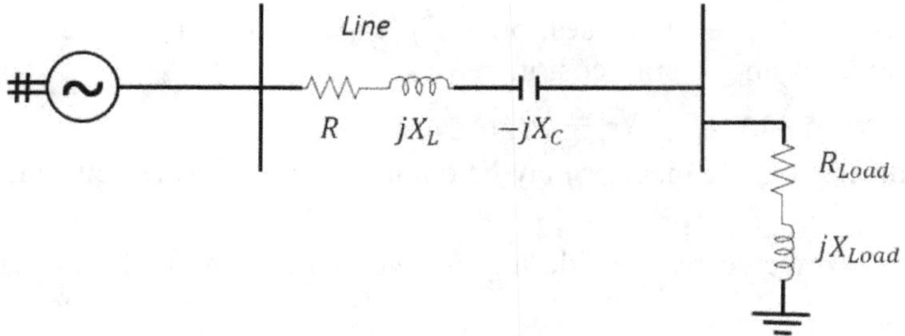

Figure 12.30 With series capacitor

If electrical system frequency equal to mechanical system oscillating frequency then a sub-synchronous resonance exists due to which shaft of the alternator damages.

The variation in receiving end voltages with and without shunt inductor at the receiving end of the transmission line is shown in the figure 12.31.

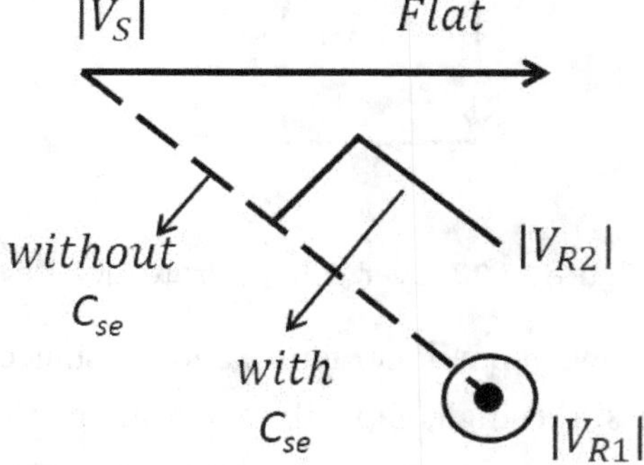

Figure 12.31 Variation in receiving end voltages with and without shunt inductor

12.5 COMPARISON BETWEEN SHUNT CAPACITOR AND SERIES CAPACITOR

(i) **with shunt capacitor:** A shunt capacitor has high voltage rating and low current rating

$$\Delta V_C = \frac{X\, Q_{sh\,cap}}{|V_S|} \qquad\qquad .....(12.31)$$

(ii) **with series capacitor:**

$$\Delta V_C = I\, X_C \sin \varnothing_R = I^2 X_C \frac{\sin \varnothing_R}{I} = \frac{Q_{se\,cap}}{I} \sin \varnothing_R \qquad .....(12.32)$$

From 12.5.1 and 12.5.2,

$$\frac{X\,Q_{sh\,cap}}{|V_S|} = \frac{Q_{se\,cap}}{I}\sin\varnothing_R$$

$$\frac{Q_{sh\,cap}}{Q_{se\,cap}} = \frac{|V_S|}{IX}\sin\varnothing_R = \frac{\sin\varnothing_R}{\left(\frac{IX}{|V_S|}\right)} = \frac{\sin\varnothing_R}{V_{Xp.u}}$$

$V_{Xp.u}$ is max 10% or 0.1

Considering $\cos\varnothing_R = 0.8, \sin\varnothing_R = 0.6$

$$\frac{Q_{sh\,cap}}{Q_{se\,cap}} = \frac{0.6}{0.1} \simeq 6$$

$$Q_{sh\,cap} > Q_{se\,cap}$$

Therefore, reactive power delivered by shunt capacitor is greater than the reactive power delivered by shunt inductor.

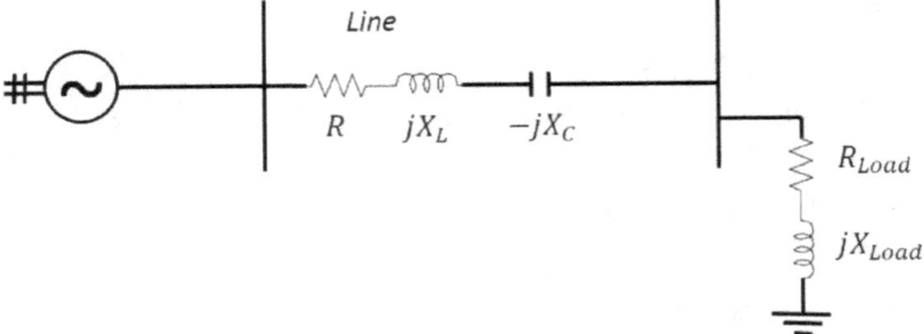

12.6 SYNCHRONOUS INDUCTOR AND SYNCHRONOUS CAPACITOR

Analysis

Consider an alternator connected to load through transmission line as shown in figure 12.32. Assume that synchronous machine connected to the receiving end of the line is operating at no-load condition i.e., $\delta = 0$.

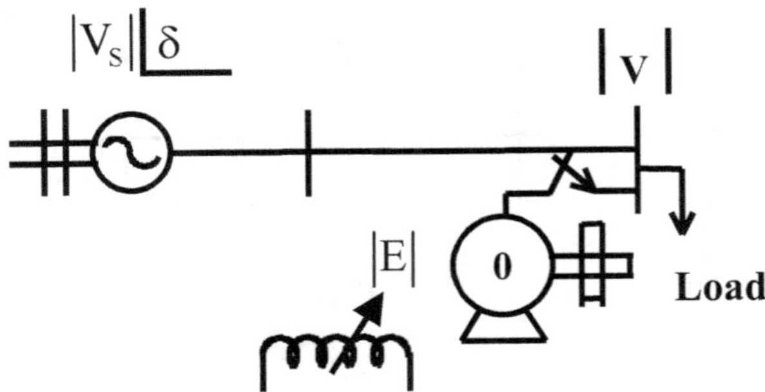

Figure 12.32 A synchronous machine at noload condition

The reactive power delivered to the receiving end of the transmission line is

$$Q_R = \frac{|V_S||V_R|}{|B|}\sin(\beta - \delta) - \frac{|A||V_R|^2}{|B|}\sin(\beta - \alpha)$$

Assuming short transmission Line i.e., $|A| = 1, \alpha = 0$ and neglecting resistance of the transmission line

Impedance of transmission line, $Z = 0 + jX_L$

Impedance of transmission line, $|B| = |Z| = X_L$

Impedance phase angle of transmission line, $\beta = \tan^{-1}\left(\frac{X_L}{0}\right)$

$$\therefore \beta = 90^0$$

$$Q_R = \frac{|V|}{X_L}[|E| - |V|] \qquad\qquad .....(12.33)$$

Case 1: Synchronous machine with over excitation

$$\Rightarrow |E| > |V|$$

$$\Rightarrow Q_R \text{ is positive}$$

An over excited Synchronous machine under no-load condition is known as "synchronous capacitor".

Case 2: Synchronous machine with under excitation

$$\Rightarrow |E| < |V|$$

$$\Rightarrow Q_R \text{ is negative}$$

An under excited Synchronous machine under no-load condition is known as "synchronous capacitor".

12.7 SYNCHRONOUS PHASE MODIFIER / PHASE ADVANCERS

Analysis

A synchronous phase modifier is an over excited synchronous motor with mechanical load on the shaft as shown in the figure 12.33.

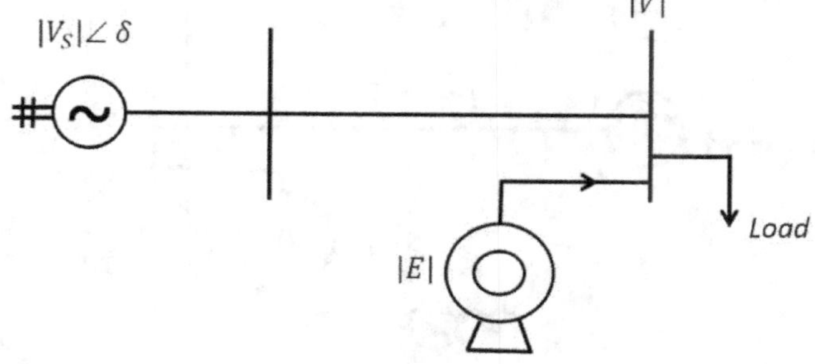

Figure 12.33 Synchronous phase modifier

Synchronous phase modifier absorbs active power and delivers reactive power.

The equivalent circuit representing power flow in a synchronous phase modifier is shown in the figure 12.34.

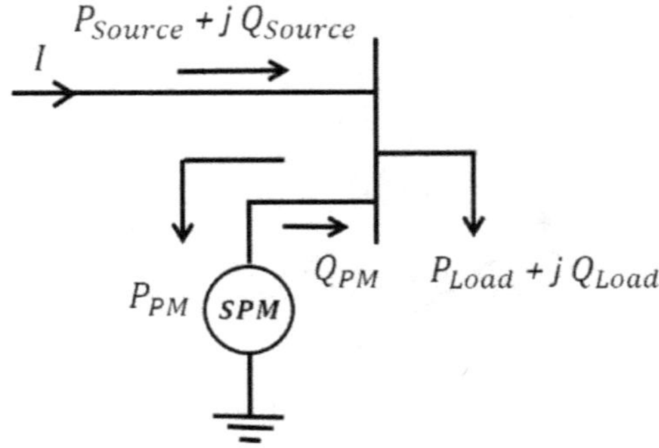

Figure 12.34 Power flow in a synchronous phase modifier

From the above figure,

Active power delivered by source, $P_{source} = P_{SPM} + P_{load}$

Reactive power delivered by source, $Q_{source} + Q_{SPM} = Q_{load}$

and $\qquad Q_{source} = Q_{load} - Q_{SPM}$

The equivalent circuit without synchronous phase modifier is shown in the figure 12.35.

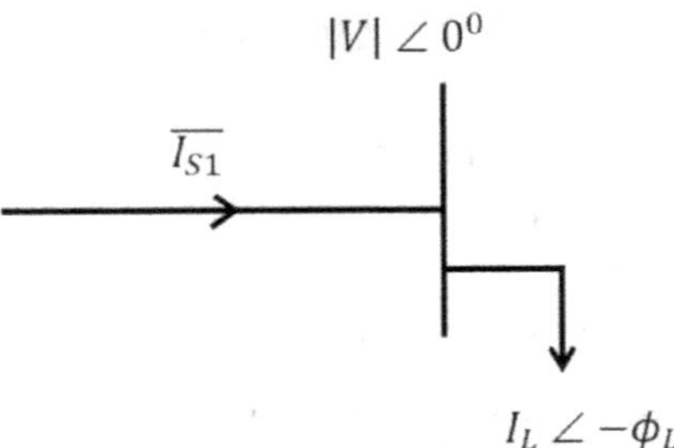

Figure 12.35 Without synchronous phase modifier

The equivalent circuit with synchronous phase modifier is shown in the figure 12.36.

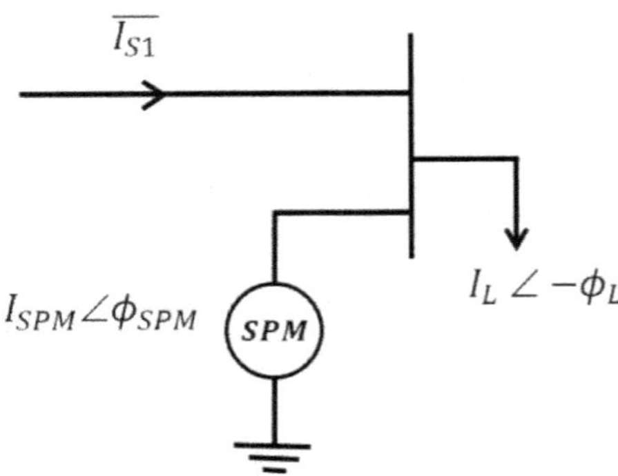

Figure 12.36 With synchronous phase modifier

The vector diagram is as follows in figure 12.37.

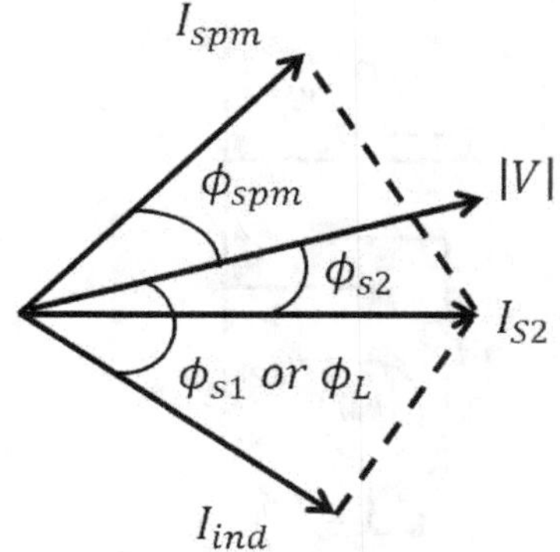

Figure 12.37 Vector diagram with and without SPM

From the above figure, power factor of the system increases.

Active power absorbed by SPM, $\quad P_{SPM} = \dfrac{|E||V|}{X_L} \sin \delta$ $\qquad$ .....(12.34)

Reactive power delivered by SPM, $Q_R = \dfrac{|V|}{X_S}\big[|E| \cos \delta - |V|\big]$ $\qquad$ .....(12.35)

If $|E| \cos \delta > |V|$ then Q_R is positive

An over excited synchronous machine operating at loaded condition is known as a "synchronous phase modifier".

Power factor at which synchronous phase modifiers operates

$$\mathrm{Cos}\ \emptyset_{SPM} = \frac{P_{SPM}}{\sqrt{P_{SPM}^2 + Q_{SPM}^2}} \text{lead}$$

For the same reactive power injected, the excitation current for SPM is greater than Synchronous condenser.

To improve the power factor from $\cos\emptyset_1$ to $\cos\emptyset_2$, reactive power delivered by SPM is $Q_{SPM} = P_1 \tan\emptyset_1 - P_2 \tan\emptyset_2$

Where P_1 and P_2 are not equal or P_2 is greater than P_1.

$$P_1 = P_{load} \text{ and } P_2 = P_{load} + P_{SPM}$$

KEY POINTS

Shunt capacitor

1.   Increases power factor
2.   Increases $|V_R|$
3.   Used during low voltages
4.   Power factor correction device

Shunt Inductor
1. Decreases $|V_R|$
2. Used during over voltages
3. Eliminates Ferranti effect

Series capacitor
1. Increases $|V_R|$
2. Power system stability device
3. Commercial known as boosters

Series Inductor
1. Equal to reactance of transmission line.
2. It acts as fault current limiter
3. Smoother reactor in power electronic circuits and HVDC system

Synchronous capacitor
1. Over excited synchronous machine at no-load condition
2. Improves power factor

Synchronous Inductor
1. Under excited synchronous machine at no-load condition
2. Eliminates Ferranti effect

Synchronous phase modifiers
1. Power factor correction device
2. Over excited synchronous machine operated at loaded condition.
3. Absorbs active power and delivers reactive power.

SOLVED PROBLEMS

Q1. A 3-Φ Induction motor rated at 400v, 50Hz coupled to a pump is running at lower power factor of 0.6. The input is 4.5 KVA and it is proposed to improve the power factor to 0.8 lag by connecting a delta connected capacitor bank. Determine the required capacitance per phase.

Solution:

Given Data:

Initial power factor, $\cos \phi_1 = 0.6$

Final power factor, $\cos \phi_2 = 0.8$

Apparent power input to induction motor, S = 4.5 KVA

Operating voltage = 400 V

Active power input to Induction motor, $P_{IM} = S_{IM} \times \cos \phi_1$
$$= 4.5 \times 0.6 = 2.7 \text{ KW}$$

Reactive power delivered per phase, $Q_{\text{shunt cap.}} = \dfrac{P}{3}[\tan\emptyset_1 - \tan\emptyset_2]$

$$= \dfrac{2.7}{3}[\tan(\cos^{-1}(0.6)) - \tan(\cos^{-1}(0.8))]$$

$$= 0.525\text{KVAR}$$

$$\Rightarrow Q_{\text{shunt cap./phase}} = \omega\, C_{\text{shunt}\frac{\text{cap}}{\text{phase}}}\, V_{ph}^2$$

$$\Rightarrow 0.525 * 10^3 = 2\pi * 50\ \ C_{\text{sh cap/phase}} \cdot (400)^2$$

$$\Rightarrow C_{\text{sh cap/phase}} = 10.44\ \mu F$$

Q2. At an Industrial substation with a 4MW Load. A Capacitor of 2MVAR is installed to maintain the load power factor at 0.97 lag. Assume that the capacitor bank is out of service. Determine the initial power factor.

Solution:

Given Data:

Final power factor, $\cos\phi_2 = 0.97$

Reactive power input to induction motor, $Q = 2$ MVAR

Active power load, $P = 4$ MW

Reactive power delivered per phase, $Q_{\text{shunt cap.}} = P[\tan\emptyset_1 - \tan\emptyset_2]$

$$2 \times 10^6 = 4 \times 10^6[\tan(\cos^{-1}\phi_1 - \tan(\cos^{-1}(0.8))]$$

$$\phi_1 = 86.8$$

$$\cos\phi_1 = 0.8$$

Q3. A 400V, 50 Hz, $3-\emptyset$ balanced source supplies power to a star connected load whose rating is $12\sqrt{3}$ KVA, 0.8 power factor lag. Determine the rating in KVAR of the delta connected capacitive reactance power bank to increase the power factor to unity.

Solution:

Given Data:

Initial power factor, $\cos\phi_1 = 0.8$

Final power factor, $\cos\phi_2 = 1$

Reactive power input, $Q = 12\sqrt{3}$ KVA $= S(\text{Sin}\,\emptyset_1 - \text{Sin}\emptyset_2)$

$$= 12\sqrt{3}(0.6 - 0) = 12.47\text{ KVAR}$$

Q4. A three phase, 11KV, 50HZ, 200KW load has a power factor of 0.8, lag. A delta connected three phase capacitor is used to improve the power factor to unity. Determine the capacitance per phase of the capacitor in μF.

Solution:

Given Data:

Initial power factor, $\cos \phi_1 = 0.8$

Final power factor, $\cos \phi_2 = 1$

Active power $= 200$ KW

Reactive power delivered by shunt capacitor, $Q_{sh.cap} = P(\tan\phi_1 - \tan\phi_2)$

$$= 200\,[\tan(\cos^{-1}0.8) - \tan(\cos^{-1}1)]$$

$$\frac{3V^2_{ph}}{X_{C_{sh.cap}}} = 150 \text{ KVAR}$$

$$3\,\omega\,C_{sh.cap|\phi}V^2_{ph} = 150$$

$$3 \times 2\pi \times 50 \times C_{sh.cap|\phi} = 11^2 \times 10^6 = 150 \times 10^3$$

Shunt capacitance connected per phase, $C_{sh.cap|\phi} = 1.3\ \mu F$

Q5. In a 400KV, power Network, 360KV is recorded at a 400KV bus. Determine the reactive power absorbed by a shunt reactor rated for 50MVAR, 400KV connected at the bus.

Solution:

Given Data:

Actual operating voltage $= 400$ KV

Recorded operating voltage $= 360$ KV

Reactive power absorbed by shunt reactor, $Q \propto V_{ph}^2$

$$Q_1 \propto V_{ph1}^2$$

$$Q_2 \propto V_{ph2}^2$$

$$\Rightarrow \frac{Q_1}{Q_2} = \left(\frac{400}{360}\right)^2$$

$$Q_2 = 40.5 \text{ MVAR}$$

Q6. The thevenin's equivalent impedance of a bus-bar in a 3-ϕ in 220KV system is 0.25 PU at a base of 250MVA. Calculate the reactive power needed in MVAR to increase the voltage by 4 KV & decrease the voltage by 2KV. Identify the equipment required in each case.

Solution:

Given Data:

Let base MVA $= 250$ MVA

Let base KV $= 220$ KV

$$\text{p. u. thevenin's impedance, } Z_{p.u} = Z_\Omega \left(\frac{MVA_b}{KV_b^2}\right)$$

$$0.25_{p.u} = Z_\Omega \left(\frac{250}{220^2}\right)$$

$$Z_\Omega = 48.4\,\Omega$$

Shunt capacitor increases voltage by 4 KV.

$$\Delta V = \frac{X.\, Q_{shunt\ cap.}}{|V_S|}$$

$$4000 = \frac{48.4}{220\times 10^3}\, Q_{shunt.cap}$$

$$Q_{shunt\ cap.} = 18.18\ \text{MVAR}$$

Shunt inductor decreases voltage by 2 KV

$$\Delta V = \frac{X.\, Q_{shunt\ ind.}}{|V_S|}$$

$$2000 = \frac{48.4}{220\times 10^3}\, Q_{shunt\ ind.}$$

$$Q_{shunt\ ind.} = 9.09\ \text{MVAR}$$

Q7. An interconnecting cable having a reactance of j0.05 p.u. links two generating stations G_1 & G_2 as shown in figure. $|V_1|$ and $|V_2|$ are 1.0 p.u. The load demands at the two buses are $S_{D1} = (15 + j5)$ p.u & $S_{D2} = (25 + j15)$ p.u. Calculate the total reactive power in p.u. at the generating station G_1 when load angle is 15^0.

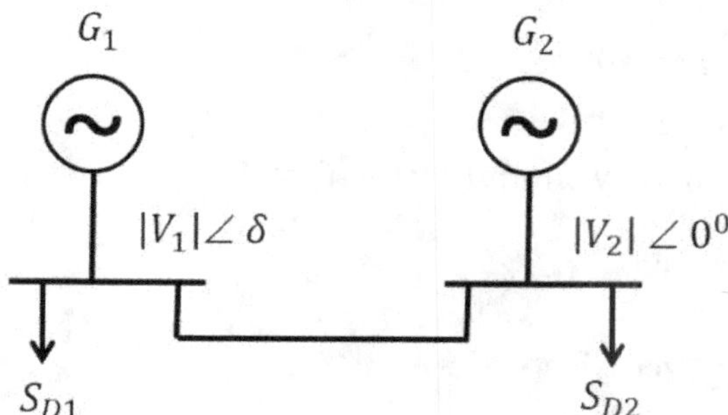

Solution:

Given Data:

Reactance of interconnecting cable = j0.05 p.u.

Complex power transmitted through the line from bus 1 to bus 2 is

$$S_{12} = V_1\, I_{12}^*$$

$$S_{12} = |V_1|\angle\delta_1 \left[\frac{|V_1|\angle\delta_1 - |V_{21}|\angle 0}{jX}\right]$$

$$= |1|\angle 15^\circ \left[\frac{|1|\angle 15^\circ - |1|\angle 0}{j0.05}\right]$$

$$= 5\ \angle 7.5^\circ$$

Reactive power transmitted through the line, $Q_{12} = 5.22\ \sin 7.5^\circ = 0.68$ p. u.

Reactive power at generating station G_1, $Q_{G1} = Q_{D1} + Q_{12} = 5 + 0.68 = 5.68$ p.u.

Q8. Consider the following transmission line diagram in which a series capacitor was placed at the midpoint. What is the point at which maximum voltage occurs?

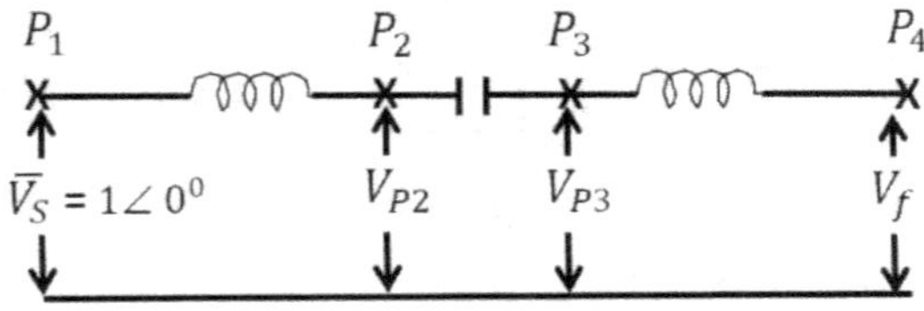

$$X_{P1P2} = j0.1 \text{ p.u.}, \quad X_{P2P3} = -j0.05 \text{ p.u.}, \quad X_{P3P1} = j0.1 \text{ p.u.}$$

Solution:

Voltage at point P1, $V_{p1} = V_S = 1\angle 0^0 \text{p.u}$

Voltage at point P2, $V_{p2} = 1\angle 0^0 - \bar{I}(j\,0.1)$

Voltage at point P3, $V_{p3} = 1\angle 0^0 - \bar{I}(j\,0.1 - j0.15)$

$$= 1\angle 0^0 - \bar{I}(j\,0.05)$$

Voltage at point P4, $V_{p4} = 1\angle 0^0 - \bar{I}(j\,0.1 - j0.15 + j0.1)$

$$= 1\angle 0^0 - \bar{I}(j\,0.05)$$

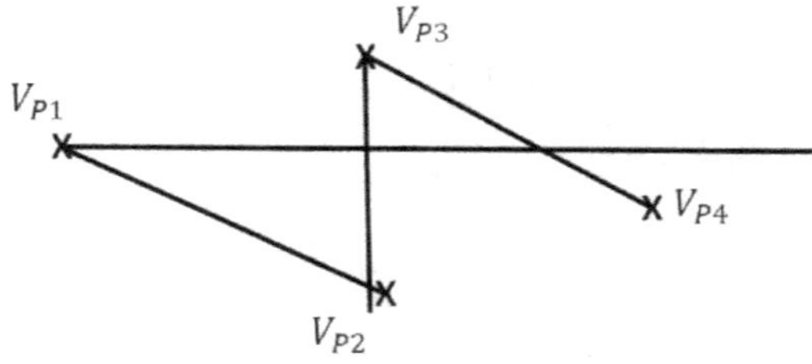

Maximum voltage occurs at point 3.

Q9. The sending end voltage of a three phase,300 km long lossless transmission line is 400 KV. It has characteristic impedance of 400 ohms and phase constant of 0.002 rad/km and is operating on no-load.

1. Find the reactance and it's rating that has to be connected in shunt at the receiving end side to make receiving ned voltage same as sending end voltage.

2. What is the value of the resistor at the receiving end per phase to make receiving end and sending end voltages magnitudes equal.

Solution:

Given data:

Sending end voltage of transmission line = 400 KV

Characteristic impedance of transmission line, $Z_c = 400\Omega$ and

Phase constant, $\beta = 0.002$ rad/km

$$A = D = \cos \beta l = \cos(0.002 * 300) = 0.825 \angle 0^0$$

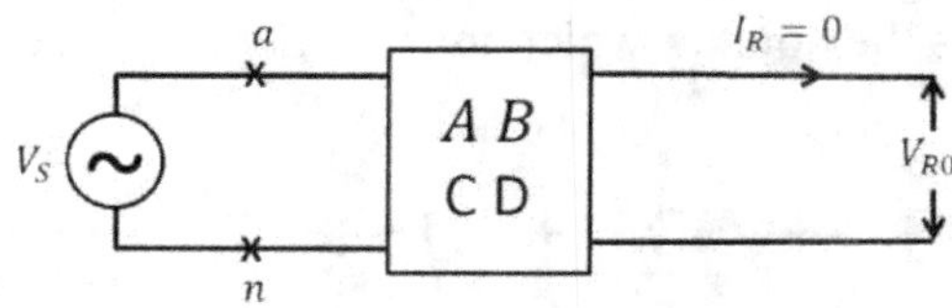

$$\text{As } |A| < 1 \Rightarrow |V_{ro}| > |V_S|$$

i.e., Ferranti effect occurs.

(i) Shunt reactor at the receiving end

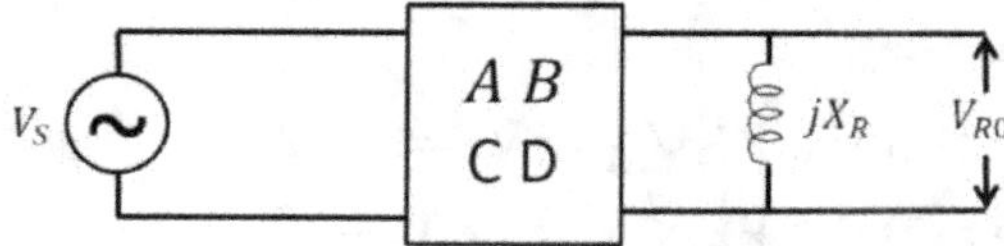

$$B = jZ_c \sin \beta l = j400 \sin(0.002 * 300) = 225.8 \angle 90^0 \text{ohms}$$

$$C = j\frac{1}{Z_c} \sin \beta l = j * \frac{1}{400} \sin(0.002 * 300) = 1.41 * 10^{-3} \angle 90^0 \text{s}$$

to get $|V_{ro}| = |V_S| \Rightarrow |A_{eq}| = 1$

equivalent ABCD parameters,

$$\begin{bmatrix} A_{eq} & B_{eq} \\ C_{eq} & E_{eq} \end{bmatrix} = \begin{bmatrix} A & B \\ C & D \end{bmatrix} \begin{bmatrix} 1 & 0 \\ \dfrac{1}{jx_r} & 1 \end{bmatrix}$$

$$A_{eq} = A * 1 + B * \frac{1}{jX_r} = 0.825 * \frac{j\,225.8}{j\,X_r} = 0.825 + \frac{225.8}{X_r}$$

$$|A_{eq}| = 1 = 0.825 + \frac{225.8}{X_r} \Rightarrow X_r = 1290 \text{ ohms}$$

i.e. 1290 ohms is connected between any one phase to neutral per phase reactance of star connected system.

Three phase reactor rating is

$$\text{Three phase } Q_{reactor,} = 3 * \frac{V_{ph}^2}{X_r} = \frac{3\left(\frac{400\text{ k}}{\sqrt{3}}\right)^2}{1290} = 124.03 \text{ MVAR}$$

(ii) For load resistance $= Z_C$

$$|V_r| = |V_s| \Rightarrow R_{load/ph} = 400 \text{ohms}$$

Q10. A three phaseInduction Motor delivers 500 HP at an efficiency of 90% when operating at 0.8 pf lagging. A loaded synchronous motor is connected in parallel with Induction Motor. If overall power factor is unity then find:

(i) reactive power to be injected by synchronous modifier

(ii) power factor of synchronous motor

(iii) KVA supplied by source without and with synchronous modifier

Solution:

Given data:

Output of induction motor = 500HP

$$= 500 * 746 = 373 \text{ KW}$$

Input to induction motor $= \dfrac{P_{out}}{X} = \dfrac{373}{0.9} = 414.4 \text{KW}$

Power factor of induction motor, $\cos \emptyset_{IM} = 0.8$ lag

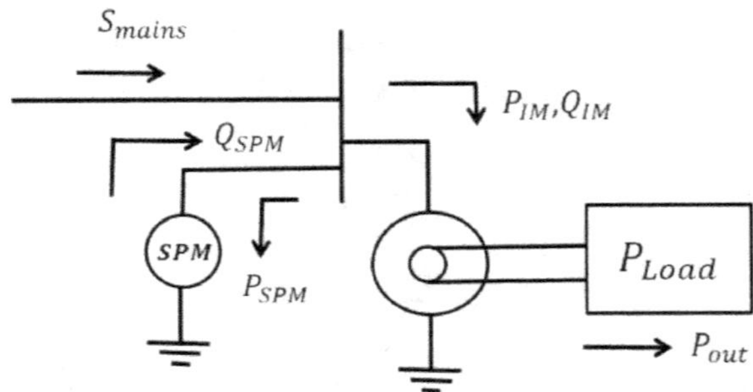

Real power drawn by synchronous modifier, $P_{SM} = 120\text{KW}$

Power balance equations, $P_{source} = P_{IM} + P_{SM} = 414.4 + 120 = 534.4\text{KW}$

$$Q_{source} + Q_{SM} = Q_{IM}$$

As overall power factor to be at unity, $Q_{source} = 0, Q_{SM} = Q_{IM}$

(i) Reactive power to be injected by induction motor

$$Q_{IM} = P_{IM}.\tan\emptyset_{IM} = 414.4 \tan(\cos^{-1}0.8) = 310.5 \text{ KVAR}$$

Reactive power to be injected by synchronous modifier

$$Q_{sm} = Q_{IM} = 310.5 \text{ KVAR}$$

(ii) Power factor of synchronous modifier, $\cos \emptyset_{sm} = \dfrac{P_{SM}}{\sqrt{P_{SM}^2 + Q_{SM}^2}}$

$$= \dfrac{120}{\sqrt{120^2 + (310.5)^2}} = 0.36\text{lead.}$$

(iii) KVA supplied by source without synchronous modifier

$$\overline{S}_{source} = P_{IM} + jQ_{IM} = 414.4 + j310.5\text{KVA}$$

$$= \sqrt{(414.4)^2 + (310.5)^2} = 517.8\text{KVA}$$

With synchronous modifier,

$$S_{source} = P_{source} + jQ_{source} = 534.4 + j0 = 534.4 \text{ KVA}$$

Q11. A three phasetransmission line is having an impedance of $(5 + j20)\dfrac{ohms}{ph}$, delivers a load of 30 MW at a p.f. of 0.8 lag and voltage of 33KV, 50 Hz.

(i) Determine the capacity of shunt condenser to be placed at receiving end to maintain sending end voltage at 33 KV.

(ii) If condenser is interpreted as a delta connected capacitor bank then what is capacitance/phase in capacitor bank?

Solution:

Given data:

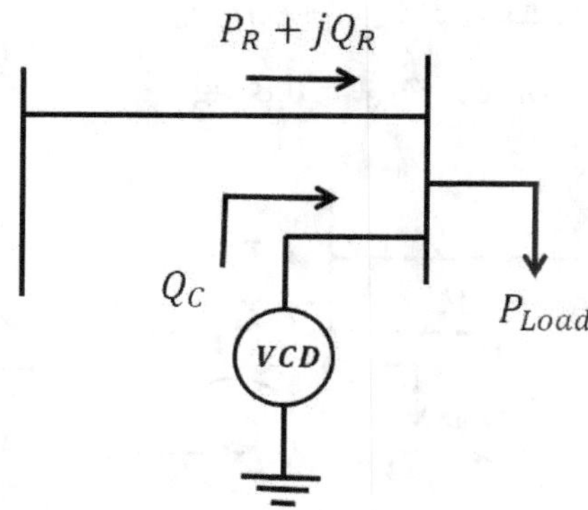

Note that voltages after synchronous condition at both ends are equal, then there is a synchronous condenser at receiving end.

Reactive power, $Q_{load} = P_{load} . \tan\varnothing = 30 \tan(\cos^{-1} 0.8) = 22.5$ MVAR

Active power, $P_{load} = 30$ MW

For power balance equation, $P_R = P_{load} = 30$ MW;

For a transmission line, $\begin{bmatrix} A & B \\ C & D \end{bmatrix} = \begin{bmatrix} 1 & Z \\ 0 & 1 \end{bmatrix} = \begin{bmatrix} 1 & 5+j20 \\ 0 & 1 \end{bmatrix} = \begin{bmatrix} 1\angle 0^0 & 20.6\angle 75.96^0 \\ 0 & 1\angle 0^0 \end{bmatrix}$

(i) Assuming a short transmission line,

$$|A| = 1; \ \alpha = 0^0; \ |B| = 20.6; \ \beta = 75.96^0$$

Active power delivered, $P_R = \dfrac{|V_s||V_r|}{|B|} \cos(\beta - \delta) - \dfrac{|A||V_r|^2}{|B|} \cos(\beta - \alpha)$

$$\Rightarrow 30 * 10^6 = \dfrac{|33*10^3||33*10^3|}{20.6} \cos(75.96 - \delta) - \dfrac{1.0 \,|33*10^3|}{20.6} \cos(75.96 - 0)$$

$$\Rightarrow \delta = 40.1^\circ$$

Reactive power delivered, $Q_R = \dfrac{|V_s||V_r|}{|B|} \sin(\beta - \delta) - \dfrac{|A||V_r|^2}{|B|} \sin(\beta - \alpha)$

$$\Rightarrow = \dfrac{|33*10^3||33*10^3|}{20.6} \sin(75.96 - 40.1) - \dfrac{1.0 \,|33*10^3|}{20.6} \sin(75.96 - 0) \ = -20.28 \text{MVAR}$$

Now, $Q_C = Q_{load} - Q_R = 22.5 - (-20.28) = 42.78$ MVAR

(ii) Capacitor bank is in delta, Δ

$$Q_{c/ph} = \dfrac{V_{ph}^2}{Xc/ph} = 2\pi f \, C/ph \, V_{ph}^2$$

$$C/ph = \frac{Q_c/ph}{V_{ph}^2(2\pi f)}$$

$$= \frac{\left(\frac{42.78}{3}\right)}{(33)^2 \, 2\pi * 50}$$

$$= 0.416\mu F$$

UNSOLVED PROBLEMS

1. A shunt reactor of 100MVAR is operating at 98% of its rated voltage & at 96% of its rated frequency. Calculate the reactive power absorbed by shunt reactor.

   Ans: **100.04 MVAR**

2. A lossless transmission line is having surge impedance loading 2280 MW. A series capacitive compensation 30% is done. Calculate the surge Impedance Loading of the compensated transmission line.

   Ans: **2725 MW**

3. A three phase 11 KV, generator feeds powers to a constant power unity power factor load of 100 MW through a three phase transmission line. The line-line voltage at the terminals of the machine is maintained constant at 11KV. The per unit positive sequence impedance of the line based on 100 MVA & 11 KV is j0.2. The line to line voltage at the load terminals is measured to be less than 11 KV. Calculate the total reactive power to be injected at the terminals of the load to increase the line to line voltage at the load terminals to 11KV.

   *Ans:*1 **0.1 MVAR**

OBJECTIVE QUESTIONS

1. An _____________ absorbs lagging reactive power.

2. A _____________ delivers leading reactive power.

3. The reactive power balance equation with a voltage control device at the receiving end of the line is _____________

4. A _____________ increases voltage at the receiving end of the line by compensating the excess inductive reactive power.

5. A _____________ decreases voltage at the receiving end of the line by compensating the excess capacitive reactive power.

6. Shunt capacitor is predominantly known as _____________

7. Series capacitor is predominantly known as _____________

8. Series capacitor is commercially known as________________.

9. An over excited synchronous machine operating at no-load condition is known as________________.

10. An under excited synchronous machine operating at no-load condition is known as________________.

11. An over excited synchronous machine operating at loaded condition is known as________________.

12. ________________is a power loss device.

13 MATLAB Programming

13.1 Y$_{BUS}$ FORMATION BY INSPECTION METHOD

MATLAB Program:

```
clc
clear all
n=input('enter no. of buses:');      % no. of buses excluding reference
nl= input('enter no. of lines:');    % no. of transmission lines
sb= input('enter starting bus of each line:');  % starting bus of a line
eb= input('enter ending bus of each line');     % ending bus of a line
zser= input('enter resistance and reactance of each line:'); % line
resistance and reactance (R, X)
yshty=input('enter shunt admittance of the
bus:');                                          % shunt admittance
i=0
k=1
while
i<nl
                                   % impedance of a line
                                   (R+jX)
zser1(i+1)=zser(k)+j*zser(k+1)
i=i+1
k=k+2
end
zser2=reshape(zser1,nl,1);
yser=ones(nl,1)./zser2;
ybus=zeros(n,n);
for
```

```
i=1:nl
ybus(sb(i),sb(i))=ybus(sb(i),sb(i))+(j*yshty(i))+yser(i);
ybus(eb(i),eb(i))=ybus(eb(i),eb(i))+(j*yshty(i))+yser(i);
ybus(sb(i),eb(i))=-yser(i);
ybus(eb(i),sb(i))=-yser(i);
end
```

Input:

```
n=5
nl=7
sb=[1 1 2 2 2 3 4]
eb=[2 3 3 4 5 4 5]
zser=[0.02 0.06 0.08 0.24 0.06 0.18 0.06 0.18 0.04 0.12 0.01 0.03 0.08
0.24] yshty=[0.03 0.025 0.02 0.02 0.015 0.01 0.025]
```

Output:

```
bus admittance matrix ybus
6.25+(-18.70)j  -5.00+(  15.00)j  -1.25+(  3.75)j  0.00+(  0.00)j  0.00+(
0.00)j  -5.00+(  15.00)j  10.83+(-32.42)j  -1.67+(  5.00)j  -1.67+(  5.00)j  -
2.50+(  7.50)j  -1.25+(  3.75)j  -1.67+(  5.00)j  12.92+(-38.70)j-10.00+(
30.00)j  0.00+(  0.00)j  0.00+(  0.00)j  -1.67+(  5.00)j-10.00+(  30.00)j
12.92+(-38.70)j  -1.25+(  3.75)j
0.00+(  0.00)j  -2.50+(  7.50)j  0.00+(  0.00)j  -1.25+(  3.75)j  3.75+(-
11.21)j
```

13.2 Y_{BUS} FORMATION BY SINGULAR TRANSFORMATION METHOD WITHOUT MUTUAL COUPLING

Steps involving singular transformation:

1. Obtain the oriented graph for the given system.

2. Get the bus incidence matrix which is the one which indicates the incidence of all the elements to nodes in connected graph. The size of this matrix is e*(n-1) where e is the number of elements in the graph and n is the number of nodes (A)

3. Get the primitive admittance matrix from the graph of size e*e. If mutual coupling between the lines is neglected then the resulting primitive matrix is a diagonal matrix(off diagonal elements are zero)([y])

4. Ybus can be obtained from the equation, Ybus = A^t * [y] *A

MATLAB Program:

```matlab
clc
clear all
n=input('enter no. of buses:');
nl= input('enter no. of lines:');
sb= input('enter starting bus of each line:');
eb= input('enter ending bus of each line');
zser= input('enter resistance and reactance of each line:');
yshty= input('enter shunt admittance of the bus:');
% no. of buses excluding reference
% no. of transmission lines
% starting bus of a line
% ending bus of a line
% line resistance and reactance (R, X)
% shunt admittance
i=0;
k=1;

while i<nl
zser1(i+1)=zser(k)+j*zser(k+1);
% impedance of a line (R+jX)i=i+1;
k=k+2;

end
zser2=reshape(zser1,nl,1);
yser=ones(nl,1)./zser2;
ypri=zeros(nl+n,nl+n);
ybus=zeros(n,n)
;
a=zeros(nl+n,n)
;
for i=1:n
a(i,i)=1;
end
for i=1:nl
a(n+i,sb(i))=1;
a(n+i,eb(i))=-
```

```
1;
ypri(n+i,n+i)=yser(i);
ypri(sb(i),sb(i))=ypri(sb(i),sb
(i))+yshty(i);
ypri(eb(i),eb(i))=ypri(sb(i),eb
(i))+yshty(i);
end
at=transpose(a)
;
ybus=at*ypri*a;
zbus=inv(ybus);
```

Input:

```
n= 3
nl=3
sb=[1 1 2 ]
eb=[2 3 3 ]
zser=[0.01 0.03 0.08 0.24 0.06
0.18 ]
yshty=[0.01 0.025 0.02]
```

Output:

```
bus      admittance
matrix
ybus:
11.29 + (-33.75)j
-10.00 + (30.00)j
-1.25 + (
3.75)j
-10.00 + (0.00)j
11.70 + (-35.00)j
-1.67 + (
5.00)j
-1.25 + (3.75)j
-1.67 + ( 5.00)j
2.94 + ( -8.75)j
```

13.3 Y_{BUS} FORMATION BY SINGULAR TRANSFORMATION METHOD WITH MUTUAL COUPLING

MATLAB program:

```
clc
clear all
n=input('enter no. of buses:');
nl= input('enter no. of lines:');
sb= input('enter starting bus of each line:');
eb= input('enter ending bus of each line');
zser= input('enter resistance and reactance of each line:');
% no. of buses excluding reference
% no. of transmission lines
% starting bus of a line
% ending bus of a line
% line resistance and reactance (R, X)nmc= input('enter no. of mutual
couplings:');
f1= input('enter first line no. of each mutual coupling:');
s1= input('enter second line no. of each mutual coupling:');
mz= input('enter mutual impedance between lines:');
i=0
k=1
while i<nl
zser1(i+1)=zser(k)+j*zser(k+1)
% impedance of a line (R+jX)
i=i+1
k=k+2
end
i=0
k=1
while i<nmc
mz1(i+1)=mz(k)+j*mz(k+1)
i=i+1
k=k+2
end
zser2=reshape(zser1,nl,1);
mz2=reshape(mz1,2,1);
zpri=zeros(nl,nl);
```

```
a=zeros(nl,n)
for i=1:nl
zpri(i,i)=zser2(i);
if sb(i)~=0
a(i,sb(i))=1;
end
if eb(i)~=0
a(i,eb(i))=-1
end
end
for i=1:nmc
zpri(f1(i),s1(i))=mz2(i);
zpri(s1(i),f1(i))=mz2(i);
end
ypri=inv(zpri)
at=transpose(a)
ybus=at*ypri*a
zbus=inv(ybus)
```

Input:

```
n=3
nl=5
sb=[0 0 2 0 1]
eb=[1 2 3 1 3]
zser=[0 0.6 0 0.5 0 0.5 0 0.4 0 0.2]
nmc=2
f1=[1 1]
s1=[2 4]
mz=[0 0.1 0 0.2 ]
```

<u>Output:</u>

Bus admittance matrix is

```
8.021 -0.208 -5.000
-0.208 4.083 -2.000
-5.000 -2.000 7.000
```

Bus impedance matrix is

```
0.271 0.126 0.230
0.126 0.344 0.189
0.230 0.189 0.361
```

13.4   FORMATION OF Y-BUS MATRIX

PROGRAM:

```
Formation of Y Bus Matrix
clc
clear all
disp('');
b = input('Enter no. of buses: ');
s = input('Enter no. of impedences: ');
for i=1:s
sb(i) = input('Enter starting bus no. ');
rb(i) = input('Enter receiving bus no. ');
imp(i) = input('Enter impedance of bus: ');
lc(i) = input('Enter line charging admittance: ');
ybus = diag(0,b-1);
end
for i=1:s
k1 = sb(i);
k2 = rb(i);
adm(i) = 1/imp(i);

ybus(k1,k1) = ybus(k1,k1) + adm(i) + lc(i); ybus(k2,k2) = ybus(k2,k2) +
adm(i) + lc(i); ybus(k1,k2) = -adm(i); ybus(k2,k1) = ybus(k1,k2);
end
ybus
```

OUTPUT:

```
Enter no. of buses: 4
Enter no. of impedances: 5
Enter starting bus no. 1
Enter receiving bus no. 2
```

```
Enter impedance of bus: 0.2+0.8j
Enter line charging admittance: 0.02j
Enter starting bus no. 2
Enter receiving bus no. 3
Enter impedance of bus: 0.3+0.9j
Enter line charging admittance: 0.03j
Enter starting bus no. 2
Enter receiving bus no. 4
Enter impedance of bus: 0.25+1j
Enter line charging admittance: 0.04j
Enter starting bus no. 3
Enter receiving bus no. 4
Enter impedance of bus: 0.2+0.8j
Enter line charging admittance: 0.02j
Enter starting bus no. 1
Enter receiving bus no. 3
Enter impedance of bus: 0.1+0.4j
Enter line charging admittance: 0.01j
ybus =
```

```
0.8824  - 3.4994i  -0.2941 + 1.1765i  -0.5882 + 2.3529i               0

-0.2941 + 1.1765i  0.8627  - 3.0276i  -0.3333 + 1.0000i  -0.2353 + 0.9412i

-0.5882 + 2.3529i  -0.3333 + 1.0000i  1.2157  - 4.4694i  -0.2941 + 1.1765i

      0            -0.2353 + 0.9412i  -0.2941 + 1.1765i  0.5294  - 2.0576i
```

13.5 FORMATION OF Y-BUS MATRIX USING BUS BUILDING ALGORITHM

PROGRAM:

Formation of Y-Bus Matrix usign bus building algorithm clc

clear all

```
y10 = 1/0.6j;
y12 = 1/0.2j;
y23 = 1/0.1j;
```

```
y13 = 1/0.3j;
y34 = 1/0.4j;
y30 = 1/0.8j;

Y11 = y10+y12+y13;
Y12 = -y12;
Y13 = -y13;
Y14 = 0;

Y21 = -y12;
Y22 = y12+y23;
Y23 = -y23;
Y24 = 0;

Y31 = -y13;
Y32 = -y23;
Y33 = y30+y13+y23+y34;
Y34 = -y34;

Y41 = 0;
Y42 = 0;
Y43 = -y34;
Y44 = y34;

Y = [Y11 Y12 Y13 Y14; Y21 Y22 Y23 Y24; Y31 Y32 Y33 Y34; Y41
Y42 Y43   Y44]
```

OUTPUT:

```
Y =

   0 -10.0000i      0     + 5.0000i      0     + 3.3333i      0

   0 + 5.0000i      0     -15.0000i      0     +10.0000i      0

   0 + 3.3333i      0     +10.0000i      0     -17.0833i      0     + 2.5000i

   0               0                     0     + 2.5000i      0     - 2.5000i
```

13.6 FORMATION OF Z-BUS MATRIX USING BUS BUILDING ALGORITHM – 1

PROGRAM:

```
%Formation of Z-Bus matrix using bus building algorithm clc
clear all
z12 = 0.2j;
z13 = 0.4j;
z23 = 0.2j;
disp('STEP1: Add an element between reference node and node(1)') zBus1 =
[z12]
disp('STEP2: Add an element between existing node(1) and new node(3)')
zBus2 = [zBus1(1,1) zBus1(1,1);
zBus1(1,1) zBus1(1,1)+z13]
disp('STEP3: Add an element between existing node(3) and reference
node')
zBus3 = [zBus2(1,1) zBus2(1,2) zBus2(1,2); zBus2(2,1) zBus2(2,2)
zBus2(2,2); zBus2(2,1) zBus2(2,2) zBus2(2,2)+z23]
disp('Using Krons reduction method')
zz11 = zBus3(1,1) - ((zBus3(1,3) * zBus3(3,1)) / zBus3(3,3)); zz12 =
zBus3(1,2)- ((zBus3(1,3) * zBus3(3,2)) / zBus3(3,3)); zz21 = zz12;
zz22 = zBus3(2,2)- ((zBus3(2,3) * zBus3(3,2)) / zBus3(3,3));
disp('RESULT:')
zBus = [zz11 zz12;
zz21 zz22]
```

OUTPUT:

```
STEP1: Add an element between reference node and node(1)
zBus1 =
0 + 0.2000i
STEP2: Add an element between existing node(1) and new node(3)
zBus2 =
0 + 0.2000i       0       + 0.2000i
0 + 0.2000i       0       + 0.6000i
STEP3: Add an element between existing node(3) and referencenode
zBus3 =

0 + 0.2000i       0       + 0.2000i       0       + 0.2000i
```

```
0 +  0.2000i        0      +  0.6000i        0      +  0.6000i

0 +  0.2000i        0      +  0.6000i        0      +  0.8000i
```

Using Krons reduction method

RESULT:

```
zBus =

              0 +  0.1500i        0      +  0.0500i

              0 +  0.0500i        0      +  0.1500i
```


13.7   FORMATION OF Z-BUS MATRIX USING BUS BUILDING ALGORITHM – 2

PROGRAM:

```
%Formation of Z-Bus matrix using bus building algorithm clc
clear all
z10 = 1.25j;
z12 = 0.25j;
z23 = 0.4j;
z24 = 0.125j;
z30 = 1.25j;
z34 = 0.2j;
disp('STEP1: Add an element between reference node(0) andnode(1)')
zBus1 = [z10]
disp('STEP2: Add an element between existing node(1) and new node(2)')
zBus2 = [zBus1(1,1)  zBus1(1,1);
zBus1(1,1)  zBus1(1,1)+z12]
disp('STEP3: Add an element between existing node(2) and new node(3)')
zBus3  =  [zBus2(1,1)  zBus2(1,2)  zBus2(1,2);  zBus2(2,1)  zBus2(2,2)
zBus2(2,2); zBus2(2,1) zBus2(2,2) zBus2(2,2)+z23]
disp('STEP4:  Add  an  element  between  existing  node(3)  and  reference
node(0)')
zBus4  =  [zBus3(1,1)  zBus3(1,2)  zBus3(1,3)  zBus3(1,3)  ;  zBus3(2,1)
zBus3(2,2)  zBus3(2,3)  zBus3(2,3);  zBus3(3,1)  zBus3(3,2)  zBus3(3,3)
zBus3(3,3); zBus3(3,1) zBus3(3,2) zBus3(3,3) zBus3(3,3)+z30]
disp('Fictitious node(0) can be eliminated')
```

```matlab
zz11 = zBus4(1,1) - ((zBus4(1,4) * zBus4(4,1)) / zBus4(4,4)); zz12 =
zBus4(1,2) - ((zBus4(1,4) * zBus4(4,2)) / zBus4(4,4)); zz13 = zBus4(1,3)
- ((zBus4(1,4) * zBus4(4,3)) / zBus4(4,4)); zz21 = zz12;
zz22 = zBus4(2,2) - ((zBus4(2,4) * zBus4(4,2)) / zBus4(4,4));
zz23 = zBus4(2,3) - ((zBus4(2,4) * zBus4(4,3)) / zBus4(4,4));
zz31 = zz13;
zz32 = zz23;
zz33 = zBus4(3,3) - ((zBus4(3,4) * zBus4(4,3)) / zBus4(4,4));
zBus5 = [zz11 zz12 zz13;
zz21 zz22 zz23;
zz31 zz32 zz33]
disp('STEP5: Add an element between existing node(3) and new node(4)')
zBus6  =  [zBus5(1,1)   zBus5(1,2)   zBus5(1,3)   zBus5(1,3);   zBus5(2,1)
zBus5(2,2)   zBus5(2,3)   zBus5(2,3);   zBus5(3,1)   zBus5(3,2)   zBus5(3,3)
zBus5(3,3); zBus5(3,1) zBus5(3,2) zBus5(3,3) zBus5(3,3)+z34]
disp('STEP6: Add an element between existing nodes (2) and (4)')
zBus7 =
[zBus6(1,1)   zBus6(1,2)   zBus6(1,3)   zBus6(1,4)   zBus6(1,2)-zBus6(1,4);
zBus6(2,1)   zBus6(2,2)   zBus6(2,3)   zBus6(2,4)   zBus6(2,2)-zBus6(2,4);
zBus6(3,1)   zBus6(3,2)   zBus6(3,3)   zBus6(3,4)   zBus6(3,2)-zBus6(3,4);
zBus6(4,1)   zBus6(4,2)   zBus6(4,3)   zBus6(4,4)   zBus6(4,2)-zBus6(4,4);
zBus6(2,1)-zBus6(4,1)    zBus6(2,2)-zBus6(4,2)    zBus6(2,3)-zBus6(4,3)
zBus6(2,4)-zBus6(4,4) z24+zBus6(2,2)+zBus6(4,4)-2*zBus6(2,4)]
disp('Fictitious node(0) can be eliminated')
zzz11 = zBus7(1,1) - ((zBus7(1,5) * zBus7(5,1)) / zBus7(5,5)); zzz12 =
zBus7(1,2)  -  ((zBus7(1,5)  *  zBus7(5,2))  /  zBus7(5,5));  zzz13  =
zBus7(1,3)  -  ((zBus7(1,5)  *  zBus7(5,3))  /  zBus7(5,5));  zzz14  =
zBus7(1,4) - ((zBus7(1,5) * zBus7(5,4)) / zBus7(5,5)); zzz21 = zzz12;
zzz22 = zBus7(2,2) - ((zBus7(2,5) * zBus7(5,2)) / zBus7(5,5));

zzz23 = zBus7(2,3) - ((zBus7(2,5) * zBus7(5,3)) / zBus7(5,5));
zzz24 = zBus7(2,4) - ((zBus7(2,5) * zBus7(5,4)) / zBus7(5,5));
zzz31 = zzz13;
zzz32 = zzz23;
zzz33 = zBus7(3,3) - ((zBus7(3,5) * zBus7(5,3)) / zBus7(5,5)); zzz34 =
zBus7(3,4) - ((zBus7(3,5) * zBus7(5,4)) / zBus7(5,5));
zzz41 = zzz14;
zzz42 = zzz24;
zzz43 = zzz34;
zzz44 = zBus7(4,4) - ((zBus7(4,5) * zBus7(5,4)) / zBus7(5,5));
```

```
disp('RESULT:')
zBus = [zzz11 zzz12 zzz13 zzz14;
zzz21 zzz22 zzz23 zzz24;
zzz31 zzz32 zzz33 zzz34;
zzz41 zzz42 zzz43 zzz44]
```

OUTPUT:

```
STEP1: Add an element between reference node(0) and node(1)
zBus1 =
0 + 1.2500i
STEP2: Add an element between existing node(1) and new node(2) zBus2 =

          0 + 1.2500i          0      + 1.2500i

          0 + 1.2500i          0      + 1.5000i


STEP3: Add an element between existing node(2) and new node(3) zBus3 =

     0 + 1.2500i       0      + 1.2500i       0      + 1.2500i
     0 + 1.2500i       0      + 1.5000i       0      + 1.5000i

     0 + 1.2500i       0      + 1.5000i       0      + 1.9000i


STEP4: Add an element between existing node(3) and reference node(0)

   zBus4 =
   0 + 1.2500i    0   + 1.2500i    0   + 1.2500i    0   + 1.2500i
   0 + 1.2500i    0   + 1.5000i    0   + 1.5000i    0   + 1.5000i
   0 + 1.2500i    0   + 1.5000i    0   + 1.9000i    0   + 1.9000i

   0 + 1.2500i    0   + 1.5000i    0   + 1.9000i    0   + 3.1500i


   Fictitious node(0) can be eliminated
   zBus5 =
   0           + 0.7540i       0      + 0.6548i       0      + 0.4960i
   0           + 0.6548i       0      + 0.7857i       0      + 0.5952i

   0           + 0.4960i       0      + 0.5952i       0      + 0.7540i
```

STEP5: Add an element between existing node(3) and new node(4) zBus6 =

```
0 + 0.7540i    0 + 0.6548i    0 + 0.4960i    0 + 0.4960i
0 + 0.6548i    0 + 0.7857i    0 + 0.5952i    0 + 0.5952i
0 + 0.4960i    0 + 0.5952i    0 + 0.7540i    0 + 0.7540i
0 + 0.4960i    0 + 0.5952i    0 + 0.7540i    0 + 0.9540i
```

STEP6: Add an element between existing nodes (2) and (4)

zBus7 =

```
    Columns 1 through 4
0 + 0.7540i    0 + 0.6548i    0 + 0.4960i    0 + 0.4960i
0 + 0.6548i    0 + 0.7857i    0 + 0.5952i    0 + 0.5952i
0 + 0.4960i    0 + 0.5952i    0 + 0.7540i    0 + 0.7540i
0 + 0.4960i    0 + 0.5952i    0 + 0.7540i    0 + 0.9540i
0 + 0.1587i    0 + 0.1905i    0 - 0.1587i    0 - 0.3587i

    Column 5
0 + 0.1587i
0 + 0.1905i
0 - 0.1587i
0 - 0.3587i
0 + 0.6742i
```

Fictitious node(0) can be eliminated

RESULT:

zBus =

```
0 + 0.7166i    0 + 0.6099i    0 + 0.5334i    0 + 0.5805i
0 + 0.6099i    0 + 0.7319i    0 + 0.6401i    0 + 0.6966i
0 + 0.5334i    0 + 0.6401i    0 + 0.7166i    0 + 0.6695i
0 + 0.5805i    0 + 0.6966i    0 + 0.6695i    0 + 0.7631i
```

13.8 DETERMINATION OF BUS CURRENTS, BUS POWER AND LINE FLOWS FOR A SPECIFIED BUS SYSTEM PROFILE

```matlab
MATLAB Program:
clc
clear all
vbus=[1.05+0j;     .98-0.06j; 1-0.05j];              % bus volatges
yline= [0 0 10 -20 10 -30 10 -20 0 0 16 -32 10 -30 16
-32 0 0 ];                                           % line G, B
n= 3                                                 % no. of buses
i=0;
k=1;
while i<9
yline1(i+1)=yline(k)+j*yline(k+1);
i=i+1
K=K+2

end
yline2=reshape(yline1,3,3
)                                                    % line admittance
Sp=zeros(n,n
)
for
i=1:n
for k=1:n
if i==k
    continu
    e
else
    I(i,k)=yline2(i,k)*(vbus(i)-vbus(k));            % line current
    I(k,i)=        -
    I(i,k)
    S(i,k)=vbus(i)*conj(I(i,k)
    );                                               % line power
    S(k,i)=vbus(k)*conj(I(k,i)
    );
    Sl(i,k)=S(i,k)+S(k,i
    );                                               % line losses
    Sp(i,i)=Sp(i,i)+S(i,
    k);                                              % bus power
```

```
end
end
end
Sbus=[Sp(1,1);           Sp(2,2);
Sp(3,3)]
%bus curentcalculation ;-
Ibus=conj((Sbus./vbus))                          % bus current
```

Input:

```
Yline= [0 0 10 -20 10 -30 10 -20 0 0 16 -32 10 -30
16 -32 0 0]
n= 3
```

Output:

```
line current matrix
is
0.00+( 0.00)j 1.90+( -0.80)j 2.00+( -1.00)j
-1.90+( 0.80)j 0.00+(   0.00)j -
0.64+(                        0.48)j
-2.00+( 1.00)j 0.64+( -0.48)j 0.00+( 0.00)j
line flow matrix is
        0.00)         0.84)
0.00+( j       2.00+( j      2.10+( 1.05)j
-1.91+( -0.67)j 0.00+(   0.00)j -0.66+( -0.43)j
-2.05+( -0.90)j 0.66+(   0.45)j 0.00+( 0.00)j
Line  losses  matrix
is
        0.00)         0.17)
0.00+( j       0.09+( j      0.05+( 0.15)j
        0.17)         0.00)
0.09+( j       0.00+( j      0.01+( 0.02)j
        0.15)         0.02)
0.05+( j       0.01+( j      0.00+( 0.00)j
Bus power matrix is
        1.89)         0.00)
4.10+( j       0.00+( j      0.00+( 0.00)j
0.00+( 0.00)j -2.57+( -1.10)j 0.00+( 0.00)j
0.00+( 0.00)j 0.00+( 0.00)j -1.39+( -0.45)j
Sbus    matrix
```

```
is
4.10+( 1.89)j -2.57+( -1.10)j -1.39+( -0.45)j
Ibus   matrix
is
3.90+( -1.80)j -2.54+(  1.28)j -1.36+( 0.52)j
```

13.9   CALCULATION OF ABCD PARAMETERS

Calculation of efficiency and regulation if sending end parameters are given:

```
clc
clear all
len=input('enter the length of the line in KM:');
r1=input('enter the resistance/phase/Km of the line:');
x1=input('enter the reactance/phase/Km of the line:');
s= input('enter the suspectance/phase/Km of the line:');
vs=input('enter the sending end voltage/ phase of the line:');
is=input('enter the sending end current of the line:');
z=r1+j*x1;
y=j*s;
Z=z*len;
Y=y*len;
if len<=80
x=1;
elseif len<=240
x=2;
else
x=3;
end
switch x
case 1
A=1;
B=Z;
C=0;
D=A;
case 2
p=input('enter 1 for PI and 2 for T');
A=1+(Y*Z)/2;
D=A;
```

```
if p==1
B=Z;
C=Y*(1+(Z*Y)/4);
else
C=Y;
B=Z*(1+(Z*Y)/4);
end
case 3
A=cosh(sqrt(Y*Z));
B=sqrt(Z/Y)*sinh(sqrt(Y*Z));
C=sqrt(Y/Z)*sinh(sqrt(Y*Z));
D=A;
end
ABCD=A*D-B*C;
vr=D*vs-B*is;
ir=-C*vs+A*is;
spower=3*vs*conj(is)
rpower=3*vr*conj(ir)
efficiency= real(rpower)/real(spower)*100;
regulation=(abs(vs)/abs(A)-abs(vr))/abs(vr)*100;
```

Input:

```
z=0.153+.38j;
y=0.0+.000003j;
vs=63508;
is=105-50.5j;
```

For short transmissionlineenter the length in KM 50

```
spower = 2.0005e+007 +9.6215e+006i
rpower = 1.9693e+007 +8.8477e+006i
abcd constants of transmission line are
A=1.000000+j0.000000
B=7.650000+j19.000000
C=0.000000+j0.000000
D=1.000000+j0.000000
sending end vs=63508.000000+j0.000000
```

```
sending end current=105.000000+j-50.500000
efficiency=98.442631
reg=2.819985
ad-bc=1.000000
```

<u>For medium transmission line with pi network</u> enter the length in KM200

```
enter 1 for PI and 2 for T 1
spower = 2.0005e+007 +9.6215e+006i
rpower = 1.8549e+007 +1.2389e+007i
abcd constants of transmission line are
A=0.977200+j0.009180
B=30.600000+j76.000000
C=-0.000003+j0.000593
D=0.977200+j0.009180
sending end vs=63508.000000+j0.000000
sending end current=105.000000+j-50.500000
efficiency=92.720923
reg=17.475816
ad-bc=1.000000
```

<u>For medium transmission line with T network</u> enter the length in KM 200

```
enter 1 for PI and 2 for T2
spower = 2.0005e+007 +9.6215e+006i
rpower = 1.8551e+007 +1.2505e+007i
abcd constants of transmission line are
A=0.977200+j0.009180
B=29.902320+j75.274054
C=0.000000+j0.000600
D=0.977200+j0.009180
sending end vs=63508.000000+j0.000000
sending end current=105.000000+j-50.500000
efficiency=92.731488
reg=17.253285
ad-bc=1.000000
```

```
For long transmission line enter the length in KM300

spower = 2.0005e+007 +9.6215e+006i
rpower = 1.7673e+007 +1.2772e+007i
abcd constants of transmission line are
A=0.949067+j0.020304
B=44.341614+j112.371757
C=-0.000006+j0.000885
D=0.949067+j0.020304
sending end vs=63508.000000+j0.000000
sending end current=105.000000+j-50.500000
efficiency=88.342221
reg=32.155708
ad-bc=1.000000
```

Calcuation of efficiency and regulation if receiving end parameters are given:

```
clc
clear all
len=input('enter the length in KM');
r1=input('enter the resistance/phase/Km of the line:');
x1=input('enter the reactance/phase/Km of the line:');
s= input('enter the suspectance/phase/Km of the line:');
vr=input('enter the receiving end voltage/ phase of the line:');
ir=input('enter the receiving end current of the line:');
z=r1+j*x1;
y=j*s;
Z=z*len;
Y=y*len;
if len<=80
x=1;
elseif len<=240
x=2;
else
x=3;
end
switch x
case 1
```

```
A=1;
B=Z;
C=0;
D=A;
case 2
p=input('enter 1 for PI and 2 for T');
A=1+(Y*Z)/2;
D=A;
if p= =1
B=Z;
C=Y*(1+(Z*Y)/4);
else
C=Y;
B=Z*(1+(Z*Y)/4);
end
case 3
A=cosh(sqrt(Y*Z));
B=sqrt(Z/Y)*sinh(sqrt(Y*Z));
C=sqrt(Y/Z)*sinh(sqrt(Y*Z));
D=A;
end
ABCD=A*D-B*C;
vs=A*vr+B*ir;
is=C*vr+D*ir;
spower=3*vs*conj(is)
rpower=3*vr*conj(ir)
efficiency= real(rpower)/real(spower)*100;
regulation=(abs(vs)/abs(A)-abs(vr))/abs(vr)*100;
```

Input:

```
z=0.153+.38j;
y=0.0+.000003j;
vr=63508;
ir=105-50.5j;
```

Short transmission line (50 km) given Vr and Ir

```
A=1.000000+j0.000000
B=7.650000+j19.000000
C=0.000000+j0.000000
D=1.000000+j0.000000
sending end vs=65270.750000+j1608.675000
sending end current=105.000000+j-50.500000
efficiency=98.466513
reg=2.806845
ad-bc=1.000000
```

Medium transmission line (200 km, PI) given Vr and Ir

```
A=0.977200+j0.009180
B=30.600000+j76.000000
C=-0.000003+j0.000593
D=0.977200+j0.009180
sending end vs=69111.017600+j7017.703440
sending end current=102.894689+j-10.714295
efficiency=94.775033
reg=11.929289
ad-bc=1.000000
```

Medium transmission line (200 km, T) given Vr and Ir

```
A=0.977200+j0.009180
B=29.902320+j75.274054
C=0.000000+j0.000600
D=0.977200+j0.009180
sending end vs=69001.100927+j6976.711950
sending end current=103.069590+j-10.279900
efficiency=94.718111
reg=11.746427
ad-bc=1.000000
```

Long transmission line given Vr and Ir

```
A=0.949067+j0.020304
B=44.341614+j112.371757
C=-0.000006+j0.000885
D=0.949067+j0.020304
sending end vs=70603.973689+j10849.218291
```

```
sending end current=100.287831+j10.388005
efficiency=92.700537
reg=18.487449
ad-bc=1.000000
```

Result:

13.10   DETERMINATION OF POWER ANGLE DIAGRAM FOR MATLAB PROGRAM

(i)   SALIENT POLE SYNCHRONOUS MACHINE.

(ii)   NON-SALIENT POLE SYNCHRONOUS MACHINE.

MATLAB PROGRAM:

```
clc
clear all
power angle curve p=input('power in MW = '); pf=input('power factor =
'); vt=input('line voltage = '); xd=input('xd in ohm = '); xq=input('xq
in ohm = '); vt_ph=vt*1000/sqrt(3); pf_a=acos(pf); q=p*tan(pf_a); i=(p-
j*q)*1000000/(3*vt_ph);   delta=0:1:180;   delta_rad=delta*(pi/180);   if
xd==xq
non   salient   syn   motor   ef=vt_ph+(j*i*xd);   excitation_emf=abs(ef)
reg=(abs(ef)-abs(vt_ph))*100/abs(vt_ph)                      power_non=abs
(ef)*vt_ph*sin(delta_rad)/xd; net_power=3*power_non/1000000;
plot(delta,net_power);
xlabel('\delta (deg)');
ylabel('3 phase power(MW)');
title('plot:power angle curve for non salient pole syn mc'); legend('non
salient power')
end
if xd~=xq
%salient pole motor
eq=vt_ph+(j*i*xq);
del=angle(eq);
theta=del+pf_a;
id_mag=abs(i)*sin(theta);
ef_mag=vt_ph*cos(del)+id_mag*xd
```

```
reg=(ef_mag-abs(vt_ph))*100/abs(vt_ph)
pp=ef_mag*vt_ph*sin(delta_rad)/xd;
reluct_power=vt_ph^2*(xd-xq)*sin(2*delta_rad)/(2*xd*xq);
net_reluct_power=3*reluct_power/1000000;
power_sal=pp+reluct_power;
net_power_sal=3*power_sal/1000000;
plot(delta,net_reluct_power);
hold on
plot(delta,net_power_sal);
xlabel('\delta (deg)');
ylabel('3 phase power(MW)');
title('plot:power   angle   curve   for   salient   pole   syn   mc');
legend('reluctpower','salient power')
end
grid;
```

Input data for salient pole synchronous machine:

```
power in MW = 48
power factor = .8
line voltage = 34.64
xd in ohm = 13.5
xq in ohm = 9.333
```

Result:

```
ef_mag = 3.0000e+004
reg =   50.0032
```

Input data for non salient pole synchronous machine:

```
power in MW = 48
power factor = .8
line voltage = 34.64
xd in ohm = 10
xq in ohm = 10
```

Result:

```
excitation_emf = 2.7203e+004
reg =   36.0171
```

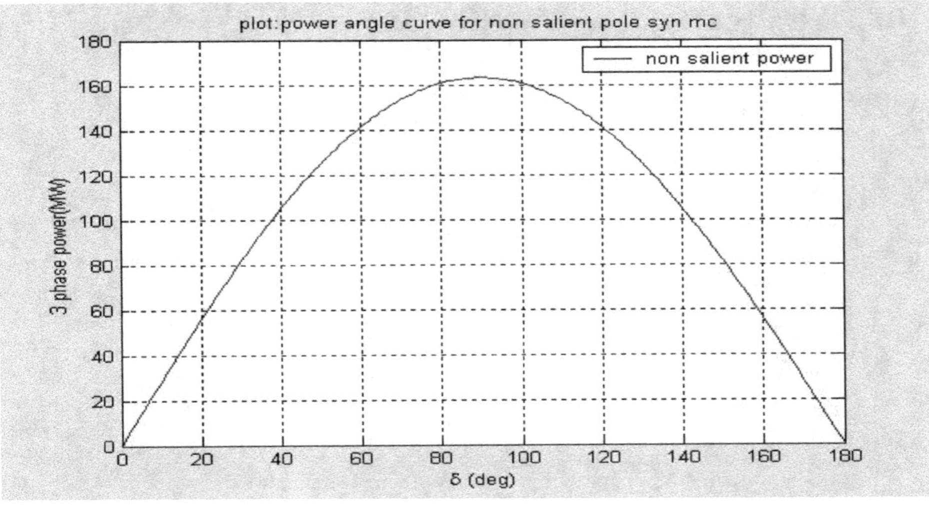

Graph 1

The above graph shows the power angle curve. The maximum power transfer occurs at delta=90°. For values of delta >90⁰ , the power output of the machine reduces successively and finally the machine may stall.The system is stable only if the displacement angle δ is in the range from -90° to +90° in which the slope dP/d δ is positive

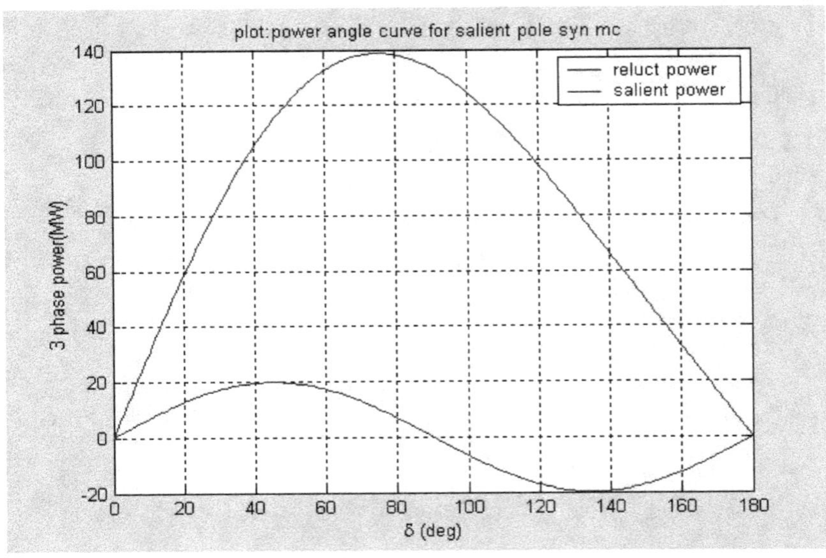

Graph 2

The graph 2 consists of a second harmonic component of power.

13.11 DETERMINATION OF SWING CURVE

MATLAB PROGRAM:

```
clc
clear all
tfc=input('enter fault clearing time=');
pi = 0.9;
e1=1.1;
e2=1.0;
m=0.016;
x0=0.45;
x1=1.25;
x2=0.55;
pm0=(e1*e2)/x0;
pm1=(e1*e2)/x1;
pm2=(e1*e2)/x2;
w=0;
d=asin(pi/pm0);
for t=0:0.05:1
dg=d*180/3.1414;
if(t<tfc)
pm=pm1;
else
pm=pm2;
end
k1=w*.05;
l1=(pi-pm*sin(d))*.05/m;
k2=(w+.5*l1)*.05;
l2=(pi-pm*sin(d+.5*k1))*.05/m;
k3=(w+.5*l2)*.05;
l3=(pi-pm*sin(d+.5*k2))*.05/m;
```

```
k4=(w+l3)*.05;

l4=(pi-pm*sin(d+k3))*.05/m;

deld=(k1+2*k2+2*k3+k4)/6;

delw=(l1+2*l2+2*l3+l4)/6;

d=d+deld;

w=w+delw;

fprintf('%8.3f \t %8.3f \n', t, dg);

end
```

Input data for swing curve:

```
Pi = 0.9;          e1=1.1;              e2=1.0
M=0.016
X0=0.45 p.u;       X1=1.25 p.u;         X2=0.55 p.u
```

Swing curve for TFC 0.3

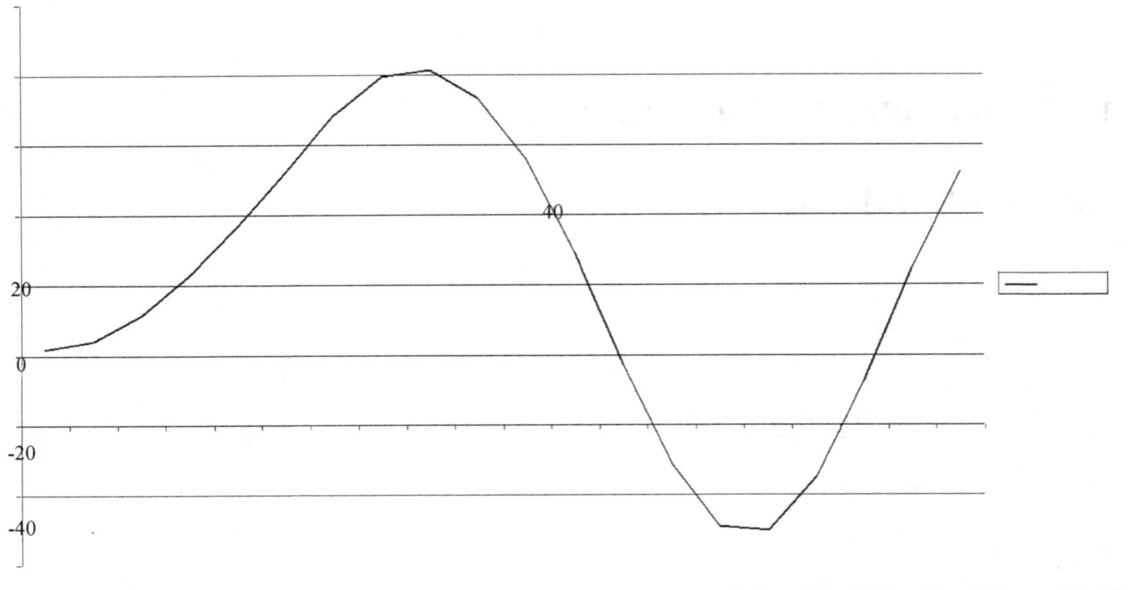

time in secs

Swing curve for TFC=0.42 secs

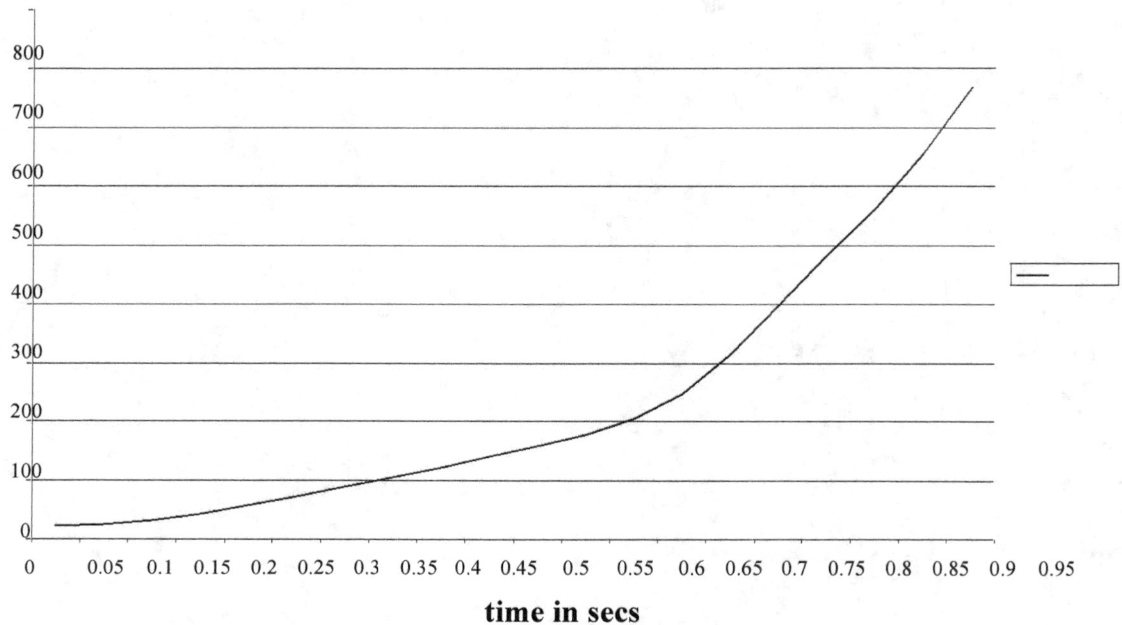

13.12   JACOBIAN MATRIX CALCULATION

MATLAB PROGRAM:

```
clc
clear all
n=4
v=[1 1 1 1]
ybus=[70-90j   -20+40j   -50+50j   0+0j;-20+40j   43.08-55.39j   0+0j   -
23.077+15.39j;  -50+50j 0+0j 75-75j -25+25j;0+0j  -23.077+15.39j  -25+25j
48.077-40.39j]
for i=1:n
for j=1:n
y(i,j)=abs(ybus(i,j))
yn(i,j)=angle(ybus(i,j))
v(i)=abs(v(i))
vn(i)=angle(v(i))
end
end
```

```
J1=zeros(n,n)
J2=zeros(n,n)
J3=zeros(n,n)
J4=zeros(n,n)
i=2
while i<=n
J2(i,i)=J2(i,i)+2*v(i)*y(i,i)*cos(yn(i,i))
J4(i,i)=J4(i,i)-2*v(i)*y(i,i)*sin(yn(i,i))
for j=1:n
if i==j
continue;
else
J1(i,i)=J1(i,i)+v(i)*v(j)*y(i,j)*sin(yn(i,j)-vn(i)+vn(j))
J1(i,j)=-1*v(i)*v(j)*y(i,j)*sin(yn(i,j)-vn(i)+vn(j))
J2(i,i)=J2(i,i)+v(j)*y(i,j)*cos(yn(i,j)-vn(i)+vn(j))
J2(i,j)=v(i)*y(i,j)*cos(yn(i,j)-vn(i)+vn(j))
J3(i,i)=J3(i,i)+v(i)*v(j)*y(i,j)*cos(yn(i,j)-vn(i)+vn(j))
J3(i,j)=-1*v(i)*v(j)*y(i,j)*cos(yn(i,j)-vn(i)+vn(j))
J4(i,i)=J4(i,i)-v(j)*y(i,j)*sin(yn(i,j)-vn(i)+vn(j))
J4(i,j)=-1*v(i)*y(i,j)*sin(yn(i,j)-vn(i)+vn(j))
end
end
i=i+1
end
J11=J1(2:n,2:n)
J22=J2(2:n,2:n)
J33=J3(2:n,2:n)
J44=J4(2:n,2:n)
Jacobian=[J11 J22;J33 J44]
```

Output:

55.39	0.00	-15.39	43.08	0.00	-23.08
0.00	75.00	-25.00	0.00	75.00	-25.00
-15.39	-25.00	40.39	-23.08	-25.00	48.08
-43.08	0.00	23.08	55.39	0.00	-15.39
0.00	-75.00	25.00	0.00	75.00	-25.00
23.08	25.00	-48.08	-15.39	-25.00	40.39

13.13 MATLAB PROGRAM TO SOLVE LOAD FLOW EQUATIONS USING GAUSS-SEIDEL METHOD WITH PQ BUSES

Algorithm:

Step 1: Read the data such as line data, specified power, specified voltages, Q limits at the generatorbuses and tolerance for convergences

Step 2: Compute Y-bus matrix.

Step 3: Initialize all the bus voltages.

Step 4: Iter=1

Ste p 5: Consider i=2, where i' is the bus number.

Step 6: check whether this is PV bus or PQ bus. If it is PQ bus goto step 8 otherwise go to next step.

Step 7: Compute Qi check for q limit violation. QGi=Qi+QLi.

If QGi>Qi max,equateQGi = Qimax. Then convert it into PQ bus.

If QGi<Qi min, equate QGi = Qi min. Then convert it into PQ bus.

Step 8: Calculate the new value of the bus voltage using gauss seidal formula.i=1 nVi=(1.0/Yii) [(Pi-j Qi)/vi0*- Σ YijVj- Σ YijVj0]

J=1   J=i+1

Adjust voltage magnitude of the bus to specify magnitude if Q limits are not violated.

Step 9: If all buses are considered go to step 10 otherwise increments the bus no. i=i+1 and Go tostep6.

Step 10: Check for convergence. If there is no convergence goes to step 11 otherwise go to step12.

Step 11: Update the bus voltage using the formula.

Vinew=Vi old+ α(vinew-Viold) (i=1,2,…..n) i≠ slackbus ,α is the acceleration factor=1.4

Step 12: Calculate the slack bus power, Q at P-V buses real and reactive give flows real and reactanceline losses and print all the results including all the bus voltages and all the bus angles.

Step 13: Stop.

Procedure:

1. Enter the command window of the MATLAB.
2. Create a new M – file by selecting File - New – M – File.
3. Type and save the program in the editor Window.
4. Execute the program by pressing Tools – Run.
5. View the results.

PROBLEM

(a) The following figure shows the one-line diagram of a simple three-bus power system with generation at bus 1. The magnitude of voltage at bus 1 is adjusted to 1.05 per unit. The scheduled loads at buses 2 and 3 are as marked on the diagram. Line impedances are marked in per unit on a 100-MVA base and the line charging susceptances are neglected. Using the Gauss-Seidel method, determine the phasor values of the voltage at the load buses 2 and 3 (P-Q buses) accurate to four decimal places and obtain full solution using MATLAB.

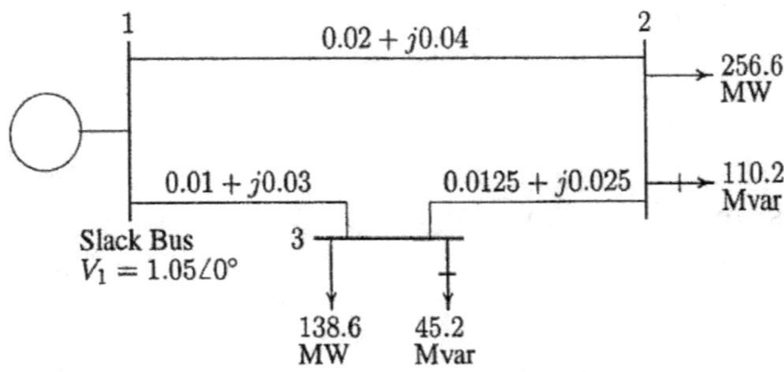

MATLAB PROGRAM:

```
clc
clear all
n=3;
V=[1.05 1 1];
Y=[20-j*50    -10+j*20     -10+j*30
-10+j*20     26-j*52      -16+j*32
-10+j*30     -16+j*32     26-j*62]
P=[inf -2.566     -1.386];
Q=[inf -1.102     -0.452];
diff=10;
iter=1;
Vprev=V;
while (diff>0.00001 | iter==1),
abs(V);
abs(Vprev);
Vprev=V;
for i=2:n
sumyv=0;
for k=1:n,
if(i~=k)
```

```
sumyv=sumyv+Y(i,k)*V(k);
end
end

V(i)=(1/Y(i,i))*((P(i)-j*Q(i))/conj(V(i))-sumyv);
end
diff=max(abs(abs(V(2:n))-abs(Vprev(2:n))));
V
iter=iter+1
end
```

Solution:

(a)   Line impedances are converted to admittances

$$y_{12} = \frac{1}{0.02 + j0.04} = 10 - j20 \qquad y_{13} = 10 - j30$$

$$y_{23} = 16 - j32$$

```
 YBUS  =   [20-j*50 -10+j*20    -10+j*30
-10+j*20            26-j*52     -16+j*32
-10+j*30            -16+j*32    26-j*62]
```

At the P-Q buses, the complex loads expressed in per units are

$$S_2^{sch} = -\frac{(256.6 + j110.2)}{100} = -2.566 - j1.102 \quad \text{pu}$$

$$S_3^{sch} = -\frac{(138.6 + j45.2)}{100} = -1.386 - j0.452 \quad \text{pu}$$

Starting from an initial estimate of

$$V_2^{(0)} = 1.0 + j0.0 \text{ and } V_3^{(0)} = 1.0 + j0.0,$$

$$V_2^{(1)} = \frac{\dfrac{P_2^{sch} - jQ_2^{sch}}{V_2^{*(0)}} + y_{12}V_1 + y_{23}V_3^{(0)}}{y_{12} + y_{23}}$$

$$= \frac{\dfrac{-2.566 + j1.102}{1.0 - j0} + (10 - j20)(1.05 + j0) + (16 - j32)(1.0 + j0)}{(26 - j52)}$$

$$= 0.9825 - j0.0310$$

$$V_3^{(1)} = \frac{\frac{P_3^{sch} - jQ_3^{sch}}{V_3^{\bullet(0)}} + y_{13}V_1 + y_{23}V_2^{(1)}}{y_{13} + y_{23}}$$

$$= \frac{\frac{-1.386 + j0.452}{1 - j0} + (10 - j30)(1.05 + j0) + (16 - j32)(0.9825 - j0.0310)}{(26 - j62)}$$

$$= 1.0011 - j0.0353$$

The process is continued and a solution is converged with an accuracy of $1*10^{\wedge 5}$ per unit in six iterations as given below.

The final solution is

$$V_2 = 0.9800 - j0.0600 = 0.98183\angle -3.5035° \quad \text{pu}$$
$$V_3 = 1.0000 - j0.0500 = 1.00125\angle -2.8624° \quad \text{pu}$$

MATLAB OUTPUT:

```
Iter=1   V =   1.0500 + 0.0000i    0.9825 - 0.0310i   1.0011 - 0.0353i
Iter=2   V =   1.0500 + 0.0000i    0.9816 - 0.0520i   1.0008 - 0.0459i
iter =3  V =   1.0500 + 0.0000i    0.9808 - 0.0578i   1.0004 - 0.0488i
iter =4  V =   1.0500 + 0.0000i    0.9803 - 0.0594i   1.0002 - 0.0497i
iter =5  V =   1.0500 + 0.0000i    0.9801 - 0.0598i   1.0001 - 0.0499i
iter =6  V =   1.0500 + 0.0000i    0.9801 - 0.0599i   1.0000 - 0.0500i
```

13.14   LOAD FLOW ANALYSIS BY GAUSS-SEIDAL METHOD

PROGRAM:

```
%Gauss Seidal Method
clc
clear all
y12 = 2-6j;
y13 = 1-3j;
y23 = 0.666-2j;
y24 = 1-3j;
y34 = 2-6j;
p2 = 0.5;
p3 = -1.0;
p4 = 0.3;
q3 = 0.5;
```

```
q4 = -0.1;
disp('STEP1: Formulate Y-Bus Matrix')
Y11 = y12+y13;
Y12 = -y12;
Y13 = -y13;
Y14 = 0;
Y21 = -y12;
Y22 = y12+y23+y24;
Y23 = -y23;
Y24 = -y24;
Y31 = -y13;
Y32 = -y23;
Y33 = y13+y23+y34;
Y34 = -y34;
Y41 = 0;
Y42 = -y24;
Y43 = -y34;
Y44 = y24+y34;
yBus = [Y11 Y12 Y13 Y14; Y21 Y22 Y23 Y24; Y31 Y32 Y33 Y34; Y41 Y42 Y43 Y44]
disp('STEP2: Initialize bus voltages')
v1 = 1.04;
v2 = 1.04;
v3=1.0;
v4 = 1.0;
disp('STEP3:   Calculate   Q2   value   for   PV   Bus')   Q2   =
v2*[(Y21*v1)+(Y22*v2)+(Y23*v3)+(Y24*v4)] Q2cal = -imag(Q2)
disp('STEP4: Calculate V2 value')
V2 = (1/Y22)*[((p2-(Q2cal*j))/v2)-(Y21*v1)-(Y23*v3)-(Y24*v4)]  delta2 =
angle(V2)
V2new = v2*(cos(delta2)+j*sin(delta2))
V3 = (1/Y33)*[((p3-(q3*j))/v3)-(Y31*v1)-(Y32*V2new)-(Y34*v4)]
V4 = (1/Y44)*[((p4-(q4*j))/v4)-(Y41*v1)-(Y42*V2new)-(Y43*V3)]
```

OUTPUT:

```
STEP1: Formulate Y-Bus Matrix
yBus =
  3.0000-9.0000i   -2.0000+6.0000i   -1.0000+3.0000i      0
```

```
-2.0000+6.0000i   3.6660-11.0000i    -0.6660+2.0000i    -1.0000+3.0000i
-1.0000+3.0000i   -0.6660+2.0000i    3.6660-11.0000i    -2.0000+6.0000i
0                 -1.0000+3.0000i    -2.0000 + 6.0000i  3.0000-9.0000i
```

```
STEP2: Initialize bus voltages
STEP3: Calculate Q2 value for PV Bus
Q2 =
0.0693 - 0.2080i
Q2cal =
0.2080
STEP4: Calculate V2 value
V2 =
1.0513 + 0.0339i
delta2 =
0.0322
V2new =
1.0395 + 0.0335i
V3 =
1.0317 - 0.0894i
V4 =
1.0343 - 0.0151i
```

13.15   NEWTON RAPHSON LOAD FLOW METHOD USING MATLAB

PROGRAM:

```
%Newton Raphson Method
clc
clear all
y12 = 0.0839+0.5183j;
hlc = 0.0636j;
disp('STEP1: Formulate Y-Bus matrix')
Y11 = 1/y12 + hlc;
Y12 = -1/y12;
Y21 = -1/y12;
Y22 = 1/y12 + hlc;
yBus = [Y11 Y12Y21 Y22]
Y11m = abs(Y11);
```

```
Y12m = abs(Y12);
Y21m = abs(Y21);
Y22m = abs(Y22);
Y11a = angle(Y11);
Y12a = angle(Y12);
Y21a = angle(Y21);
Y22a = angle(Y22);
disp('STEP2: Initialize bus voltages')
v1 = 1.05
v2 = 1.0
d1 = 0
d2 = 0
X0 = [d2;v2]
disp('STEP3: Calculate P2cal, Q2cal, dP2, dQ2')
P2cal = v2*((v1*Y21m*cos(Y12a+d2-d1))+(v2*Y22m*cos(Y22a+d2-d1)))
P2spec = -0.3
dP2 = P2spec-P2cal
Q2cal = -v2*((v1*Y21m*sin(Y12a+d1-d2))+(v2*Y22m*sin(Y22a+d2-d1)))
disp('Check for q limit')
disp('Q2cal < Q2min')
Q2spec = -0.1
dQ2 = Q2spec-Q2cal
disp('STEP4: Form Jacobian matrix')
J11 = (v2*v1*Y21m*sin(Y12a+d1-d2))+(v2*v2*Y22m*0);
J12 = (v1*Y21m*cos(Y12a+d1-d2))+(2*v2*Y22m*cos(Y22a));
J21 = (v2*v1*Y21m*cos(Y12a+d1-d2))-(v2*v2*Y22m*0);
J22 = (-v1*Y21m*sin(Y12a+d1-d2))-(2*v2*Y22m*sin(Y22a));
Jmatrix = [J11 J12;J21 J22]
Jinv = inv(Jmatrix)
disp('STEP5: Compute dX')
pq = [dP2;dQ2];
dX = Jinv*pq
delta2 = dX(1,1);
dV2 = dX(2,1);
Xnew = X0+dX
v2new = v2+dV2
Delta2new = d2+delta2
```

OUTPUT:

```
STEP1: Formulate Y-Bus matrix
yBus =
0.3043      - 1.8165i   -0.3043    + 1.8801i

-0.3043     + 1.8801i    0.3043    - 1.8165i


STEP2: Initialize bus voltages
v1 =1.0500
v2 =1
d1 =0
d2 =0
X0 =0
1
STEP3: Calculate P2cal, Q2cal, dP2, dQ2 P2cal =-0.0152
P2spec =-0.3000
dP2 =-0.2848
Q2cal =-0.1576
Check for q limit
Q2cal < Q2min
Q2spec =-0.1000
dQ2 =0.0576
STEP4: Form Jacobian matrix
Jmatrix =
1.9741 0.2891
-0.31961.6589
Jinv =
0.4927 -0.0859
0.0949 0.5863
STEP5: Compute dX
dX =-0.1452
0.0067
Xnew =-0.1452
1.0067
v2new =1.0067
Delta2new =-0.1452
```

13.16   FAST DECOUPLED POWER FLOW MODEL USING MATLAB

PROGRAM:

```matlab
%Fast decoupled power flow method
clc
clear all
y11 = 0.0839+0.5183j;
hlc = 0.0636*j;
disp('STEP 1: Form Y-Bus matrix')
Y11 =  (1/y11)+hlc;
Y12 = -(1/y11);
Y21 = -(1/y11);
Y22 =(1/y11)+hlc;
ybus=[Y11 Y12;Y21 Y22]
Y11m = abs(Y11);
Y12m = abs(Y12);
Y21m = abs(Y21);
Y22m = abs(Y22);
Y11a = angle(Y11);
Y12a = angle(Y12);
Y21a = angle(Y21);
Y22a = angle(Y22);
disp('STEP 2: Initialize bus voltages')
v1 = 1.05
v2 = 1.02
d1 = 0
d2 = 0
disp('STEP 3: Check for Q-limit violation')
Q2cal = -((v2*v1*Y21m*sin(Y21a-d2+d1))+(v2*v2*Y22m*sin(Y22a)))
Q2min =10/100
Q2spec=0.1;
disp('Q2cal < Q2min, therefore Q2cal is with in the limit')
disp('Bus 2 act as pvbus')
v2 = 1;
disp('STEP 4: Calculate dP2 and dQ2')
P2cal = ((v2*v1*Y21m*cos(Y21a-d2+d1))+(v2*v2*Y22m*cos(Y22a)))
P2spec= 60/100
dP2  = P2spec-P2cal
```

```
dQ2  = Q2spec-Q2cal
disp('STEP 5: Bus susceptance matrix')
B1 = [-1.817]
B1_inv = inv(B1)
B2 = [-1.817]
B2_inv = inv(B1)
disp('STEP 6: calculate d2new and V2new')
del  = -B1_inv*(dP2/v2);
d2new = d2+del
dV2  = -B2_inv*(dQ2/v2);
V2new = v2+dV2
```

OUTPUT:

```
STEP 1: Form Y-Bus matrix


        ybus =
          0.3043 - 1.8165i      -0.3043 + 1.8801i

          -0.3043 + 1.8801i      0.3043 - 1.8165i

        STEP 2: Initialize bus voltages
        v1 =       1.0500
        v2 =       1.0200
        d1 =       0

        d2 =       0

        STEP 3: Check for Q-limit violation
        Q2cal =   -0.1237

        Q2min =   0.1000


Q2cal < Q2min, therefore Q2cal is within the limit Bus 2 act as pv bus


STEP 4: Calculate dP2 and dQ2


        P2cal =    -0.0152
        P2spec =   0.6000
        dP2 =      0.6152

        dQ2 =      0.2237
```

```
STEP 5: Bus susceptance matrix
B1 = -1.8170


B1_inv =     -0.5504
B2 = -1.8170
B2_inv =     -0.5504


STEP 6: calculate d2new and V2new
        d2new =    0.3386

        V2new =    1.1231
.
```

13.17   LG, LL, LLG FAULT ANALYSIS

INPUT DATA (IMPEDANCES IN PER- UNIT)

```
BUS  CODE     Z0                          Z1
 0    1      0.0      0.40     0.0    0.25
 0    2      0.0      0.10     0.0    0.25
 1    2      0.0      0.30     0.0    0.125
 1    3      0.0      0.35     0.0    0.15
 2    3      0.0      0.7125   0.0    0.25
```

RESULTS

The following results were obtained after simulation of the codes using a MATLAB environment:

The complex bus impedance matrix

```
Zbus1=
        0              +0 + 0.1050i  0            +
        0.1450i                      0.1300i
        0              +0 + 0.1450i  0            +
        0.1050i                      0.1200i
        0              +0 + 0.1200i  0            +
        0.1300i                      0.2200i
Zbus0=
        0              +0 + 0.0545i  0            +
        0.1820i                      0.1400i
        0              +0 + 0.0864i  0            +
        0.0545i                      0.0650i
        0              +0 + 0.0650i  0            +
        0.1400i                      0.3500i
```

Line-to-ground fault analysis

Single line to-ground fault at bus No. 1 Total fault current = 6.3559 per unit

Bus Voltages during the fault in per unit

```
Bus Voltage Magnitude
No. Phase a Phase b  Phase c
1   0.0000  1.0414   1.0414
2   0.4396  0.9510   0.9510
3   0.1525  1.0108   1.0108
```

Line currents for fault at bus No. 1

```
From To    Line CurrentMagnitude
Bus  Bus   Phase a  Phase b  Phase c
1    F     6.3559   0.0000   0.0000
2    1     2.2564   0.2225   0.2225
2    3     0.6780   0.0424   0.0424
3    1     0.6780   0.0424   0.0424
```

Single line to-ground fault at bus No. 2 Total fault current=7.9708 per unit

Bus Voltages during the fault in per unit

```
Bus Voltage Magnitude
No. Phase a Phase b  Phase c
1   0.2972  0.9401   0.9401
2   0.0000  0.9319   0.9319
3   0.1896  0.9355   0.9355
```

Line currents for fault at bus No. 2

```
From To    Line CurrentMagnitude
Bus  Bus   Phase a  Phase b  Phase c
1    2     1.9827   0.5679   0.5679
1    3     0.6111   0.1860   0.1860
2    F     7.9708   0.0000   0.0000
3    2     0.6111   0.1860   0.1860
```

Single line to-ground fault at bus No. 3 Total fault current=   3.7975 per unit
Bus Voltages during the fault in per unit

```
       Bus Voltage Magnitude
       No. Phase a Phase b  Phase c
       1   0.4937  1.0064   1.0064
       2   0.6139  0.9671   0.9671
       3   0.0000  1.0916   1.0916
```

Line currents for fault at bus No. 3

```
       From To    Line CurrentMagnitude
       Bus  Bus   Phase a Phase b  Phase c
       1    3     2.2785  0.0000   0.0000
       2    1     0.5190  0.2152   0.2152
       2    3     1.5190  0.0000   0.0000
       3    F     3.7975  0.0000   0.0000
```

MATLAB CODES
Z_{bus} IMPEDANCE MATRIX

```
% This program forms the complex bus impedance matrix by the method
% of building algorithm. Bus zero is taken as reference.
function [Zbus] = zbuild(linedata)
nl  = linedata(:,1);  nr  = linedata(:,2);  R  = linedata(:,3);  X  =
linedata(:,4);
nbr=length(linedata(:,1)); nbus = max(max(nl), max(nr)); for k=1:nbr
if R(k) == inf | X(k) ==inf
R(k) = 99999999; X(k) = 99999999;
else, end
end
ZB = R + j*X;
Zbus = zeros(nbus, nbus); tree=0; %%%%new
% Adding a branch from a new bus to reference bus 0 for I =1:nbr
ntree(I) =1;
if nl(I) == 0 | nr(I) == 0
if nl(I)==0 n = nr(I); elseif nr(I) == 0 n = nl(I); end
if  abs(Zbus(n,  n)) == 0  Zbus(n,n)  = ZB(I);tree=tree+1;  %%new  else
```

```
Zbus(n,n) = Zbus(n,n)*ZB(I)/(Zbus(n,n) + ZB(I));
end
ntree(I) = 2;
else,end
end
% Adding a branch from new bus to an existing bus while tree <nbus %%%
new
for n = 1:nbusnadd = 1;
if abs(Zbus(n,n)) == 0 for I = 1:nbr
if nadd == 1;
if nl(I) == n | nr(I) == n
if nl(I)==nk = nr(I); elseif nr(I) == n k = nl(I); end
if abs(Zbus(k,k)) ~= 0 for m = 1:nbus
if m ~= n
Zbus(m,n) = Zbus(m,k);
Zbus(n,m) = Zbus(m,k); else, end
end
Zbus(n,n) = Zbus(k,k) + ZB(I); tree=tree+1; %%new nadd = 2; ntree(I) =
2;
else, end else, end
else, end end
else, end end
end %%%%%%new
% Adding a link between two old buses for n = 1:nbus
for I = 1:nbr
if ntree(I) == 1
if nl(I) == n | nr(I) == n
if nl(I)==n
k = nr(I); elseif nr(I) == n k = nl(I);
end
DM = Zbus(n,n) + Zbus(k,k) + ZB(I) - 2*Zbus(n,k); for jj = 1:nbus
AP = Zbus(jj,n) - Zbus(jj,k); for kk = 1:nbus
AT = Zbus(n,kk) - Zbus(k, kk); DELZ(jj,kk) = AP*AT/DM;
end end
Zbus = Zbus - DELZ; ntree(I) = 2;
else,end
else,end
end end
```

Double- Line- Ground fault

```
% The program dlgfault is designed for the double line-to-ground
% fault analysis of a power system network. The program requires
% the positive-, negative- or zero-sequence bus impedance matrices,
% Zbus1 Zbus2,and Zbus0. The bus impedances matrices may be defined
% by the user, obtained by the inversion of Ybus or it may be
% determined either  from the  function Zbus =zbuild(zdata)
% or the function Zbus = zbuildpi(linedata,  gendata,  yload).
% The program prompts the user to enter the faulted bus number
% and the fault impedance Zf. The prefault bus voltages are
% defined by the reserved Vector V. The array V may be defined or
% it is returned from the power flow programs lfgauss, lfnewton,
% decouple or perturb. If V does not exist the prefault bus voltages
% are automatically set to 1.0 per unit. The program obtains the
% total fault current, bus voltages and line currents during the fault.

function dlgfault(zdata0,  Zbus0,  zdata1,  Zbus1,  zdata2,  Zbus2,  V)  if
exist('zdata2')  ~=1
zdata2=zdata1; else, end
if exist('Zbus2')  ~=  1 Zbus2=Zbus1;
else, end
nl = zdata1(:,1); nr = zdata1(:,2);
nl0  =  zdata0(:,1);  nr0  =  zdata0(:,2);  nbr=length(zdata1(:,1));  nbus  =
max(max(nl),  max(nr));  nbr0=length(zdata0(:,1));
R0  =   zdata0(:,3); X0   =zdata0(:,4);
R1  =   zdata1(:,3); X1   =zdata1(:,4);
R2  =   zdata2(:,3); X2   =zdata2(:,4);
for k = 1:nbr0
if R0(k)  ==  inf  |  X0(k)  ==  inf
R0(k)  = 99999999; X0(k)  = 999999999;
else, end end
ZB1 = R1 + j*X1; ZB0 = R0 + j*X0; ZB2 = R2 + j*X2;
if exist('V')  == 1
if length(V)  == nbus V0 = V;
else, end
else, V0 = ones(nbus, 1) + j*zeros(nbus, 1); end
fprintf('\nDouble line-to-ground fault analysis \n') ff = 999;
while ff > 0
```

```
nf = input('Enter Faulted Bus No. -> '); while nf<= 0 | nf>nbus
fprintf('Faulted bus No. must be between 1 & %g \n', nbus) nf =
input('Enter Faulted Bus No. -> ');
end
fprintf('\nEnter Fault Impedance Zf = R + j*X in ')
Zf = input('complex form (for bolted fault enter 0). Zf = '); fprintf('
\n')
fprintf('Double line-to-ground fault at bus No. %g\n', nf)
a=cos(2*pi/3)+j*sin(2*pi/3);
sctm = [1 1 1; 1 a^2 a; 1 a a^2];
Z11 = Zbus2(nf, nf)*(Zbus0(nf, nf)+ 3*Zf)/(Zbus2(nf, nf)+Zbus0(nf,
nf)+3*Zf); Ia1 = V0(nf)/(Zbus1(nf,nf)+Z11);
Ia2 =-(V0(nf) - Zbus1(nf, nf)*Ia1)/Zbus2(nf,nf);
Ia0 =-(V0(nf) - Zbus1(nf, nf)*Ia1)/(Zbus0(nf,nf)+3*Zf); I012=[Ia0; Ia1;
Ia2];
Ifabc = sctm*I012; Ifabcm=abs(Ifabc); Ift = Ifabc(2)+Ifabc(3);
Iftm = abs(Ift);
fprintf('Total fault current = %9.4f per unit\n\n', Iftm) fprintf('Bus
Voltages during the fault in per unit \n\n') fprintf('      Bus   ------
-Voltage Magnitude------- \n') fprintf(' No.   Phasea        Phaseb
  Phase c\n')
for n = 1:nbus
Vf0(n)= 0 - Zbus0(n, nf)*Ia0; Vf1(n)= V0(n) - Zbus1(n, nf)*Ia1; Vf2(n)=
0 - Zbus2(n, nf)*Ia2;
Vabc = sctm*[Vf0(n); Vf1(n); Vf2(n)];
Va(n)=Vabc(1); Vb(n)=Vabc(2); Vc(n)=Vabc(3); fprintf(' %5g',n)
fprintf(' %11.4f', abs(Va(n))),fprintf(' %11.4f', abs(Vb(n))) fprintf('
%11.4f\n', abs(Vc(n)))
end
fprintf(' \n')
fprintf('Line currents for fault at bus No. %g\n\n', nf)
fprintf('   From       To         -----Line Current Magnitude---- \n')
fprintf(' Bus  Bus   Phasea     Phaseb       Phase  c     \n')   for
n=1:nbus
for I = 1:nbr
if nl(I) == n | nr(I) == n
ifnl(I)==n  k = nr(I); elseif nr(I) == n k = nl(I); end
```

```
if k ~= 0
Ink1(n, k) = (Vf1(n) - Vf1(k))/ZB1(I);
Ink2(n, k) = (Vf2(n) - Vf2(k))/ZB2(I);
else, end else, end
end
for I = 1:nbr0
if nl0(I) == n | nr0(I) == n
ifnl0(I)==n k = nr0(I); elseif nr0(I) == n k = nl0(I); end
if k ~= 0
Ink0(n, k) = (Vf0(n) - Vf0(k))/ZB0(I);
else, end else, end
end
for I = 1:nbr
if nl(I) == n | nr(I) == n
ifnl(I)==n  k = nr(I); elseif nr(I) == n k = nl(I); end
if k ~= 0
Inkabc  =  sctm*[Ink0(n,  k);  Ink1(n,  k);  Ink2(n,  k)];  Inkabcm  =
abs(Inkabc); th=angle(Inkabc);
if real(Inkabc(2)) < 0
fprintf('%7g', n), fprintf('%10g', k),
fprintf(' %11.4f', abs(Inkabc(1))),fprintf(' %11.4f', abs(Inkabc(2)))
fprintf(' %11.4f\n',abs(Inkabc(3)))
elseif  real(Inkabc(2))  ==0  &imag(Inkabc(2))  >  0  fprintf('%7g',  n),
fprintf('%10g',k),
fprintf(' %11.4f', abs(Inkabc(1))),fprintf(' %11.4f', abs(Inkabc(2)))
fprintf(' %11.4f\n',abs(Inkabc(3)))
else, end else, end
else, end end
if n==nf
fprintf('%7g',n),fprintf('    F'),
fprintf(' %11.4f', Ifabcm(1)),fprintf(' %11.4f', Ifabcm(2)) fprintf('
%11.4f\n', Ifabcm(3))
else, end end resp=0;
while strcmp(resp, 'n')~=1 &strcmp(resp, 'N')~=1 &strcmp(resp, 'y')~=1
&strcmp(resp, 'Y')~=1
resp = input('Another fault location? Enter ''y'' or ''n'' within single
```

```
quote -> ');
if  strcmp(resp,  'n')~=1  &strcmp(resp,  'N')~=1  &strcmp(resp,  'y')~=1
&strcmp(resp, 'Y')~=1
fprintf('\n Incorrect reply, try again \n\n'), end end
if resp == 'y' | resp == 'Y' nf = 999;
else ff = 0; end
end % end for while
```

Line- Ground fault

```
% The program lgfault is designed for the single line-to-ground
% fault analysis of a power system network. The program requires
% the positive-, negative- and zero-sequence bus impedancematrices,
% Zbus1 Zbus2,and Zbus0.The bus impedances matrices may be defined
% by the user, obtained by the inversion of Ybus or it may be
% determined  either  from the  function Zbus =zbuild(zdata)
% or the function Zbus = zbuildpi(linedata,  gendata,  yload).
% The program prompts the user to enter the faulted bus number
% and the fault impedance Zf. The prefault bus voltages are
% defined by the reserved Vector V. The array V may be defined or
% it is returned from the power flow programs lfgauss, lfnewton,
% decouple or perturb. If V does not exist the prefault bus voltages
% are automatically set to 1.0 per unit. The program obtains the
% total fault current, bus voltages and line currents during the fault.

function lgfault(zdata0,  Zbus0,  zdata1,  Zbus1,  zdata2,  Zbus2,  V)  if
exist('zdata2')  ~=1
zdata2=zdata1; else, end
if exist('Zbus2') ~= 1 Zbus2=Zbus1;
else, end
nl = zdata1(:,1); nr = zdata1(:,2);
nl0 = zdata0(:,1);  nr0 = zdata0(:,2);  nbr=length(zdata1(:,1)); nbus =
max(max(nl), max(nr)); nbr0=length(zdata0(:,1));
R0 =  zdata0(:,3); X0  =zdata0(:,4);
R1 =  zdata1(:,3); X1  =zdata1(:,4);
R2 = zdata1(:,3); X2 = zdata1(:,4);
for k=1:nbr0
if R0(k)==inf | X0(k) ==inf
R0(k) = 99999999; X0(k) = 99999999;
```

```
else, end end
ZB1 = R1 + j*X1; ZB0 = R0 + j*X0; ZB2 = R2 + j*X2;
if exist('V') == 1
if length(V) == nbus V0 = V;
else, end
else, V0 = ones(nbus, 1) + j*zeros(nbus, 1); end
fprintf('\nLine-to-ground fault analysis \n') ff = 999;
while ff > 0
nf = input('Enter Faulted Bus No. -> '); while nf<= 0 | nf>nbus
fprintf('Faulted bus No. must be between 1 & %g \n', nbus) nf =
input('Enter Faulted Bus No. -> ');
end
fprintf('\nEnter Fault Impedance Zf = R + j*X in ')
Zf = input('complex form (for bolted fault enter 0). Zf = '); fprintf('
\n')
fprintf('Single  line  to-ground  fault  at  bus  No.  %g\n',  nf)
a=cos(2*pi/3)+j*sin(2*pi/3);
sctm = [1 1 1; 1 a^2 a; 1 a a^2];
Ia0 = V0(nf)/(Zbus1(nf,nf)+Zbus2(nf, nf)+ Zbus0(nf, nf)+3*Zf); Ia1=Ia0;
Ia2=Ia0; I012=[Ia0; Ia1;Ia2];
Ifabc = sctm*I012; Ifabcm =abs(Ifabc);
fprintf('Total  fault  current  =  %9.4f  per  unit\n\n',  Ifabcm(1))
fprintf('Bus Voltages during the fault in per unit \n\n') fprintf('
   Bus -------Voltage Magnitude------- \n') fprintf('No.  Phasea
   Phaseb  Phase c\n')
for n = 1:nbus
Vf0(n)= 0 - Zbus0(n, nf)*Ia0; Vf1(n)= V0(n) - Zbus1(n, nf)*Ia1; Vf2(n)=
0 - Zbus2(n, nf)*Ia2;
Vabc = sctm*[Vf0(n); Vf1(n); Vf2(n)];
Va(n)=Vabc(1); Vb(n)=Vabc(2); Vc(n)=Vabc(3); fprintf(' %5g',n)
fprintf(' %11.4f', abs(Va(n))),fprintf(' %11.4f', abs(Vb(n))) fprintf('
%11.4f\n', abs(Vc(n)))
end
fprintf(' \n')
fprintf('Line currents for fault at bus No. %g\n\n', nf)
fprintf('   From      To          -----Line Current Magnitude---- \n')
fprintf('  Bus   Bus   Phasea      Phaseb       Phase   c    \n')  for
n=1:nbus
for I = 1:nbr
```

```
if nl(I) == n | nr(I) == n
ifnl(I)==n  k = nr(I); elseif nr(I) == n k = nl(I); end
if k ~= 0
Ink1(n, k) = (Vf1(n) - Vf1(k))/ZB1(I);
Ink2(n, k) = (Vf2(n) - Vf2(k))/ZB2(I);
else, end else, end
end
for I = 1:nbr0
if nl0(I) == n | nr0(I) == n
ifnl0(I)==n k = nr0(I); elseif nr0(I) == n k = nl0(I); end
if k ~= 0
Ink0(n, k) = (Vf0(n) - Vf0(k))/ZB0(I);
else, end else, end
end
for I = 1:nbr
if nl(I) == n | nr(I) == n
ifnl(I)==n  k = nr(I); elseif nr(I) == n k = nl(I); end
if k ~= 0
Inkabc  =  sctm*[Ink0(n,  k);  Ink1(n,  k);  Ink2(n,  k)]; Inkabcm =
abs(Inkabc); th=angle(Inkabc);
if real(Inkabc(1)) > 0
fprintf('%7g', n), fprintf('%10g', k),
fprintf('  %11.4f',  abs(Inkabc(1))),fprintf('  %11.4f',  abs(Inkabc(2)))
fprintf('  %11.4f\n',abs(Inkabc(3)))
elseif  real(Inkabc(1))  ==0  &imag(Inkabc(1))  <  0  fprintf('%7g',  n),
fprintf('%10g',k),
fprintf('  %11.4f',  abs(Inkabc(1))),fprintf('  %11.4f',  abs(Inkabc(2)))
fprintf('  %11.4f\n',abs(Inkabc(3)))
else, end else, end
else, end end
if n==nf
fprintf('%7g',n),fprintf('     F'),
fprintf('  %11.4f',  Ifabcm(1)),fprintf('  %11.4f',  Ifabcm(2)) fprintf('
%11.4f\n', Ifabcm(3))
else, end
end resp=0;
while strcmp(resp, 'n')~=1 &strcmp(resp, 'N')~=1 &strcmp(resp, 'y')~=1
&strcmp(resp, 'Y')~=1
resp = input('Another fault location? Enter ''y'' or ''n'' within single
```

```
quote -> ');
if  strcmp(resp,  'n')~=1  &strcmp(resp,  'N')~=1  &strcmp(resp,  'y')~=1
&strcmp(resp, 'Y')~=1
fprintf('\n Incorrect reply, try again \n\n'), end end
if resp == 'y' | resp == 'Y' nf = 999;
else ff = 0; end
end % end for while
%Ink0
%Ink1
%Ink2
```

Line- line fault

```
% The program llfault is designed for the line-to-line
% fault analysis of a power system network. The program requires
% the positive- and negative-sequence bus impedance matrices,
% Zbus function Zbus = zbuildpi(linedata, gendata, yload).
% The program prompts the user to enter the faulted bus number
% and the fault impedance Zf. The prefault bus voltages are
% defined by the reserved Vector V. The array V may be defined or
% it is returned from the power flow programs lfgauss, lfnewton,
% decouple or perturb. If V does not exist the prefault bus voltages
% are automatically set to 1.0 per unit. The program obtains the
% total fault current, bus voltages and line currents during the fault.

function llfault(zdata1, Zbus1, zdata2, Zbus2, V) if exist('zdata2') ~=
1
zdata2=zdata1; else, end
if exist('Zbus2') ~= 1 Zbus2=Zbus1;
else,end
nl = zdata1(:,1); nr = zdata1(:,2);
R1 = zdata1(:,3); X1 = zdata1(:,4);
R2 = zdata2(:,3); X2 = zdata2(:,4); ZB1 = R1 + j*X1; ZB2 = R2 + j*X2;
nbr=length(zdata1(:,1)); nbus = max(max(nl), max(nr));
if exist('V') == 1
if length(V) == nbus V0 = V;
else, end
else, V0 = ones(nbus, 1) + j*zeros(nbus, 1); end
fprintf('\nLine-to-line fault analysis \n') ff = 999;
```

```
while ff > 0
nf = input('Enter Faulted Bus No. -> '); while nf<= 0 | nf>nbus
fprintf('Faulted bus No. must be between 1 & %g \n', nbus) nf =
input('Enter Faulted Bus No. -> ');
end
fprintf('\nEnter Fault Impedance Zf = R + j*X in ')
Zf = input('complex form (for bolted fault enter 0). Zf = '); fprintf('
\n')
fprintf('Line-to-line    fault    at    bus    No.    %g\n',    nf)
a=cos(2*pi/3)+j*sin(2*pi/3);
sctm = [1 1 1; 1 a^2 a; 1 a a^2]; Ia0=0;
Ia1  =  V0(nf)/(Zbus1(nf,nf)+Zbus2(nf,  nf)+Zf);  Ia2=-Ia1;  I012=[Ia0;
Ia1;Ia2];
Ifabc = sctm*I012; Ifabcm  =abs(Ifabc);
fprintf('Total  fault  current  =  %9.4f  per  unit\n\n',  Ifabcm(2))
fprintf('Bus Voltages during the fault in per unit \n\n') fprintf('
   Bus -------Voltage Magnitude-------  \n') fprintf('No.    Phasea
   Phaseb    Phase c\n')
for n = 1:nbus Vf0(n)= 0;
Vf1(n)= V0(n) - Zbus1(n, nf)*Ia1; Vf2(n)= 0 - Zbus2(n, nf)*Ia2;
Vabc = sctm*[Vf0(n); Vf1(n); Vf2(n)];
Va(n)=Vabc(1); Vb(n)=Vabc(2); Vc(n)=Vabc(3); fprintf(' %5g',n)
fprintf(' %11.4f', abs(Va(n))),fprintf(' %11.4f', abs(Vb(n))) fprintf('
%11.4f\n', abs(Vc(n)))
end
fprintf(' \n')
fprintf('Line currents for fault at bus No. %g\n\n', nf)
      fprintf From  To    -----Line Current Magnitude--
      ('                    -- \n')
      fprintf Bus    Bus   Phasea Phaseb    Phase c\n')
      ('
for n= 1:nbus for I = 1:nbr
if nl(I) == n | nr(I) == n
ifnl(I)==n
k = nr(I);
elseif nr(I) == n k =nl(I);
end
if k ~= 0 Ink0(n, k) = 0;
Ink1(n, k) = (Vf1(n) - Vf1(k))/ZB1(I);
```

```
Ink2(n, k) = (Vf2(n) - Vf2(k))/ZB2(I);
Inkabc   =   sctm*[Ink0(n,   k);   Ink1(n,   k);   Ink2(n,   k)];   Inkabcm   =
abs(Inkabc); th=angle(Inkabc);
if real(Inkabc(2)) < 0
fprintf('%7g', n), fprintf('%10g', k),
fprintf('  %11.4f',  abs(Inkabc(1))),fprintf('  %11.4f',  abs(Inkabc(2)))
fprintf('  %11.4f\n',abs(Inkabc(3)))
elseif  real(Inkabc(2))  ==0  &imag(Inkabc(2))  >  0  fprintf('%7g',  n),
fprintf('%10g',k),
fprintf('  %11.4f',  abs(Inkabc(1))),fprintf('  %11.4f',  abs(Inkabc(2)))
fprintf('  %11.4f\n',abs(Inkabc(3)))
else, end else, end
else, end end
if n==nf
fprintf('%7g',n),fprintf('     F'),
fprintf('  %11.4f',  Ifabcm(1)),fprintf('  %11.4f',  Ifabcm(2))  fprintf('
%11.4f\n', Ifabcm(3))
else, end end resp=0;
while strcmp(resp, 'n')~=1 &strcmp(resp, 'N')~=1 &strcmp(resp, 'y')~=1
&strcmp(resp, 'Y')~=1
resp = input('Another fault location? Enter ''y'' or ''n'' within single
quote -> ');
if strcmp(resp, 'n')~=1 &strcmp(resp, 'N')~=1 &strcmp(resp, 'y')~=1
&strcmp(resp, 'Y')~=1
fprintf('\n Incorrect reply, try again \n\n'), end end
if resp == 'y' | resp == 'Y' nf = 999;
else ff = 0; end
end % end for while
```

Balanced Three- phase fault

```
% The program symfaultisdesigned for the balancedthree-phase
% fault analysis of a power system network. The program requires
% the bus impedance matrix Zbus. Zbus may be defined by the
% user, obtained by the inversion of Ybus or it may be
% determined  either  from  the  function Zbus =zbuild(zdata)
% or the function Zbus = zbuildpi(linedata,  gendata,  yload).
% The program prompts the user to enter the faulted bus number
% and the fault impedance Zf. The prefault bus voltages are
% defined by the reserved Vector V. The array V may be defined or
```

```
% it is returned from the power flow programs lfgauss, lfnewton,
% decouple or perturb. If V does not exist the prefault bus voltages
% are automatically set to 1.0 per unit. The program obtains the
% total fault current, the postfault bus voltages and line
currents.functionsymfault(zdata, Zbus, V)

nl = zdata(:,1); nr = zdata(:,2); R = zdata(:,3); X = zdata(:,4);
nc = length(zdata(1,:)); if nc> 4
BC = zdata(:,5);
elseif nc ==4, BC = zeros(length(zdata(:,1)), 1); end
ZB = R + j*X;
nbr=length(zdata(:,1)); nbus = max(max(nl), max(nr)); if exist('V') == 1
if length(V) == nbus V0 = V;
else, end
else, V0 = ones(nbus, 1) + j*zeros(nbus, 1); end
fprintf('\Three-phase balanced fault analysis \n') ff = 999;
while ff > 0
nf = input('Enter Faulted Bus No. -> '); while nf<= 0 | nf>nbus
fprintf('Faulted bus No. must be between 1 & %g \n', nbus) nf =
input('Enter Faulted Bus No. -> ');
end
fprintf('\nEnter Fault Impedance Zf = R + j*X in ')
Zf = input('complex form (for bolted fault enter 0). Zf = ');
fprintf(' \n')
fprintf('Balanced three-phase fault at bus No. %g\n', nf)
If = V0(nf)/(Zf + Zbus(nf, nf));
Ifm = abs(If); Ifmang=angle(If)*180/pi;
fprintf('Total fault current = %8.4f per unit \n\n',Ifm)
%fprintf(' p.u. \n\n', Ifm)
fprintf('Bus Voltages during fault in per unit \n\n') fprintf('   Bus
   Voltage  Angle\n')
fprintf('  No.   Magnitude   degrees\n')
for n = 1:nbus if n==nf
Vf(nf)  =   V0(nf)*Zf/(Zf  +  Zbus(nf,nf));  Vfm  =  abs(Vf(nf));
angv=angle(Vf(nf))*180/pi; else, Vf(n) = V0(n) - V0(n)*Zbus(n,nf)/(Zf +
Zbus(nf,nf));
Vfm = abs(Vf(n)); angv=angle(Vf(n))*180/pi;
end
```

```
fprintf(' %4g', n), fprintf('%13.4f', Vfm),fprintf('%13.4f\n', angv) end
fprintf(' \n')
fprintf('Line currents for fault at bus No. %g\n\n', nf)
fprintf('   From        To        Current      Angle\n')
fprintf('   Bus   Bus   Magnitudedegrees\n')
for n= 1:nbus
%Ign=0;
for I = 1:nbr
if nl(I) == n | nr(I) == n
ifnl(I)==n   k = nr(I); elseif nr(I) == n k = nl(I); end
if k==0
Ink = (V0(n) - Vf(n))/ZB(I);
Inkm = abs(Ink); th=angle(Ink);
%if th<= 0
if real(Ink) > 0
fprintf('   G    '),    fprintf('%7g',n),    fprintf('%12.4f',    Inkm)
fprintf('%12.4f\n',th*180/pi)
elseif real(Ink) ==0 &imag(Ink) < 0
fprintf('   G    '),    fprintf('%7g',n),    fprintf('%12.4f',    Inkm)
fprintf('%12.4f\n',th*180/pi)
else, end Ign=Ink; elseif k ~= 0
Ink = (Vf(n) - Vf(k))/ZB(I)+BC(I)*Vf(n);
%Ink = (Vf(n) - Vf(k))/ZB(I);
Inkm = abs(Ink); th=angle(Ink);
%Ign=Ign+Ink;
%if th<= 0
if real(Ink) > 0
fprintf('%7g', n), fprintf('%10g', k),
fprintf('%12.4f', Inkm), fprintf('%12.4f\n', th*180/pi) elseif real(Ink)
==0 &imag(Ink) < 0
fprintf('%7g', n), fprintf('%10g', k),
fprintf('%12.4f', Inkm), fprintf('%12.4f\n', th*180/pi) else, end
else, end else, end
end
if n==nf
fprintf('%7g',n),fprintf('    F'), f
printf('%12.4f', Ifm) fprintf('%12.4f\n',Ifmang)
else, end end
```

```
resp=0;
while strcmp(resp, 'n')~=1 &strcmp(resp, 'N')~=1 &strcmp(resp, 'y')~=1
&strcmp(resp, 'Y')~=1
resp = input('Another fault location? Enter ''y'' or ''n'' within single
quote -> ');
if strcmp(resp, 'n')~=1 &strcmp(resp, 'N')~=1 &strcmp(resp, 'y')~=1
&strcmp(resp, 'Y')~=1
fprintf('\n Incorrect reply, try again \n\n'), end end
if resp == 'y' | resp == 'Y' nf = 999;
else ff = 0; end
end % end for while
```

NETWORK CODE

```
zdata1=[  1    0 0.25
0
0         2   0    0.25
1         2   0    0.125
1         3   0    0.15
2         3   0    0.25];


zdata0=  1    0 0.40
[0
0         2   0    0.1
1         2   0    0.3
1         3   0    0.35
2         3   0    0.7125]
                   ;
```

```
zdata2=zdata1;  Zbus1=zbuild(zdata1)  Zbus0=zbuild(zdata0)  Zbus2=Zbus1;
symfault(zdata1,Zbus1)
lgfault(zdata0, Zbus0, zdata1, Zbus1, zdata2, Zbus2) llfault(zdata1,
Zbus1, zdata2, Zbus2) dlgfault(zdata0,Zbus0,zdata1,Zbus1,zdata2,Zbus2)
```

Line-to-line fault analysis

Line-to-line fault at bus No. 1

Total faultcurrent=5.9726 perunit

Bus Voltages during the fault in per unit

```
Bus Voltage Magnitude
No. Phase a Phase b  Phase c
1   1.0000  0.5000   0.5000
2   1.0000  0.5541   0.5541
3   1.0000  0.5080   0.5080
```

Line currents for fault at bus No. 1

```
From To    Line CurrentMagnitude
Bus  Bus   Phase a  Phase b  Phase c
1    F     0.0000   5.9726   5.9726
2    1     0.0000   1.9112   1.9112
2    3     0.0000   0.5973   0.5973
3    1     0.0000   0.5973   0.5973
```

Line-to-line fault at bus No. 2
Total faultcurrent=5.9726 perunit
Bus Voltages during the fault in per unit

```
Bus  Voltage Magnitude
No.  Phase a Phase b  Phase c
1    1.0000 0.5541   0.5541
2    1.0000 0.5000   0.5000
3    1.0000 0.5218   0.5218
```

Line currents for fault at bus No. 2

```
From To    Line Current Magnitude
Bus  Bus   PhaseaPhasebPhasec
1    2     0.0000 1.9112    1.9112
1    3     0.0000 0.5973    0.5973
2    F     0.0000 5.9726  5.9726
3    2     0.0000 0.5973  0.5973
```

Line-to-line fault at bus No. 3
Total fault current=3.9365 per unit

Bus Voltages during the fault in per unit

```
     Bus Voltage Magnitude
     No. Phase a Phase b  Phase c
     1    1.0000  0.6128   0.6128
     2    1.0000  0.6364   0.6364
     3    1.0000  0.5000   0.5000
```

Line currents for fault at bus No. 3

```
     From To    Line CurrentMagnitude
     Bus  Bus   Phase a  Phase b  Phase c
     1    3     0.0000   2.3619   2.3619
     2    1     0.0000   0.3149   0.3149
     2    3     0.0000   1.5746   1.5746
     3    F     0.0000   3.9365   3.9365
```

Double line-to-ground fault analysis

Double line-to-ground fault at bus No. 1 Total fault current=5.8939 per unit

Bus Voltages during the fault in per unit

```
     Bus Voltage Magnitude
     No. Phase a Phase b  Phase c
     1    1.0727  0.0000   0.0000
     2    0.9008  0.3756   0.3756
     3    1.0196  0.1322   0.1322
```

Line currents for fault at bus No. 1

```
     From To    Line CurrentMagnitude
     Bus  Bus   Phase a  Phase b  Phase c
     1    F     0.0000   6.6601   6.6601
     2    1     0.2063   2.2302   2.2302
     2    3     0.0393   0.6843   0.6843
     3    1     0.0393   0.6843   0.6843
```

Double line-to-ground fault at bus No. 2

Total fault current=9.4414 per unit

Bus Voltages during the fault in per unit

```
Bus Voltage Magnitude
No. Phase a Phase b Phase c
1   0.8411  0.2894  0.2894
2   0.8155  0.0000  0.0000
3   0.8269  0.1835  0.1835
```

Line currents for fault at bus No. 2

```
From To    Line CurrentMagnitude
Bus  Bus   Phase a Phase b Phase c
1    2     0.6727  2.0868  2.0868
1    3     0.2203  0.6482  0.6482
2    F     0.0000  7.6129  7.6129
3    2     0.2203  0.6482  0.6482
```

Double line-to-ground fault at bus No. 3 Total fault current=   3.2609 per unit

Bus Voltages during the fault in per unit

```
Bus VoltageMagnitude
No. Phase a Phase b Phase c
1   1.0109  0.4498  0.4498
2   0.9402  0.5362  0.5362
3   1.1413  0.0000  0.0000
```

Line currents for fault at bus No. 3

```
From To    Line CurrentMagnitude
Bus  Bus   Phase a Phase b Phase c
1    3     0.0000  2.5565  2.5565
2    1     0.1848  0.4456  0.4456
2    3     0.0000  1.7043  1.7043
3    F     0.0000  4.2608  4.2608
```

Balanced three-phase fault analysis

Balanced three-phase fault at bus No. 1 Total fault current = 6.8966 per unit

Bus Voltages during fault in per unit

```
    Bus Voltage      Angle
    No. Magnitude   degrees
    1   0.0000         0.0000
    2   0.2759         0.0000
    3   0.1034         0.0000
```

Line currents for fault at bus No. 1

```
    From To   Current    Angle
    Bus  Bus  Magnitude  degrees
    G    1    4.0000     -90.0000
    1    F    6.8966     -90.0000
    G    2    2.8966     -90.0000
    2    1    2.2069     -90.0000
    2    3    0.6897     -90.0000
    3    1    0.6897     -90.0000
```

Balanced three-phase fault at bus No. 2 Total fault current = 6.8966 per unit

Bus Voltages during fault in per unit

```
    Bus Voltage   Angle
    No. Magnitude degrees
    1   0.2759    0.0000
    2   0.0000    0.0000
    3   0.1724    0.0000
```

Line currents for fault at bus No. 2

```
From To Current     Angle
BusBusMagnitudedegrees
                 G   1    2.8966    -90.0000
                 1   2    2.2069    -90.0000
```

```
    1   3    0.6897    -90.0000
    G   2    4.0000    -90.0000
    2   F    6.8966    -90.0000
    3   2    0.6897    -90.0000
```

Balanced three-phase fault at bus No. 3 Total fault current = 4.5455 per unit

Bus Voltages during fault in per unit

```
    Bus   Voltage     Angle
    No.   Magnitude degrees
    1     0.4091     0.0000
    2     0.4545     0.0000
    3     0.0000     0.0000
```

Line currents for fault at bus No. 3

```
    From ToB Current   Angledeg
    Bus  us  Magnitude rees
    G    1   2.3636    -90.0000
    1    3   2.7273    -90.0000
    G    2   2.1818    -90.0000
    2    1   0.3636    -90.0000
    2    3   1.8182    -90.0000
    3    F   4.5455    -90.0000
```

13.18   SOLUTION OF SWING EQUATION BY POINT BY POINT METHOD

```
Problem: - f = 50Hz generator 50 MVA supplying 50 MW with inertia
constant 'H' = 2.7
```

```
MJ/MVA at rated speed. E = 1.05 pu, V = 1 pu, X1 = X2 = 0.4 pu. three
phase fault at line 2.
```

```
(a)  plot swing curve for a sustained fault up to a time of 5 secs.
```

```
(b)  plot swing curve if fault is cleared by isolating line in 0.1
     seconds.
```

```
(c)  Find the critical clearing angle.
```

MATLAB PROGRAM:-

```
MVA base = 50
GIVEN:-
E = 50; V =1; Xd = 0.2; X1 =0.4; X2 = 0.4;H = 2.7;
prefault condition
del = 0:pi/10:pi;
del1 =del;
del2 = del;
M = 2.7/(180*50);                       % angular momentum = H/180*f
Peo = (1.05/0.4)*sin(del);              % Initial power curve
Po = 1 ;                                % power output in pu = 50 MW/50 MVA
delo = asind(0.4/1.05);                 % initial load angle in degrees //Pe
= (E*V/X) sin(delo)
During fault
Pe2 = 1.05*sin(del1);                   % Power curve during fault
Post fault condition
Pe3 = (1.05/0.6)*sin(del2);             % Power curve after clearing fault
plot(del,Peo);
set(gca,'XTick',0:pi/10:pi);
set(gca,'XTickLabel',{'0','','','','','pi/2','','','','','pi'});
title('Power Curve');
xlabel('Load angle');
ylabel('Genpower');
text((2/3)*pi,(1.05/0.4)*sin((2/3)*pi),'\leftarrowintial
curve','HorizontalAlignment','left');
text(pi/2,2.75,'2.625*sin\delta','HorizontalAlignment','center');
hold all
plot(del1,Pe2);
text((2/3)*pi,1.05*sin((2/3)*pi),'\leftarrow                    during
fault','HorizontalAlignment','left');
text(pi/2,1.80,'1.05*sin\delta','HorizontalAlignment','center');
plot(del2,Pe3);
text((2/3)*pi,(1.05/0.6)*sin((2/3)*pi),'\leftarrow              fault
cleared','HorizontalAlignment','left');
text(pi/2,1.1,'1.75*sin\delta','HorizontalAlignment','center');
hold off
t = 0.05;        % time step preferably 0.05 seconds
t1 = 0:t:0.5;
```

```
Similarly develop for
(a) sustained fault at t = 0
for discontinuity at t = 0 , we take the average of accelerating power
before and after the fault
at t = 0-, Pa1 = 0
at t = 0+. Pa2 = Pi - Pe2
at t =  0 ,Pa =Pa1+Pa2/2
(b) Fault cleared in 0.10 seconds ,2nd step ---- 3rd element [1]0 [2]
0.05,[3]0.10
(c) critical clearing angle
delo = degtorad(delo);         % initial load angle in rad
delm = pi - asin(1/1.75);    % angle of max swing
c1 = ((delm-delo)-(1.05*cos(delo))+(1.75*cos(delm)))/(1.75-1.05);
cclang = acos(c1);                              % critical clearing
angle in rad
cclang = radtodeg(cclang);                % critical clearing angle in
degree
cclang = int16(cclang);                    % converting to integer
fprintf('\n\n\t\t Critical Clearing angle is %d degree ',cclang);
```

Solutions

CHAPTER 1

1. Series, parallel
2. Wave length
3. Physical length of the line, operating voltage and effect of capacitance.
4. Short
5. Open circuited or lightly loaded
6. Greater
7. Characteristic impedance, Zc
8. Surge impedance, Zs
9. 400 ohms
10. 40 ohms
11. Characteristic Impedance Loading.
12. Surge Impedance Loading.
13. Characteristic Impedance Loading.
14. Surge Impedance Loading.
15. Open circuited
16. Short circuited
17. Open and short circuit impedances

CHAPTER 2

5. Absolute system
6. Per unit system

7. Per unit system

8. Per unit system

9. One

10. Per unit system

11. Absolute value.

12. Reference value.

13. Alternator, alternator

14. Per unit

15. Equal

16. Three phase

17. Line to Line

18. Base MVA and square of the base KV

CHAPTER 6

45. PQ

46. Taylor's series of expansion

47. Jacobian matrix

48. Newton Raphson method

49. Linear, quadratic

50. Reactive

51. Gauss Seidel, Newton Raphson

CHAPTER 8

1. Symmetrical components

2. Two

3. Internal, external

4. Positive

5. Negative

6. Zero

7. Operator–'k' or 'a' or 'α'

8. $1 + k + k^2 = 0$

9. $E_{R1} = E_R,\ E_{R2} = 0$ and $E_{R0} = 0$

10. $3I_{R0}$

11. Ground
12. Neutral
13. ground and neutral
14. In. Xn
15. (a) Solid neutral (b) Reactance neutral (c) Isolated neutral
16. Series
17. Shunt
18. Solid
19. Reactance
20. Isolated
21. Delta
22. $X_{Ro}+3X_n$
23. X_{R1}
24. X_{R2}
25. Sub–transient reactance, Xd"
26. Transient reactance, Xd'
27. Steady state reactance,Xd
28. Xd" <Xd' <Xd
29. Reactance diagram
30. 2 to 3.5
31. Negative and zero sequence
32. Three
33. Balanced
34. Equal
35. Zero sequence
36. Zero sequence
37. Electrical fault
38. Symmetrical
39. DC off set
40. Fault level
41. Fault calculation

CHAPTER 9

1. Open circuit fault
2. Short circuit fault
3. Sub-transient reactance

4. Zero sequence current
5. Negative sequence current
6. Positive sequence current
7. Faults involved with ground
8. Zero
9. L-G fault
10. L-L-G fault
11. L-L fault
12. Series
13. Anti-parallel
14. $3I_{R0}$
15. Ground
16. Neutral
17. L-L-G fault
18. Involved with ground
19. Not involved with ground
20. Nature of fault
21. Negative and zero sequence currents
22. Unbalanced
23. Sequence networks=3
24. LG fault
25. LG fault
26. Negative and zero
27. LLG fault

CHAPTER 10

1. LLL fault
2. Sequence voltage components
3. Positive, terminals short circuited
4. LLLG
5. $X_{R1}=X_{1eq}$
6. Twice the maximum of symmetrical short circuit current.
7. MVA in 3phase
8. Z_{bus}

9. Rated voltage and rated symmetrical breaking current.

10. 33.3A

11. Sequence network do not have mutual coupling

12. Positive and negative sequence

13. Grounded star–delta

14. Zero sequence, positive or negative sequence impedances

15. Zero sequence

16. Zero and negative sequence voltages

17. Per phase

18. LLL

19. LG fault

CHAPTER 11

1. 1.732 p.u.

2. 58.04°

3. 1.8 p.u.

4. 2.2p.u.

5. 8

6. 15 MJ/MVA

7. 400 MJ

8. 2p.u.

9. 121 MW

10. $2.5\ P_{max}$

11. 20 p.u.

12. Angle and area stability

13. Area stability

14. Short

15. Bundled conductors

16. Increases

17. Less

18. One and two

CHAPTER 12

1. Inductor
2. Capacitor
3. $Q_R + Q_C = Q_D$
4. Shunt capacitor
5. Shunt inductor
6. Power factor improvement device
7. Power system stability device
8. Booster
9. Synchronous capacitor
10. Synchronous capacitor
11. Synchronous phase modifier
12. Synchronous phase modifier